TECHNOLOGY OF THE GODS

The Incredible Sciences of the Ancients

David Hatcher Childress

Other Books by David Hatcher Childress:

VIMANA
ANCIENT TECHNOLOGY IN PERU & BOLIVIA
THE MYSTERY OF THE OLMECS
PIRATES AND THE LOST TEMPLAR FLEET
TECHNOLOGY OF THE GODS
A HITCHHIKER'S GUIDE TO ARMAGEDDON
LOST CONTINENTS & THE HOLLOW EARTH
ATLANTIS & THE POWER SYSTEM OF THE GODS
THE FANTASTIC INVENTIONS OF NIKOLA TESLA
LOST CITIES OF NORTH & CENTRAL AMERICA
LOST CITIES OF CHINA, CENTRAL ASIA & INDIA
LOST CITIES & ANCIENT MYSTERIES OF AFRICA & ARABIA
LOST CITIES & ANCIENT MYSTERIES OF SOUTH AMERICA
LOST CITIES OF ANCIENT LEMURIA & THE PACIFIC
LOST CITIES OF ATLANTIS, ANCIENT EUROPE & THE MEDITERRANEAN
LOST CITIES & ANCIENT MYSTERIES OF THE SOUTHWEST
YETIS, SASQUATCH AND HAIRY GIANTS

With Brien Foerster
THE ENIGMA OF CRANIAL DEFORMATION

With Steven Mehler
THE CRYSTAL SKULLS

TECHNOLOGY OF THE GODS

Adventures Unlimited Press

Technology of the Gods

ISBN 10: 0-932813-73-9
ISBN 13: 978-0-932813-73-2

Published by:
Adventures Unlimited Press
One Adventure Place Box 74
Kempton, Illinois 60946 USA
auphq@frontiernet.net

AdventuresUnlimitedPress.com

TECHNOLOGY OF THE GODS

The Incredible Sciences of the Ancients

Thanks to the many people who have helped me in researching and finishing this book including Jennifer Bolm, Christopher Dunn, Andrew Tomas, Ivan T. Sanderson, Charles Berlitz, J. Manson Valentine, Alfred Bielek, Rth Hover-McKinley, Flavia Anderson, Jerry Ziegler, John Michell, Harry Osoff, Chas Berlin, William Corliss, the A.R.E. of Virginia Beach and plenty of others.

Dedicated to all scientist-philosophers everywhere who continue to study, learn and grow. May they take us to infinity and beyond.

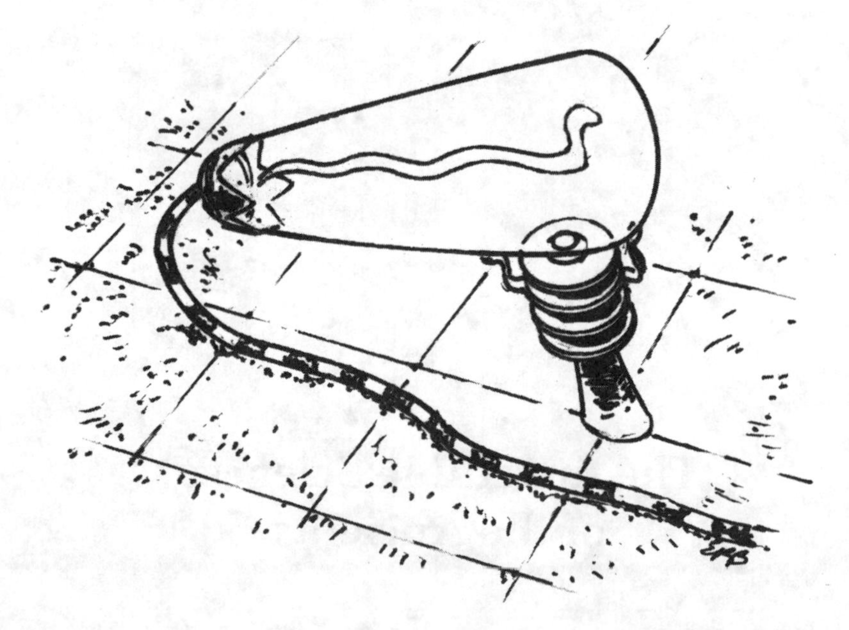

TABLE OF CONTENTS

And here, ...my dear Watson, we come into those
realms of conjecture where the most logical minds may be at fault;
each may form its own hypothesis upon the present evidence,
and yours is as likely to be correct as mine.
—Sherlock Holmes, *The Adventure of the Empty House*

From the moment I picked up your book until I laid it down
I was convulsed with laughter. Some day I intend reading it.
—Groucho Marx

All are architects of fate,
Working in these walls of time:
Some with massive deeds and great;
Some with lesser rhyme.
—Longfellow, *The Builders*

Preface

Believe me, that was a happy age,
before the days of architects, before the days of builders.
—Seneca (5 BC-65AD), *Epistle 90*

Welcome to the controversial and fascinating world of ancient technology. In this book we will explore the many bits of evidence that lead us to the astounding conclusion that ancient man was virtually as sophisticated as we are today—at least someone, from somewhere, was here using high technology. This technology included everything from electricity to heavy machinery and aircraft.

The topics of ancient flight, ancient atomic wars, ancient electricity and such, will seem odd to many people, especially "highly educated" readers. To many, these topics seem too incredible to even discuss; yet, as we shall see, there is a lot of evidence pointing to a technologically advanced past. Every culture in the world seems to have legends of ancient flight and a golden civilization before our own. Separating fact from fiction is the difficult part. A coherent time-line of the ancient past would also be helpful. Hopefully, new dating techniques such as the chlorine 23 method will accurately pinpoint when rock was hewn from its quarry and erected. Since the most ancient man-made artifacts are monuments in stone, this will give us information to accurately date the megalith masterminds and the dawn of their vanished civilization.

As a reporter, I am interested in the strange and unusual. I am also interested in the facts. In this book, I have tried to include those stories, artifacts, and places that seemed the most important and could be largely verified. Admittedly, there is much speculation in this book, and I invite readers to speculate as they will. Like all "scientists" the readers of this book will sift through the information presented, take that which seems reasonable to them, and file it in with their existing "computer files" to be accessed again later and modified as necessary. Other information they will discard and ignore.

Throughout this book, a number of ancient texts are mentioned. Spelling varies in some cases such as *Rg Veda* and *Rig Veda.* We have used the more common and easy spelling whenever possible, but we have left the original author's spelling in the extensive sources quoted in this book. We have tried to include most of the sourcebooks mentioned within the text in the bibliography, but in some cases this was impossible. Books that are mentioned by other authors within the text are referenced to the quoter's book.

Special thanks to the Sanskrit scholar Ramachandra Dikshitar, the Oxford professor who wrote *War in Ancient India.* In a special chapter in the Oxford edition of his book, he waxed poetic over his country's contribution to aviation—inventing it! Said the proud historian back in 1944: "No question can be more interesting in the present circumstances of the world than India's contribution to the science of aeronautics. There are numerous illustrations in our vast Puranic and epic literature to show how well and wonderfully the ancient Indians conquered the air. To glibly characterize everything found in this literature as imaginary and summarily dismiss it as unreal has been the practice of both Western and Eastern scholars until very recently. The very idea indeed was ridiculed and people went so far as to assert that it was physically impossible for man to use flying machines. But today what with balloons, aeroplanes and other flying machines, a great change has come over our ideas on the subject."[100]

Sadly, Dr. Dikshitar was indeed ridiculed by his fellow Oxford scholars at times, but the texts speak for themselves. What was a scientific-minded scholar to do—ignore the evidence? Most did, in fact.

In approaching the subject of advanced ancient technology I have decided to start with simple but necessary technology, such as irrigation, water and sewage, and then move to the basic combination for advanced technology: metallurgy and electricity. With it established in the reader's mind that the ancients may well have had complicated metal machines—and electricity—I then move on to the fantastic possibilities of ancient flight, atomic warfare and the idea of a world-wide power system.

Hey, it's a wild ride through ancient history, but it's my story, and I'm sticking to it. Sometimes truth is stranger than fiction.

1.
The Enigma of Ancient Technology

As we acquire knowledge,
things do not become more comprehensible,
but more mysterious.
—Will Durant

I think, myself, that in 1903, we passed through
the remains of a powdered world—
left over from an ancient inter-planetary dispute,
brooding in space....
—Charles Fort

Was the Science of Egypt Inherited from an Earlier Culture?

In my searches for lost cities and mysteries of the past, I have often found clues to the technology of the ancients. These clues can be in the form of depictions of ancient devices in rock paintings or carvings (such as the electric devices at the Temple of Hathor in Egypt) or as small models of devices (such as the miniature solid gold airplanes at the Bogota Gold Museum) or in the stories from ancient texts (such as the Ramayana or even the Bible).

In this book I would like to recap some of the evidence for ancient technology and for advanced ancient cultures in the past. What is amazing about the modern world versus the ancient one is that in the modern world, the average citizen has access to advanced technology such as electricity, a personal vehicle, telephone, fax, and computer technology. In the ancient world, high technology was largely de-

nied the masses. In fact it was often used in temples and ceremonies to gain power over people by amazing or terrifying them; this was part of worship and mystery.

The well-known author and presenter of the television documentary *Mystery of the Sphinx,* John Anthony West, says:

> Egyptian science, medicine, mathematics and astronomy were all of an exponentially higher order of refinement and sophistication than modern scholars will acknowledge. The whole of Egyptian civilization was based upon a complete and precise understanding of universal laws. And this profound understanding manifested itself in a consistent, coherent and interrelated system that fused science, art and religion into a single organic *Unity*. In other words, it was exactly the opposite of what we find in the world today.
>
> Moreover, every aspect of Egyptian knowledge seems to have been complete at the very beginning. The sciences, artistic and architectural techniques and the hieroglyphic system show virtually no signs of a period of 'development'; indeed, many of the achievements of the earliest dynasties were never surpassed, or even equaled later on. This astonishing fact is readily admitted by orthodox Egyptologists, but the magnitude of the mystery it poses is skillfully understated, while its many implications go unmentioned.
>
> How does a civilization spring full-blown into being? Look at a 1905 automobile and compare it to a modern one. There is no mistaking the process of 'development,' but in Egypt there are no parallels. Everything is right there at the start.
>
> The answer to the mystery is of course obvious, but because it is repellent to the prevailing cast of modern thinking, it is seldom seriously considered. *Egyptian civilization was not a 'development,' but a legacy.*[108]

In his highly rated NBC special of November, 1993, *Mystery of the Sphinx,* West and his researchers sought to prove that the Sphinx had been severely waterworn and was over 10,000 years old!

> *Why don't you write books people can read?*
> —Nora Joyce (to her husband, James)

The Destruction of Knowledge

As our technology has gotten more advanced, we have become able to look into the future and into outer space with a view different from that of scientists and thinkers earlier in this century. Similarly, we are now able to look at the past with greater insight and technological know-how. Just as our minds have been able to imagine a future different from that which our grandfathers could envision, we are also able to see a past different from that of the scientists and experts of the turn of the century.

Just as our scope of the universe has been pushed back to the farthest reaches of space, we are now in a position to push back to the farthest reaches of history. And many researchers are doing just that.

Atlantis, with its advanced culture, is named in ancient texts. To begin with, it is mentioned in Plato's dialogues (taken from ancient Egyptian records according to the text), and nearly every ancient culture in the world has myths and legends of an ancient world-before and the cataclysm that destroyed it.

The Mayans, Aztecs and Hopis believed in the destruction of four or more worlds before our own. The destruction of Atlantis may not even be the most recent cataclysm to befall the earth.

The most widely-known books in the world such as the *Bible,* the *Mahabharata,* the *Koran* and even the *Tao Te Ching* all speak of cataclysms and ancient civilizations that were destroyed. Ancient civilizations and stories about them filled thousands, even hundreds of thousands of volumes of books that were kept in libraries around the world in ancient times. Many ancient libraries were so huge that they were famous among local historians. The library at Alexandria is a well-known example.

Sadly, it is a fact that throughout history, huge archives and libraries have been purposely destroyed. According to the famous astronomer Carl Sagan, a book entitled *The True History of Mankind Over the Last 100,000 Years* once existed and was housed in the great library in Alexandria, Egypt. Unfortunately, this book, along with thousands of others, was burned by fanatical Christians in the third century AD Any volumes which they might have missed were burned by the Moslems to heat baths a few hundred years later.

All ancient Chinese texts were ordered destroyed in 212 BC by Emperor Chi Huang Ti, the builder of the famous Great Wall. Vast amounts of ancient texts—virtually everything pertaining to history, philosophy, and science—were seized and burnt. Whole libraries, including the royal library, were destroyed. Some of the works of Confucius and Mencius were included in this destruction of knowledge.

Fortunately, some books survived because people hid them in various underground caves, and many works were hidden in Taoist temples where they are even now religiously kept and preserved.

The Spanish conquistadors destroyed every Mayan codex that they found. Out of many thousands of Mayan books found by the Spanish, only three or four are known to exist today. Like the fanatical Christian sects of the third century and Emperor Chi Huang Ti in the second century BC, the conquistadors wanted to erase all knowledge of the past and the records preserved it.

Europe and the Mediterranean were plunged into the infamous Dark Ages when the Christian church first split after a series of church councils, beginning with the council of Nicaea in 325 AD The last Patriarch of the early Christian church, Nestorius, was deposed at the Council of Ephesus in 431 AD He was banished to Libya and the Nestorian church moved eastward. The dispute concerned the early Christian doctrine of reincarnation and the idea that Christ had a dual nature: Jesus was a Master while Christ was the Archangel Melchizedek.

In the aftermath of this struggle, all books in the Byzantine Empire were ordered destroyed except for a newly edited version of the Bible that the Catholic church was issuing. The library at Alexandria was destroyed at this time, and the great mathematician and philosopher Hypatia was dragged from her chariot by a mob and torn to pieces. The mob went on to burn the library. Thus the suppression of science and knowledge, particularly of the ancient past, began in earnest.

Knowledge has been suppressed throughout the last 2,000 years. It is sometimes said that history is written by the winners of wars, rather than the losers; given the amount of known war-oriented political propaganda still popular as "history" in this century, we should consider much of ancient history in this light.

Given this suppression, it is astonishing that the few ancient texts that have survived do indeed speak of advanced civilizations and the cataclysms that destroyed them. Similarly, they talk of wise people who lived in harmony with the earth and the natural workings of all things. But at some time in the distant past, man fell out of harmony with nature, and a catastrophe struck the whole earth.

We can see here a startling parallel between the ancient Atlantis "myth" and the situation that modern man now finds himself in. Will modern man survive his own technology and tribalism, or will he destroy himself in the natural workings of his destructive practices and disharmony with the earth?

I have had a good many more uplifting thoughts,
creative and expansive visions while soaking in comfortable baths
in well-equipped American bathrooms
than I have ever had in any cathedral.
—Edmund Wilson

Ancient Sanitation: Bathrooms of the Gods

It has been said that the mark of any advanced civilization is the level of their sanitation and plumbing. Nice bathrooms and toilets are important conveniences. Plumbing and sanitation are themselves the offshoots of the science of irrigation, something that developed at least 25,000 years ago.

Over three thousand years ago the Nabateans, an Arab people, maintained six flourishing cities in Israel's desolate Negev area, including the famous Petra. By utilizing an ingenious system of terraces and walls, these engineer-farmers managed to till the soil on an average rainfall of four inches a year. "The more one examines the Nabateans' elaborate systems, the more impressed he must be with the precision and scope of their work... They anticipated and solved every problem in a manner which we can hardly improve upon today." (*Scientific American,* April, 1956)

Some 3,000 years ago the ancient Persians discovered a method of digging underground aqueducts that would bring mountain ground water to their arid plains. Still extant and functional, the system of irrigation provides 75 percent of the water used in Iran today. (*Scientific American,* April, 1968)

For centuries, sanitary conditions in Europe were deplorable. The casual treatment of human waste sustained the horrible plagues that nearly decimated the continent on several occasions. But more than 5,000 years ago in the Tigris Valley near Baghdad, Tell Asmar had homes and temples with elaborate arrangements for sanitation. One excavated temple had six toilets and five bathrooms, with most of the plumbing equipment "connected to drains which discharged into a main sewer, one meter high and 50 meters long... In tracing one drain, the investigators came upon a line of earth ware pipes. One end of each section was about eight inches in diameter while the other end was reduced to seven inches, so that the pipes could be coupled into each other just as is done with drain pipes in the 20th Century (*Scientific American*, July, 1935).

Ancient man made tunnels through mountains for irrigation purposes and sometimes built massive dams or other large hydraulic engineering feats. The great dam built by the Queen of Sheba at Marib in Yemen is a good example. Huge hydraulic works of ancient man, hitherto unknown, are just coming to light. The Sri Lankan archaeologist A. D. Fernanado in an article in the *Journal of the Sri Lanka Branch of the Royal Asiatic Society* (1982)[144] relates the incredible discoveries made when Sri Lankan engineers wanted to place a dam at Maduru Oya, thereby drowning a large valley. As the bulldozers set to work they began to scrape against bricks which already lay in the ground. To everybody's amazement, it turned out that prehistoric engineers had made the same calculations and had built a dam at the very same spot!

Norwegian archaeologists visited the site and reported that the grandeur of these prehistoric megalithic waterworks would have impressed a Pharaoh. Thor Heyerdahl says that much of the water system was constructed out of 15 ton blocks of stone 33 feet high and arranged in the shape of square tunnels and brick walls. The dams had sluices measuring more than 6 miles in length to control the water flow to a series of artificial lakes. Millions of tons of water had been regulated by this huge and sophisticated dam.[144]

It was thought by historians a hundred years ago that since nomadic tribes had no formal bathrooms or sewage disposal, everyone must have lived that way. Nomadic tribes often packed up their tents and moved on when the garbage and sewage was just too much to take. Cities, however, are more difficult to just up and move. It was thought by early British archaeologists that ancient man did not use sophisticated sewage and water systems, mainly simply allowing rain water to eventually wash away sewage into a nearby stream or river.

However, many of the bathrooms of the ancient world were very luxurious, with colorful ceramic basins and bathtubs, just as we have today. Reginald Reynolds, in his witty book on ancient sanitation, *Cleanliness and Godliness*,[130] argues that ancients knew about sewage disposal, but had two or more clearly separated systems:

> It is the opinion of Mr. Ernest Mackay, the eminent archaeologist, that these drains were not used for the disposal of sewage, and as evidence of this he cites us the *Charaka-Samhita*, a work of the second

century AD, as nearly as can be computed, in which it is said that latrines are intended only for the sick and infirm; and for the rest that a man should proceed a bow-shot from his house to do his business... The waters were suffered on occasion to run down the walls of the house, which he says would have been a noxious custom if these waters had contained sewage. But he omits to remind us that the contrary proposition, which is the absence of sewage disposal, would have been, in a city, even more noxious than an open drain; and since there were both open water-chutes and closed drains in these houses, it were more reasonable to suppose that these two systems served separate purposes, the one to carry off rain water and bath water, the other to dispose of sewage... But this at least is beyond dispute concerning those who inhabited The Mound of the Dead, [Mohnenjo-Daro in Pakistan] that they were possessed of well-appointed bathrooms, together with the system of drainage that I have described ...as part of the Sanitary Vanguard of Mankind.

Sir G. Maspero, sometime Director-General of the Service of Antiquities in Egypt, spoke highly of the excellent hygienic and sanitary arrangements known in ancient Egypt, particularly of the elaborate bath discovered at Tell el-Amarna, in the house of a high official of the Eighteenth Dynasty. And he observes that in the midst of their paved streets they provided a channel of stone to carry off water and drainage. And in this same bathroom at Tell el-Amarna there was discovered a well-preserved closet behind a screen wall, this closet being provided with a seat of limestone, elegantly shaped.[130]

It was the opinion of Herodotus that the Egyptians were of all nations the most healthy; for he found them distinguished from other peoples by the singularity of their institutions and their manners. Reynolds tells us that "the Egyptians also (like the Pythagoreans, who imitated them) abstained from beans, which they held to be unclean, for which I can give no good explanation; though some say that Pythagoras was in this matter misunderstood by Aristotle." People were already avoiding beans, five thousand years ago.

As to Egyptian toilets, Reynolds says that they preferred the "composting" type of toilets:

> [T]hey used earth rather than water for the most part, we have yet to see whether we are wiser than the Pharaohs; for sanitation is not to be confused with any popular notion or prevalent system, but must be considered in relation to the best and most expeditious disposal of sewage, the abatement of nuisances and causes of infection, the fertility of the soil, and many other matters, such as the climate and the means at human disposal. But we know from these general observations that the priest-physicians who directed the public hygiene of Egypt found cleanliness next to godliness, and were concerned to keep at least the upper-class quarters of their cities wholesome.
>
> The Egyptians knew even the art of making drains of hammered copper, of which one was found, fully 450 yards in length, at the temple of Sahara, though this drain served only for rain water. And that their supply of water was considered a matter of great importance, for the attention of the highest officer of the State, we know from an inscription concerning the duties of the Vizier under the Eighteenth Dynasty. For this inscription says of the Vizier, "It is he who dispatches the official staff to attend to the water supply in the whole land'; and again the inscription reads that 'It is he who inspects the water supply on the first of every ten-day period.'[130]

William Corliss reports in his *Science Frontiers* newsletter (No. 123, May-June 1999) that the ancient Egyptians not only had advanced toilets and bathrooms, but also applied cosmetics copiously to themselves. Upper-class women, as well as many men, favored green, white, and black makeup. These cosmetic powders, dating from 2000 BC, have been exceptionally well preserved in their original vials made of alabaster, wood or ceramic.

A team of French chemists led by P. Walter was not surprised when their analyses of these powders found crushed galena and cerussite (two ores of lead). However, they nearly dropped their test tubes when they also found chemical compounds that are extremely rare in nature; specifically, laurionite (PbOHCl) and phosgenite ($Pb_2Cl_2CO_3$). In fact, these compounds are so rare naturally that the Egyptian powders must be artificial.

P. Walter et. al. wrote: "Taken together, these results indicate that laurionite and phosgenite must have been synthesized in Ancient Egypt using wet

An Egyptian temple flame.

chemistry. The Egyptians manufactured artificial lead-based compounds, and added them to the cosmetic product. The underlying chemical reactions are simple, but the whole process, including many repetitive operations, must have been quite difficult to achieve.

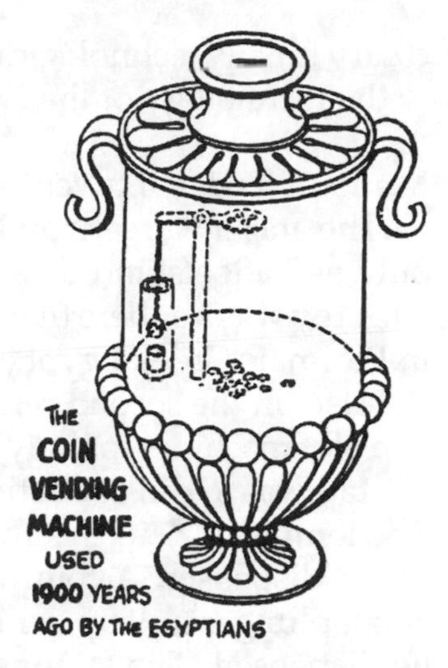

"It had been recognized earlier that the Egyptian chemists had used fire-based technology 500 years earlier in 2500 BC to manufacture blue pigment. Wet chemistry represented another forward technological step." (Nature, No. 397, 1999)

Corliss comments that, "Uncowed by the successes of the ancient Egyptian chemists, those in Nissan have synthesized artificial bird droppings for use in testing automobile paints. The real stuff, you see, is inconsistent from batch to batch."

Quality bathrooms need good soap, and the very word "soap" comes down to us from the ancient Egyptian word *swab*. In 1931, the British Egyptologist Dr. Rendel Harris maintained that the words swab and swabber derive from the Egyptian language and are of the greatest antiquity. *Wdb,* he says, among the Ancient Egyptians signified pure, and from this word he would trace the name of the *Wahabis,* who are the Puritans of Islam today. Also he says the letter *S,* expresses causation, so that *ankh* being the word meaning life, *S-ankh* signifies to make alive. From this he concludes that, if *wdb* be pure, *S-wdb* would be to make pure; that is to say, to cleanse or to swab. And since it was among the opinions strongly held by Dr. Harris that the Egyptians were a great seafaring people, he contends that this word swab came into our language through the ancient intercourse of seamen, whose language may be in parts older than any now spoken in Europe. For the seafaring use of the word he invokes Shakespeare, who wrote of:

The master, the swabber, the boatswain and I...

The seafaring skills of the Egyptians were considerable, and it is beyond dispute that they had great fleets of ships. Apparently swabbing the deck comes from ancient Egyptian and our word soap is dervived from *swab:* "that which makes clean."[130]

Proper hygiene, sanitary water, soaps and sewage disposal are all necessary for any technological civilization to advance forward. When it comes to the technology of the Gods, cleanliness is next to godliness.

Many of Today's Inventions Were Yesterday's Inventions

The ancient Greeks built steam boilers that worked, but they used them only as gadgets and toys rather than as practical sources of power. One such toy was a sphere that was spun by two steam jets, and was "invented" in Ptolmaic Greek Egypt circa 200 BC.

Also, in the second century before the Christian era, Egyptian temples had slot machines for holy water. The quantity of water which flowed from the tap was in direct relation to the weight of the coin thrown into the slot. The temple of Zeus in Athens had a similar automatically-controlled holy water dispenser. A coin was dropped into a sealed vessel which made a small plunger pull up, which then allowed a measured quantity of fluid to be dispensed. The famous Greek/Egyptian inventor Hero of Alexandria invented one such well-known device in 120 BC. It is evident from this example that temples and priests were involved in technology at an early time.

Many of the common inventions of the modern world—steam engines, clocks, vending machines, hydraulic pumps, etc.—were all known in the ancient world. Finely made instruments and tools such as the Antikythera device (to be discussed later) were common in ancient times, but archaeologists are always surprised to discover them!

Gynecology was an unknown science until the latter half of the nineteenth century. Then, according to the October 20, 1900, issue of *Scientific American*, excavations at Pompeii revealed gynecology to be but a "reinvention in the world of surgery." Instruments buried in the Temple of Vestal Virgins since the eruption of Vesuvius in AD. 79 were found to demonstrate that "gynecology was a science flourishing in its perfection long before that date... in every instance the instruments are almost in their minutest particulars exact duplicates of those in use by the most approved modern science of today... The workmanship is as fine as anything to be produced in this line in the twentieth century. The instruments are hand wrought, the screws as threadlike and capable of delicate manipulation as anything to be found in today's achievements."

Shipwrecks around the Mediterranean give us an idea of the machines that the ancient Greeks, Romans and other Mediterranean maritime cultures possessed. The

July 27, 1959, issue of *Chemical Engineering* had an article about an 80-pound valve that had been salvaged from one of Emperor Caligula's yachts. The valve was made of a zinc-free, lead-rich, anticorrosion, antifriction bronze.

Said the article, "The Caligula valve was found submerged at the bottom of Lake Lemi in Rome. Although 19 centuries old, it still exhibits highly polished surfaces and retains its plug tightly." While modern fashion and sexual trends may simply mimic those of ancient times, scientists are too often surprised at the high level of technical and scientific knowledge of ancient man.

Becoming familiar with ancient science is a good start for the layman, and two good books that are easily obtainable are *Technology in the Ancient World*[54] by Henry Hodges and *Engineering In the Ancient World* by J. Landels.[116] With these books, the science of classical times can be seen to be very similar to our own.

If we human beings want to feel humility, there is no need to look at the starred infinity above. It suffices to turn our gaze upon the world cultures that existed thousands of years before us, achieved greatness before us, and perished before us.
—C.W. Ceram, *Gods, Graves, & Scholars*

The Amazing Inventions of China

Many ancient inventions are said to have originated in China, though many of them probably came from even earlier cultures.

The Chinese possessed geared machinery from an early age—some think as far back as the early centuries BC, if not before. While modern historians prefer to trace China back only to the the Chou dynasty in 1122 BC, the Chinese themselves trace their history back to the semi-mythical "Five Monarchs."

Extant Chinese texts state that the first of the dynasties was that of the "Five Monarchs," in which there were, confusingly, nine rulers whose combined reigns lasted from 2852 to 2206 BC. Confucius ascribed to one king, Yao, whose reign started around 2357 BC, "..kindliness, wisdom, and sense of duty." He was succeeded by Shon, who built a vast network of roads, passes, and bridges through the enormous land, and many scholars attribute the building of the Silk Road to him.

All ancient Chinese texts, especially those of Lao Tzu and Confucius, as well as the *I Ching*, speak of the ancients and the glory of their civilization. They were presumably speaking about the people living at least at the time of the "Five Monarchs" and probably before. The legendary Chi-Kung people of this early period were said to have "flying carriages."

As noted above, just before he died in 212 BC, Emperor Chin Shih Huang Ti ordered that all the books and literature relating to ancient China be destroyed. Vast amounts of ancient texts—virtually everything pertaining to history, astronomy, philosophy, and science—were seized and burnt. Whole libraries, including the royal library, were destroyed. Some of the works of Confucius and Mencius were included in this destruction of knowledge.

Fortunately, some books survived, as people hid them, and many works were hidden in Taoist temples where they are even now religiously kept and preserved. They are on no account shown to anyone, but kept hidden away as they have been for thousands of years. The persecution and closing of religious temples by the Communists indicated that the lamas still had cause to keep their ancient books hidden.

Doubtless, there was a great deal of lost history relating to the early days of China and its technology. What caused the emperor Chin to want to destroy any record of the past just prior to his death? Was he such a megalomaniac that he wanted history to start with him, or was he influenced by the same evil forces that inspired Genghis Khan and Hitler to the same sort of book burning?

We have heard that in the remote past
kings had titles but no posthumous appelations.
In recent times kings not only had titles
but after their death were awarded names
based on their conduct.
That means sons passed judgment
on their fathers, subjects on their sovereign.
This cannot be allowed.
Posthumous titles are herewith abolished.
We are the First Emperor, and our successors
shall be known as the Second Emperor,
the Third Emperor, and so on,
for endless generations.
–Chin Shih Huang Ti *edict*, 212 BC

Despite their sometimes despotic rulers, invention and inovation thrived in ancient China and Central Asia. It was really the Chinese who invented movable type; the inventor was a fellow named Bi Sheng who introduced the technology in 1045 AD, four hundred years before Gutenberg first printed the Bible. The Chinese are also credited with inventing writing paper, wrapping paper, paper napkins, playing cards and paper money! Toilet paper was another one of the spin-offs of their paper industry, over 2,000 years ago. Probably, all of these inventions had existed in their past, and more.

The Chinese were well aware of earthquakes and geological changes; they developed earthquake-resistant houses as long as seven thousand years ago. The world's first-known seismograph for detecting and recording far away earthquakes was invented by Zhang Heng in 132 AD This ingenious device stood about eight feet tall and featured eight bronze dragons holding balls between their jaws. When tilted by a distant earthquake, an internal pendulum opened the jaw of the dragon facing the source of the tremor and the ball dropped into the mouth of a bronze frog waiting below each dragon.

The first mechanical clock is attributed to two Chinese inventors around 725 AD, and gunpowder was known in China at least as early as the ninth century, if not much earlier. Used there only for fireworks and enjoyment, after it was first brought to Europe in the thirteenth century, it fueled the first cannons, which were made by the Dutch and Germans.

The Chinese have always had great scope and vision regarding their projects; not only was the Great Wall a colossal endeavor, but the Grand Canal of China, which connects the Yellow River with the Yangtze, is twenty times longer than the Panama Canal—yet the Chinese constructed it without modern equipment starting over 1,300 years ago! There are other mammoth projects that still are largely unknown or waiting to be discovered, such as the largest pyramid in the world, near Xian. Even the Chinese version of the typewriter, called the Hoang typewriter, has 5,700 characters on a keyboard two feet wide and seventeen inches high!

In *The Genius of China: 3,000 Years of Science, Discovery and Invention,*[99] the author Robert Temple (distilling the book from Joseph Needham's works at Cambridge University) says that the Chinese knew and used poison gas and tear gas in the 4th century BC, 2300 years before the West got around to it! The Chinese were making cast iron in the 4th century BC (1700 years before the West), and they manufactured steel from cast iron in the 2nd century BC (2000 years before the West). The first suspension bridge was built in China in the 1st century (at least 1800 years before the West), and the Chinese invented matches in 577, a thousand years before the West.

Says Needham in the introduction of the book about the advanced level of civilization in China: "First, why should they have been so far in advance of other civilizations; and second, why aren't they now centuries ahead of the rest of the world?" Perhaps China inherited its knowledge from an older civilization. Its discoveries, like ours, are just the re-discovery of ancient technology from the roller-coaster ride of history.

a game or ceremony

Says author Andrew Tomas in his book *We Are Not the First*, "Cybernetics is an old science. In China it was known as the art of Khwai-shuh, by means of which a statue was brought to life to serve its maker. The description of a mechanical man is contained in the story of Emperor Ta-chouan. The empress found the robot so irresistible that the jealous ruler of the Celestial Empire gave orders to the constructor to break it up in spite of all the admiration that he himself had for the walking robot.

One of the world's first calculating machines was, of course, the 2,600-year-old Chinese abacus. It is only recently that modern calculators made faster calculations than the simple but efficient abacus.

This would seem fantastic.
One would think that modern engineers had exploited these forces, to the nth degree, but the truth is; that outside the common ram, or turbine, the ancients can teach us a thing or two.
Jules Verne—In reply to a statement that the exploitation of natural forces had been exhausted

The Marvelous Chinese Clocks

The marvelous clocks of Ancient China are a good example of how complicated ancient machinery could be. Although mechanical clocks have been around for thousands of years, the problem of accuracy over extended periods of weeks and months was difficult to solve. The Chinese solved this problem with a device called an escapement. This mechanism makes it possible to closely regulate the speed of a clock and to drive it with a comparatively small power source.

The first known clock with an escapement was built about 724 AD by Lyang Lingdzan, though it seems that the technology was known before then. This apparatus included a celestial sphere that turned with the heavens, a model sun and moon that went around the sphere as the real ones seem to do about the earth, and jacks that struck bells and beat drums to mark the passage of time.

The bell of Lyang's clock marked the Chinese "hour" or shî which is twice the length of one of ours. The drum sounded a shorter period, the ko. This is 1/100 of a solar day, or 14 minutes and 24 seconds on our time scale. Like more westerly peoples, the Chinese originally divided day and night into intervals, which stretched and shrank with the seasons. Later, about 1100 AD, the Chinese adopted a system of equal, permanent periods that stayed the same regardless of the wanderings of dawn and sunset. This change made clock-making easier.

In Lyang's clock, "Water, flowing [into scoops] turned a wheel automatically, rotating it one complete revolution in one day and night." The machinery of the clock included "wheels and shafts, hooks, pins and interlocking rods, mopping devices and locks checking mutually."[27]

The words "pins and interlocking rods" describe the escapement, which was needed to make the wheel revolve so slowly. The escapement was presumably a simple system of tripping lugs that held the water wheel against rotation until one scoop had been filled and then allowed it to move only far enough to bring the next scoop into the filling position. Lyang's clock kept better time than anything seen before, although it would no doubt have seemed impossibly inaccurate by our standards.

After Lyang's time, corrosion of the parts of bronze and iron put the clock out of action, and it was retired to a museum. Later mechanicians built grander clocks. In 976 AD, Jang Sz-hsun built a clock that occupied a pagodalike tower over 30 feet high. This had nineteen jacks, which not only rang bells and beat drums but also popped out of little doors holding signs to show the time. Other parts showed the movements of the heavens, the sun, the moon, and the planets. To keep his clock from being stopped by the freezing of the water in winter, Jang rebuilt it to use mercury instead of water as the working fluid.

According to L. Sprague de Camp in his book *The Ancient Engineers*,[27] the grandest of these imperial water clocks was built by Su Sung in 1090. Su Sung's memorial to the emperor Shen Dzung describes his clock, with diagrams, so that if anybody wished to do so he could reconstruct the clock with fair accuracy today.

At this time the Sung dynasty ruled most of China, though a nomad tribe, the Kitans, had conquered some of the northern provinces. Su Sung had a long career in the imperial bureaucracy. His eventual list of titles included: Official of the Second Titular Rank, President of the Ministry of Personnel, Imperial Tutor to the Crown Prince, Grand Protector of the Army, and Kai-gwo Marquis of Wu-gung.

When Su was sent on a mission to the Kitan court to congratulate the khan on having passed the winter solstice, he found that he had arrived a day too early. The Sung astronomers had erred by a quarter-hour in calculating the exact time of the solstice. Su saved his sovereign's face, and probaly his own, by lecturing on the difficulty of exactly calculating such events.

But, when Su returned to the Sung capital of Kaifeng, he urged the Emperor to let him build a clock accurate enough to avoid such contretemps. Receiving approval, Su, like any competent engineer, built a couple of wooden pilot models, one small and one full-sized, to get the bugs out of the design before tackling the final clock.

The finished machine occupied a tower at least 35 feet high, counting the penthouse on top. Water, flowing through a series of vessels, filled the thirty-six scoops of a water wheel, one after the other. An escapement allowed the wheel to rotate, one scoop-interval at a time. The wheel revolved once in nine hours, while the water fell from the scoops into a basin below the wheel.

The wheel turned a wooden shaft in iron bearings. This shaft, by means of a crown gear, turned a long vertical shaft that worked all the rest of the machinery, to which it was connected by gearing. The machinery included an armillary sphere (a set of graduated intersecting rings corresponding to the horizon, the ecliptic, and the meridian) in the penthouse on top. There was also a celestial sphere, with pearls for stars in the top story, and five large horizontal wheels bearing jacks.[27]

Altogether, Su's clock must have been an impressive spectacle, what with the continual splashing, the clatter of the escapement, the creak of the shafts in their bearings, and the frequent outbursts of drums, bells, and gongs. One shortcoming of the clock was that it was not so placed that a natural source of water could power it. Hence it had to be "wound up" from time to time. This was accomplished by hand-turned water wheels, which raised the water from the basin, into which it cascaded from the scoops of the main water wheel, to the reservoir above this wheel.

About 1126 AD a Tatar people, the Jurchens, whose kings reigned under the dynastic name of Gin, conquered the Kitan lands and some Sung provinces as well. After capturing Kaifeng, they carried away, to their own capital of Peking, Su's clock together with mechanics to run it. The captive horologists built a new tower and succeeded in making the clock run, after they had adjusted its astronomical parts for the change of latitude.

After a few years, though, the parts wore out, the clock stopped, and lightning wrecked the upper part of the tower. The Gin emperors left the remains of the clock behind when they fled before the Mongols in the 1260; and it disappeared.

Meanwhile the Sung emperors wanted another imperial clock. But Su Sung was dead, and nobody could be found who knew the subject well enough to build such a complicated mechanism.

Similar clocks continued to be built under the Mongol or Yuan dynasty. The last Yuan emperor made a hobby of mechanical engineering and took part in the construction of tail-wagging dragons and other automata. But when the Ming overthrew the Yuan in 1368, all the clocks, mechanical dragons, and other machines built for the Mongol emperors were scrapped as "useless extravagances."[27, 99]

The modern clock device, from which such devices as grandfather clocks and pocket watches are derived, is generally traced to the year 1364 when Giovanni di Dondi, of an Italian clock-making family, published a description of a weight-powered, escapement-regulated clock which, except for improvements in detail, is essentially a modern clock. Dondi became famous, and astronomers came from foreign lands to look at his marvelous clock. Galileo later substituted a pendulum for Dondi's crown-shaped balance wheel, but in watches and small clocks we still use Dondi's device.

Some time around 1502, Peter Henlein of Nuremberg invented the spring-driven watch, so called because it was originally used by watchmen. Henlein's "Nuremberg egg" was slightly larger than a modern alarm clock, had a singe hand, and hung from a chain around the neck.

Early watches gave their owners much difficulty; as Maximilian I of Bavaria used to say: "if you want troubles, buy a watch." Watches, and clocks in general have probably been giving mankind trouble for thousands of years.

The Curious Crystal Skull

Part of the enigma of ancient technology are oddball objects or devices which are clearly artificial, yet how they could have been made baffles scientists. One such curious object is the famous Mitchell-Hedges crystal skull found in the ruins of the ancient city Lubaantun in what is today Belize. Lubaantun means "place of fallen stones" in the local Mayan dialect, and the actual name of the city remains unknown. Lubaantun was first reported to the British colonial government late in the last century by the inhabitants of the Toledo settlement near Punta Gorda and in 1903 the Governor of the colony commissioned Thomas Gann to investigate it. Gann explored and excavated the main structures around the central plaza and concluded that the site's population must have been large. His report was published in England in 1904.

In 1915 R.E. Merwin of Harvard University investigated the site, locating many more structures, recognizing a ball court and drawing the first plan. His excavation of the ball court revealed three carved stone markers, each depicting two men playing the ball game. Curiously these are the only carved stones found at Lubaantun.

It wasn't until 1924 that F.A. "Mike" Mitchell-Hedges arrived at Lubaantun to help Thomas Gann with the excavation of the city. In 1927, while digging in a collapsed altar and adjoining wall, Mitchell-Hedges' adopted daughter, Anna, discovered the life-size crystal skull on her seventeenth birthday. Three months later a matching jawbone was discovered 25 feet from the altar. And thus, one of the world's strangest ancient objects came to popular attention.

The age of the skull is unknown. Rock crystal cannot be dated by conventional means. Hewlett-Packard laboratories, which studied the skull, estimated that its completion would have required a minimum of 300 years' work by a series of extremely gifted artisans. On the hardness scale, rock crystal ranks only slightly below diamonds.

The mystery of the skull deepened when it was discovered that the jawbone was carved from the same piece of crystal and that when the two pieces were attached, the cranium rocked on the jawbone base, giving the

Mike Mitchell-Hedges, Lady Richmond Brown, and Thomas Gann at Lubaantun in 1927.

impression that the skull was talking by opening and closing its mouth. In this manner, the skull could have been manipulated as a temple oracle by priests.

Mayan relief of a crystal skull.

Even more incredible properties are ascribed to the skull. The frontal lobe is said to cloud over sometimes, turning milky white. At other times the skull is said to emit an aura of light, "strong with a faint trace of the color of hay, similar to a ring around the moon."[94]

According to Frank Dorland, a crystalographer for Hewlett-Packard who studied the skull for years, the skull's eyes would sometimes flicker as if they were alive, and observers have reported strange sounds, odors, and various light effects emanating from the skull. Bizarre photographs have been taken of "pictures" which sometimes form within the skull. An example of such "pictures" forming within the skull are images of flying discs (UFO-flying saucers) and of what appears to be the Caracol observatory at the Toltec-Mayan site of Chichen Itza. In the past few years the skull has become quite famous because it has been displayed at Psychic Fairs in the USA and Canada. The skull now resides with Anna "Sammy" Mitchell-Hedges in Kitchener, Ontario or at her other home in the south of England.

F.A. Mitchell-Hedges was a fascinating person, and in some ways, his life was very much a model for an Indiana Jones sort of character. Born in 1882, "Mike" Mitchell-Hedges was destined for a life of adventure. He chronicles many of his adventures in his book *Danger My Ally*,[93] which was published in 1954. Mitchell-Hedges came to Canada and the United States in 1899, met with J.P. Morgan, won a fortune in a card game, and took off for Mexico. He was then captured and held prisoner by Pancho Villa, and later rode with Villa in Northern Mexico.

He eventually ended up in Central America. With his girlfriend, the wealthy Lady Richmond Brown (who was married at the time), he cruised the Caribbean, exploring the Bay Islands off Honduras, the San Blas Islands off Panama and the area around Jamaica.

He believed that artifacts which he had found in the Bay Islands pointed to a high civilization that was now beneath the water, and equated it with Atlantis. Mitchell-Hedges had a penchant for mystical sciences and secret societies, and championed the cause of lost civilizations and Atlantis. Eventually, he ended up in Lubaantun and the crystal was "discovered" in 1927.

Curiously, he devotes only three paragraphs to the famous crystal skull ancient societies were primitive, yet, we know that ancient steam engines, cogwheeled clocks and crystal skulls existed. What other secrets of high technology does the past have for us?

"The Skull of Doom is made of pure rock crystal and according to scientists it must have taken over 150 years, generation after generation working all the days of their lives, patiently rubbing down with sand an immense block of rock crystal until finally the perfect Skull emerged.

"It is at least 3,600 years old and according to legend was used by the High Priest of the Maya when performing esoteric rites. It is said that when he willed death with the help of the skull, death invariably followed. It has been described as the embodiment of all evil. I do not wish to try and explain this phenomena."[93]

Today, the skull continues to amaze audiences around the world and is frequently on television. Quartz crystals are today used in the highest forms of technology as well, as in computers and LCD watches.

The Crystal Skull, like other objects, is apparently a high-tech object from the past. The enigma of ancient technology is that we believe that the ancient societies were primitive, yet, we know that ancient steam engines, cogwheeled clocks and crystal skulls existed. What other secrets of high technology does the past have for us?

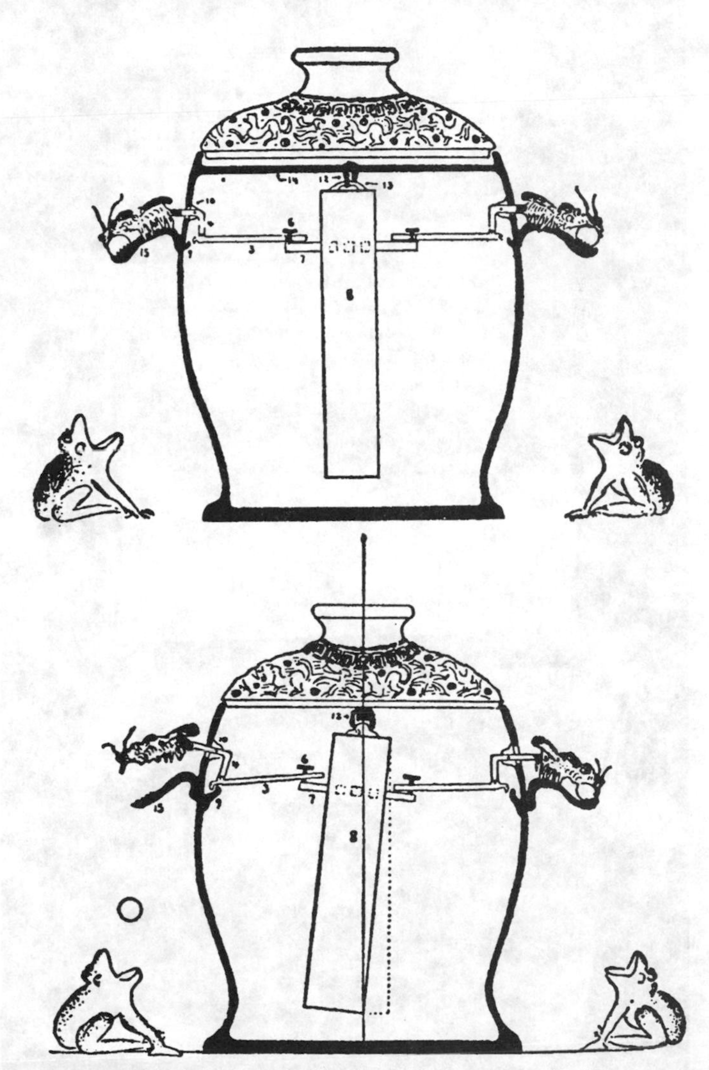

Earthquake detection device from China, circa 200 AD.

A scroll found in a secret library at Dunhuang in the Gobi Desert in the year 1900 by Sir Aurel Stein, who was working for the British Museum. It is in an unknown language. Most Ancient books were ordered destroyed in China.

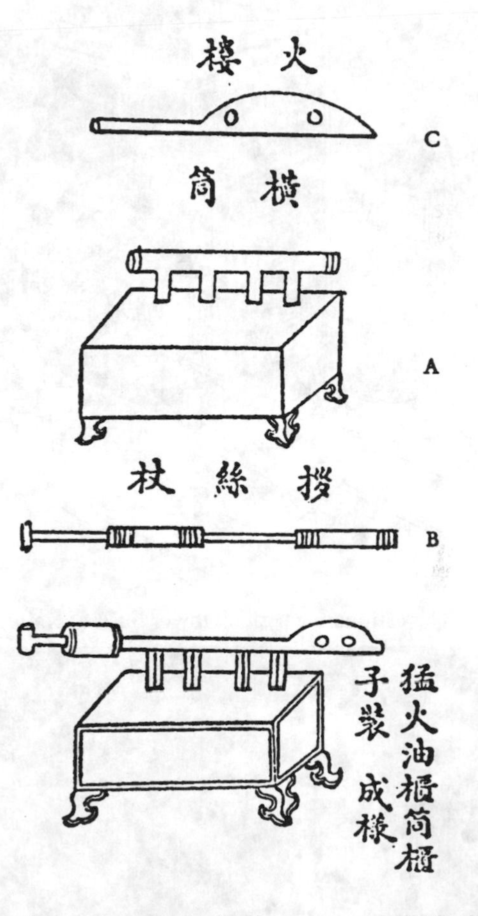

Chinese flame thrower device from circa 1040 AD. It used refined petroleum that was pumped out of a rectangular tank.

The astronomical clock tower built at Kaifeng.

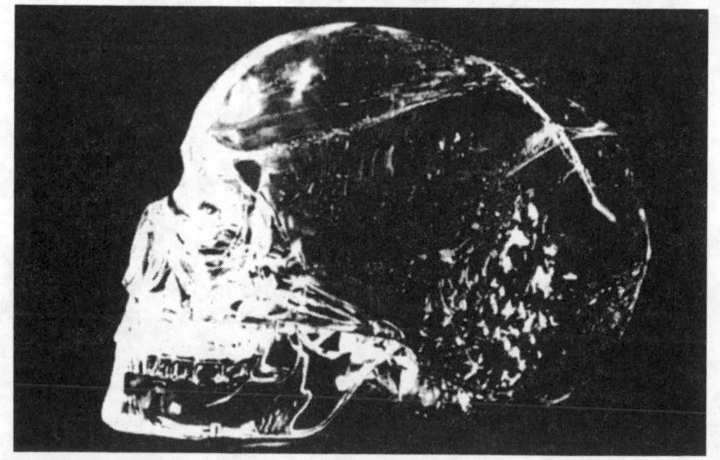

The enigmatic Mitchell-Hedges crystal skull. A relic from Atlantis?

2.
The Megalith Masterminds

Facts do not cease to exist because they are ignored.
—Aldous Huxley

Truth is one, but error proliferates.
Man tracks it down and cuts it up into
little pieces hoping to turn it into grains of truth.
—René Dauman, *The Way of the Truth*

Megalithomania

Legends of resplendent ancient civilizations and their cataclysmic destruction are part of nearly every culture in the world. The modern skeptic asks, "Well, if highly advanced ancient societies existed in the past then where is the evidence of their machinery and such; and shouldn't there be the remains of these peoples' cities?" The answer is that such evidence does exist, and hundreds of ruined cities have been found both above and below water.

The idea that man was primitive in the past and that the present represents the highest civilization our planet has achieved is fairly well accepted in the West, while other cultures view history as cyclical and see our current society as a decline from a former golden age. Out of the past we find megalithic cities built to last for thousands of years. How primitive must we suppose these people were?

Throughout the world, there exists a type of megalithic construction that is called "Atlantean" by those researchers who believe in advanced civilizations of the past. This is typically a type of construction which used gigantic blocks of stone, often crystalline granite. Huge blocks are fitted together without mortar in a polygonal style which tends to interlock the heavy blocks in a jigsaw fashion. These interlocked polygonal walls resist earthquake damage by moving with the shock wave of the quake. They momentarily jumble themselves and move freely but then fall back into place. These interlocked jigsaw walls will not collapse with an earthquake shock wave as with brick wall construction.

Such "Atlantean style" construction can be found all over the world. Classical examples of such construction are at Mycenae in the Peloponnese and the temples of Malta, along with the gigantic megalithic walls of Tiahuanaco, Ollantaytambo, Monte Alban and Stonehenge, as well as the pre-Egyptian structures of the Osirion at Abydos and the Valley Temple of the Sphinx.

Atlantean architecture is often circular, and uses the most exact rock cutting techniques to fit blocks together. Atlantean-type architecture often uses "keystone cuts"—identical shapes are cut into rock on both sides of a joint and the space is fitted with a metal clamp. These keystone cuts are typically an hourglass shape or a double-T shape. The clamps that went inside them may have been copper, bronze, silver, electrum (a mixture of silver and gold) or some other metal. In nearly every case where keystone cuts can be found, the metal clamp has already been removed—many thousands of years ago!

Many well-known, and not-so-well-known, ruins of the world contain the remains of even earlier cities within them. Such sites as Ba'albek in Lebanon, Cuzco in Peru, the Acropolis of Athens, Lixus in Morocco, Cadiz in Spain and even the Temple Mount of Jerusalem are built on gigantic remains of earlier ruins. Some modern cities, Cuzco is a good example, contain three or more levels of occupation, including modern occupants. Some archaeologists think the civilization that these earlier buildings are from is the "mythical" civilization of Atlantis.

Where then is Atlantis? Atlantis is all around us, asserted British scholar John Michell in his book *The View Over Atlantis.*[73] Michell further showed that amazing ancient ruins are a worldwide phenomenon in his *Megalithomania.*[77] Many authors have attempted to show how the worldwide distribution of megaliths points to an advanced civilization in an antediluvian sense, including such scholarly works as *Megaliths & Masterminds*[67] by Peter Lancaster Brown.

Their thesis was that the ancient world was remarkably advanced for its so-called Stone Age inheritance, and they argued that an advanced civilization called "Atlantis" preceded the dawn of history. The prehistoric civilization not only ranged worldwide but built impressive monuments and buildings as well.

The idea that man has only recently invented such things as electricity, generators, steam and combustion engines, or even powered flight does not necessarily hold true for a world that rides the rollercoaster of history.

Indeed, when we see how quickly inventions are absorbed into today's society we can imagine how quickly a highly scientific civilization may have arisen in remote antiquity. Just as today there are still primitive tribes in New Guinea and South America who live a Stone Age existence, so could Atlantis have existed during a period where other areas of the world lived in various states of development.

The ancient world of Atlantis may have been a lot like the modern world of today—juxtaposed between various factions in the government and military, while international discontent rises in various client colonies of an economic system set up by large business interests. According to the mythos that has built up around Atlantis, it was destroyed because of the wars that it had fought around the world. Today the world is again teetering on the brink of wholesale Armageddon because of political, religious and ethnic differences. Does modern man have something to gain by studying the past? Students of Atlantis believe so.

The Osirian Civilization

The Osirian Civilization, according to esoteric tradition, was an advanced civilization contemporary with Atlantis. In the world of about 15,000 years ago, there were a number of highly developed and sophisticated civilizations on our planet, each said to have a high degree of technology. Among these fabled civilizations was Atlantis, while another highly developed civilization existed in India. This civilization is often called the Rama Empire.

What is theorized is a past quite different from that which we have learned in school. It is a past with magnificent cities, ancient roads and trade routes, busy ports and adventurous traders and mariners. Much of the ancient world was civilized, and such areas of the world as ancient India, China, Peru, Mexico and Osiris were thriving commercial

Isis, Osiris, and Horus

centers with many important cities. Many of these cities are permanently lost forever, but others have been or will be discovered!

Horus.

It is said that at the time of Atlantis and Rama, the Mediterranean was a large and fertile valley, rather than a sea as it is today. The Nile River came out of Africa, as it does today, and was called the River Styx. However, instead of flowing into the Mediterranean Sea at the Nile Delta in northern Egypt, it continued into the valley, and then turned westward to flow into a series of lakes, to the south of Crete. The river flowed out between Malta and Sicily, south of Sardinia and then into the Atlantic at Gibraltar (the Pillars of Hercules). This huge, fertile valley, along with the Sahara (then a vast fertile plain), was known in ancient times as the Osirian civilization.

The Osirian civilization could also be called "Pre-Dynastic Egypt," the ancient Egypt that built the Sphinx and pre-Egyptian megaliths such as the Osirion at Abydos. In this outline of ancient history, it was the Osirian Empire that was invaded by Atlantis, and devastating wars raged throughout the world toward the end of the period of Atlantis' warlike imperial expansion.

Solon relates in Plato's dialogues that Atlantis, just near the cataclysmic end, invaded ancient Greece. This ancient Greece was one that the "ancient" Greeks knew nothing about. This "unknown ancient Greece," we shall see, is closely connected to the Osiris-Isis civilization.

The story of Osiris himself, as related by the Greek historian Plutarch, is revealing in technology. According to Egyptian mythology Osiris was born of the Earth and Sky, was the first king of Egypt and the instrument of its civilization. Osiris allegedly traveled throughout the world, teaching the arts of civilization after the flood. He weaned the inhabitants of Egypt from their barbarous ways, taught agriculture, formulated laws and taught the worship of the gods. Having accomplished this, he set off to impart his knowledge to the rest of the world.

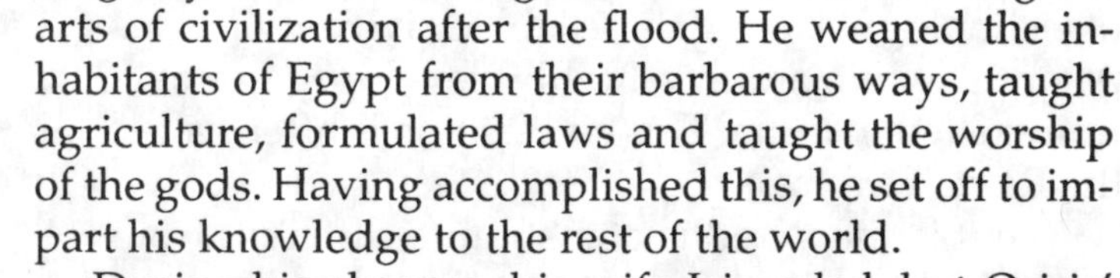

During his absence, his wife Isis ruled, but Osiris' brother and her brother-in-law, Typhon (also known as Set, and known to us as Satan) was always ready to disrupt her work. When Osiris returned from civilizing the world (or attempting to, at least), Set/Typhon/Satan decided he would kill Osiris and take Isis for him-

Osiris

self. He collected 72 conspirators to his plot and had a beautiful chest made to the exact measurements of Osiris. He threw a banquet and declared that he would give the chest to whomever could lie comfortably within it. When Osiris got in, the conspirators rushed to the chest and fastened the lid with nails. They then poured lead over the box and dumped it into the river where it was carried out to sea. When Isis heard of Osiris' death, she immediately set out to find her beloved.

A Winged Isis.

The box with Osiris in it came aground at Byblos in present day Lebanon, not too far from the massive slabs at Ba'albek. A tree grew around the box where it landed, and the king of Byblos had it cut down and used it as a pillar in his palace, Osiris still being inside. Isis eventually located Osiris and brought him back to Egypt, where Typhon (Set/Satan) broke into the box, chopped Osiris into fourteen different pieces, and scattered him about the countryside.

The loving Isis went looking for the pieces of her husband, and each time she found a piece she buried it—which is why there are temples dedicated to Osiris all over Egypt, and apparently other parts of the eastern Mediterranean as well. In another version, she only pretends to bury the pieces, in an attempt to fool Set/Typhon, and puts Osiris back together, bringing him back to life. Eventually she found all the pieces, except the phallus, and Osiris, one way or another, returned from the underworld and encouraged his son Horus (the familiar hawk-headed god) to avenge his death. Scenes in Egyptian temples frequently depict the hawk-headed Horus spearing a great serpent, Typhon or Set, in a scene that is identical to that of St. George and the dragon, though depicted thousands of years earlier.

In the happy ending, Isis and Osiris get back together, and have another child, Harpocrates. However, he is born prematurely and is lame in the lower legs as a result.[147,148]

There are many important themes in the legend of Osiris, including resurrection and the vanquishing of evil by good, and perhaps a key to the ancient Osirian civilization. Were the 14 scattered pieces of Osiris an allusion to 14 sacred sites built by the Osirians throughout the Mediterranean?

I have already mentioned the theory that the Mediterranean was once a fertile valley with many cities, farms and temples. Perhaps some of the 14 sites lie still undetected underwater and others are known but their full importance has not been identified. I believe that the early megalithic construction at Ba'albek, Jerusalem, Giza and the Osirion at Abydos would count as known sites among the number.

A key to the megalithic society of Osiris can be found in the curious buried ruins of the Osirion (the megalithic, pre-dynastic ruins at Abydos in southern Egypt). The British archaeologist Naville noted in a *London Illustrated News* article in 1914 that "here and there on the huge granite blocks was a thick knob... which was used for moving the stones. The blocks are very large—a length of fifteen feet is by no means rare; and the whole structure has decidedly the character of the primitive construction which in Greece is called cyclopean. An Egyptian example of which is at Ghizeh, the so-called temple of the Sphinx."[58]

Naville is directly relating the Osirion to the gigantic and prehistoric construction in Greece and also to the temple of the Sphinx. Other such sites around the former Osirian Empire are on the island of Malta, in Lebanon, in Israel, the Balearic Islands, and other areas of the Mediterranean. (In fact, virtually every Mediterranean island of any size has prehistoric megaliths on it.) Furthermore, the knobs, which may or may not be for moving the stones, are the same sort of knobs that occur on the gigantic stones that are used on the massive walls to be found in the vicinity of Cuzco, Peru.

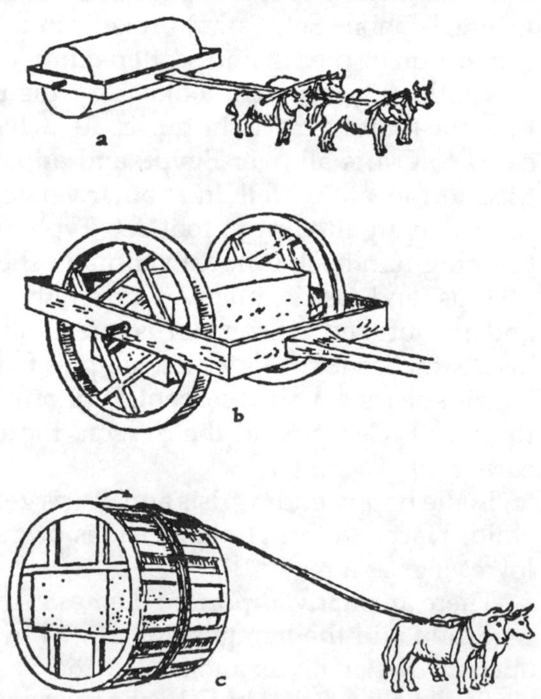

Various possible methods for moving large stones.

The lack of inscriptions indicates that the Osirion, like the valley temple of the Sphinx, was built before the use of hieroglyphics in Egypt! We know this because the Egyptians always engraved hierglyphics and decorations into any of their architecture. The only exceptions are those buildings, such as Great

Pyramid, the Osirion, and the Valley Temple of the Sphinx, that are thought by many archeologists now to be older than other structures. The Osirion is evidently a relic from the civilization of Osiris itself.

The present and the past are
perhaps both present in the future
and the future is contained in the past.
—T.S. Eliot

Ba'albek and Osiris

One of the most astonishing ancient ruins in the world is the megalithic base of Ba'albek, the pre-Roman ruins upon which a Roman-era temple sits.

The archaeological site of Ba'albek is 44 miles east of Beirut and consists of a number of ruins and catacombs. 2,500 feet long on each side, it is one of the largest stone structures in the world. A portion of it consists of gigantic cut stone blocks from some ancient time, forming a platform with a Roman temple built on top of it. The Roman temple to Jupiter and Venus was built on top of the earlier temples dedicated to the corresponding ancient deities—Ba'al and his partner, the goddess Astarte.[94]

The temples to Ba'al and Astarte may initially have been built as part of a prehistoric Sun Temple, and even then on the ruins of the more ancient structure, its purpose unknown. According to an article by Jim Theisen in the *INFO Journal*,[63] the Greeks called the temple "Heliopolis" which means "Sun Temple" or "Sun City." Even so, the original purpose of the gigantic platform may have been something else entirely.

Ba'albek is a good example of what happens to large, well-made ancient walls—they are used again and again by other builders who erect a new city or temple on top of the older one, using the handy stones that are already to be found at the site. Often, the original stones are so colossal that they could not be moved and placed elsewhere anyway. This is exactly what has been going on at many sites, in the Old World and in the Americas. Examples of very ancient stonework (3,000 to 6,000 years old) mixed with more recent ancient stonework (500 to 2,500 years ago) can be seen at Monte Alban in Mexico and at such Andean sites as Chavin, Cuzco and Ollantaytambo.

At Ba'albek, the Roman architecture (largely destroyed by an earthquake in 1759) does not pose any archaeological

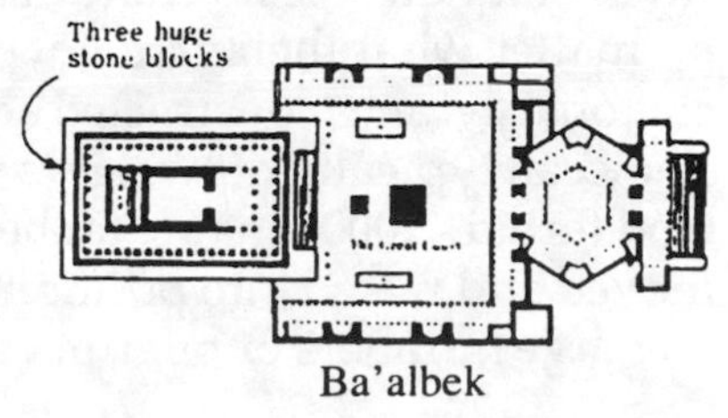

Ba'albek

problems, but the massive cut stone blocks beneath it certainly do. One part of the enclosure wall, called the Trilithon, is composed of three blocks of hewn stone that are the largest stone blocks ever used in construction on this planet, so far as is known (underwater ruins may reveal larger constructions). This is an engineering feat that has never been equaled in history.

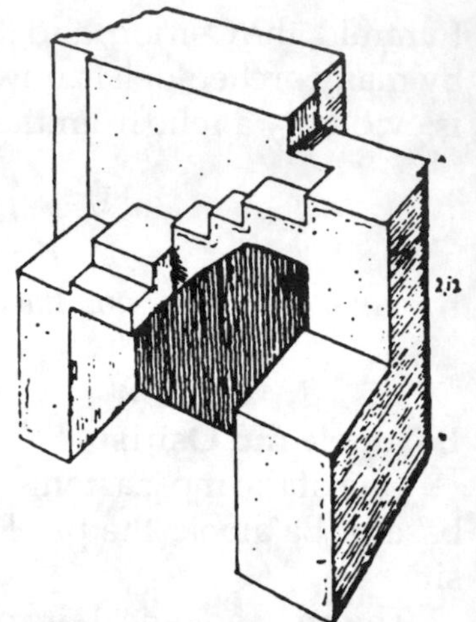

One stone in the interior staircase at Ba'albek.

The weight and even size of the stones is open to controversy. According to the author Rene Noorbergen in his fascinating book *Secrets of the Lost Races,*[32] the individual stones are 82 feet long and 15 feet thick and are estimated to weigh between 1,200 and 1,500 tons each (a ton is 2,000 pounds, which would make the blocks weigh an estimated 2,400,000 to 3,000,000 pounds each). While Noorbergen's size may be incorrect, his weight is probably closer to the truth. Even conservative estimates say that the stones weigh at least 750 tons each, which would be one and a half million pounds.[32]

It is an amazing feat of construction, for the blocks have been raised more than 20 feet in order to lie on top of smaller blocks. The colossal stones are fitted together perfectly, and not even a knife blade can be fitted between them.[32] Even the blocks on the level below the Trilithons are incredibly heavy. At 13 feet in length, they probably weigh about 50 tons each, an extremely large-sized bunch of stones by any other estimate, except when compared to the Trilithons. Yet, even the Trilithons are not the largest of the stones!

The largest hewn block, 13 feet by 14 feet and nearly 70 feet long and weighing at least 1,000 tons (both Noorbergen and Berlitz give the weight of this stone at 2,000 tons[32]), lies in the nearby quarry which is half a mile away. 1,000 tons is an incredible two million pounds! The stone is called *Hadjar el Gouble,* Arabic for "Stone of the South." Noorbergen is correct in saying that there is no crane in the world that could lift any of these stones, no matter what their actual weight is. The largest cranes in the world are stationary cranes constructed at dams to lift huge concrete blocks into place. They can typically lift weights up to several hundred tons. 1,000 tons, and God forbid, 2,000 tons are far beyond their capacity. How these blocks were moved and raised into position is beyond the comprehension of engineers.

Large numbers of pilgrims came from Mesopotamia as well as the Nile

Valley to the Temple of Ba'al–Astarte. The site is mentioned in the Bible in the *Book of Kings*. There is a vast underground network of passages beneath the acropolis. Their function is unknown, but they were possibly used to shelter pilgrims, probably at a later period.

Who built the massive platform of Ba'albek? How did they do it? According to ancient Arab writings, the first Ba'al–Astarte temple, including the massive stone blocks, was built a short time after the Flood, at the order of the legendary King Nimrod, by a "tribe of giants."[32]

But, it could be older, because history shows that some rulers liked to claim monuments that were built by others. The mythical king Nimrod, a figure so old in history that he is lost to us, may have been laying claim to the stones of Ba'albek circa 6000 BC, when the building itself had been constructed in 12000 BC or more, before the flood.

Ancient astronaut theorists have frequently suggested that Ba'albek was built by extraterrestrials. Charles Berlitz says that a Soviet scientist named Dr. Agrest suggests that the stones were originally part of a landing and takeoff platform for extraterrestrial spacecraft.[3] Author and Sumerian scholar Zechariah Sitchin believes, in a like manner, that Ba'albek is a launching pad for rockets.

Like Buddha seeking the "middle path," I seek a middle ground in this intriguing mystery of the past. While ancient astronauts may well have visited earth in the past, it seems unlikely that they would have arrived here in rockets. They would have mastered the art of anti-gravity and their spaceships would be electric solid-state models, at the very least. Such aircraft could land and take off in a pleasant grassy field, and would not need a gigantic platform.

What then was Ba'albek and who built it? The theory of Ba'albek being some remnant of the Osirian Empire, along with some of the other megalithic sites in the Mediterranean, fits in well with the Arab legend mentioned previously: that the massive stone blocks were built a short time after the Flood, at the order of King Nimrod.[94]

The largest stones of Ba'albek.

Yet, even if Ba'albek is a remnant from the Osirian civilization, how were such huge blocks transported and lifted? One clue is the massive block that still remains at the quarry a half mile away. This stone was apparently meant to take its place on the platform with the other stones, but for some reason never made it to the site. According to the *INFO* article,[32] the largest stones used in the Great Pyramid of Egypt only weigh about 400,000 pounds (these are several large granite blocks in the interior of the pyramid). The authors point out that until NASA moved the gigantic Saturn V rocket to its launch pad on a huge tracked vehicle, no man had transported such a weight as the blocks at Ba'albek.

In his book *Baalbek*[154] archaeologist Friedrich Ragette attempts to explain how Ba'albek was built and how the stones were moved into place. Explaining Ba'albek is no easy task, Ragette admits, but he does his best.

Ragette first explains that there are two quarries, one about 2 km. north of Ba'albek and a closer quarry where the largest stone block in the world still lies. He then makes this interesting remark about the quarries: "After the block was separated on its vertical side, a groove was cut along its outer base and the piece was felled like a tree on to a layer of earth by means of wedging action from behind. It seems that the Romans also employed a sort of quarrying machine. This we can deduce from the pattern of concentric circular blows shown on some blocks. They are bigger than any man could have produced manually, and we can assume that the cutting tool was fixed to an adjustable lever which would hit the block with great force. Swinging radii of up to 4 m (13 ft) have been observed."[154]

Ragette goes on to theorize that moving an 800-ton stone on rollers would be possible: "[I]f we assume that the block rested on neatly cut cylindrical timber rollers of 30 cm (12 in) diameter at half-metre distances, each roller would carry 20 tons. If the contact surface of the roller with the ground were 10 cm (4 in) wide, the pressure would be 5 kg/cm^2 (71 lbs/in^2), which requires a solid stone paving on the ramp. The theoretical force necessary to move that block horizontally would be 80 tons. Another possibility is that the whole block was encased in a cylindrical wrapping of timber and iron braces." Ragette dismisses this second idea as unlikely and cumbersome. "Also there remains the question of how the block would have been unwrapped and put in place, which brings us to the even more perplexing problem of lifting great weights."[154]

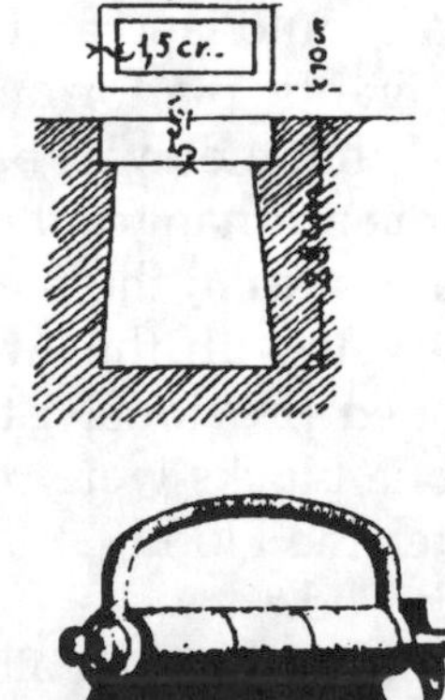

"Lewis" stone cuts & pieces.

There is, however, no evidence of an ancient road, which would have to have been paved, says

Ragette. According to the *INFO* article, "one sees no evidence of a road connecting the quarry and Temple. Even if a road existed, logs employed as rollers would have been crushed to a pulp. But, obviously, someone way back then knew how to transport million-pound stones."[32]

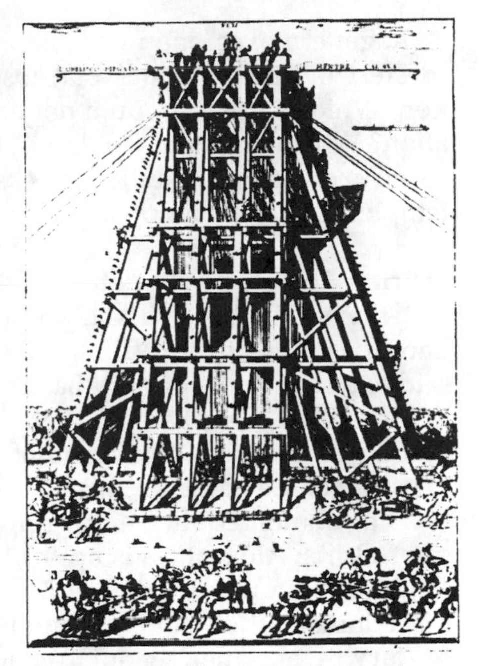
Erecting an obelisk during Roman times.

There is not a contractor today that would attempt to move or lift these stones. It is simply beyond our modern machine technology. I find it interesting that there is no discernible road between the quarry and the massive Sun Temple. This indicates one or both of two possibilities: the building of the lower platform occurred at such an ancient time in antiquity that the road is long gone, or a road was never needed for transporting the block. As the *INFO* article points out, a road would have been of little use anyway.

Ragette cannot solve the problem of lifting such a block into place, saying that it is impossible to lift a huge block such as this completely off the ground by the use of levers. He says that we know that the stone had to be lifted so that the log rollers could be removed from underneath the block and then the block lowered into place. In order to fit perfectly, the stone probably had to be lifted and lowered into place several times at least.

His suggestion is that a giant lifting frame was built around the block and then at least 160 "Lewis" stones—wedge-shaped keystones with metal loops—were inserted into the top of the block. Then a system of pulleys and tackles were used with thousands of manual workers to raise and lower the gigantic blocks a few inches.

Ragette makes no suggestion as to why the Romans, or anyone else, would go to such immense trouble, attempting a virtually impossible engineering feat, in order to lay the foundation for a temple to Jupiter. If they had cut the stone into, say, 100 pieces, they would still be of unusually large size, larger than a man, but at least could have been stacked into a wall much more easily. One is left with the unsettling thought that the reason they used these huge stones was because they *could* use them—and do it relatively easily, though today we have no idea how.

Ragette makes one final interesting comment on Ba'albek: "The real mystery of Baalbek is the total absence of written records on its construction. Which emperor would not have wanted to share the fame of its creation? Which architect would not have thought of proudly inscribing his name in one of the countless blocks of stone? Yet, nobody lays claim to the temples. It is as if Heliopolitan Jupiter alone takes all the credit."[154]

Osirian Remains in Egypt

Other vestiges of Osiris still exist in the eastern Mediterranean. The foundation ashlars of the Wailing Wall at Jerusalem are also gigantic blocks said to be similar to those at Ba'albek. Megalithic ruins found under water at Alexandria, Egypt are also believed to predate the dynastic Egypt of the Pharaohs. It is from the legend of Osiris and the many "tombs of Osiris" that we get the name for this lost civilization from the time of Atlantis.

The submerged megalithic ruins at Alexandria are another clue to ancient Osiris. Alexandria is not really an Egyptian city, it is Greek. As one might easily guess, Alexandria is named after Alexander the Great, the Macedonian king who first conquered the city-states of Greece in the 3rd century BC and then set out to conquer the rest of the world, starting with Persia. Persia was Egypt's traditional enemy, and so Egypt fell willingly into Alexander's hands. He went to Memphis near modern-day Cairo and then descended the Nile to the small Egyptian town of Rhakotis. Here he ordered his architects to build a great port city, what was to be Alexandria.

Alexander then went to the temple of Ammon in the Siwa Oasis where he was hailed as the reincarnation of a god, which is to say, some great figure from ancient Osiris or Atlantis. Which god, we do not know. He hurried on to conquer the rest of Persia and then India. Eight years after leaving Alexandria, he returned to it, in a coffin. He never saw the city, though his bones are said to rest there to this day (though no one has ever found the tomb).

Of all the mysteries of Alexandria, however, none is more intriguing than the megalithic ruins which lie to the west of the Pharos lighthouse near the promontory of Ras El Tin. Discovered at the turn of the last century by the French archaeologist M. Jondet and discussed in his paper "Les Ports submerges de l'ancienne Isle de Pharos,"[152] the prehistoric port is a large section of massive stones that today is completely submerged. Near it was the legendary Temple of Poseidon, a building now lost, but known to us

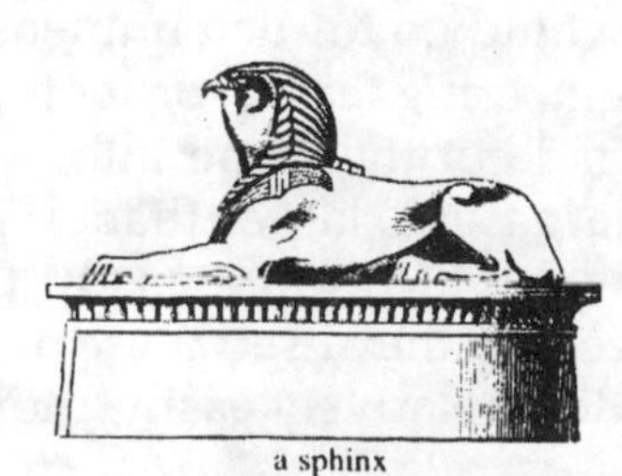
a sphinx

in literature.[152]

The Theosophical Society, upon learning of the submerged harbor of megaliths, quickly ascribed it to Atlantis. M. Jondet theorized that it might be of Minoan origin, part of a port for Cretan ships. E.M. Forster theorizes, in his excellent guide to Alexandria,[152] that it may be of ancient Egyptian origin, built by Ramses II circa 1300 BC. Most of it lies in 4 to 25 feet of water and stretches for 70 yards from east to west, curving slightly to the south.

Probably the true origin of the massive, submerged harbor, which was definitely at least partially above water at one time, is a blend of M. Jondet's theory of Minoan builders and the Theosophical Society's belief that it is from Atlantis.

In theory, with the Mediterranean slowly filling up with water, the sea would have stabilized after a few hundred years, and then the remnants of the Osirians, using a technology and science similar to that of Atlantis, built what structures and ports they could. Later, in another tectonic shift, the port area (probably used by pre-dynastic Egyptians) was submerged, and was then essentially useless.

It is interesting to note, with regard to this theory, that a temple to Poseidon was located at the tip of Ras El Tin. Atlantis was known to the ancients as Poseid, and "Poseidonis" or "Poseidon" was a legendary king of Atlantis. Similarly, Poseidonis and Osiris are thought to be the same person. The main temple at Rhakotis, the Egyptian town which Alexander found at the ancient harbor, was naturally dedicated to Osiris.

What we are learning about the megalithic masterminds is that their buildings occur all over the world, and many of them are underwater and difficult to reach!

The opposite of a correct statement is a false statement.
But the opposite of profound truth may be another profound truth.
—Niels Bohr

The Sunken Temples of Carnac

Located on the south coast of Brittany, France, the site of Carnac holds the greatest concentration of megaliths in the world. Conservative estimates claim that megaliths were being erected here by 5000 BC, nearly seven thousand years ago. They may be much older.

The Grand Menhir Brisé at Er Grah, Brittany, is said to be the largest menhir in the world and is situated on a promontory near the water. The problems of moving such a huge stone are illustrated in an article published in the *Journal for the History of Astronomy* (No. 2, pages 147-160, 1971) entitled "The Astronomical Significance of the Large Carnac Menhirs." The astronomers, Mr. and Mrs. Thom, maintain that the megalith was a lunar sighting stone. They write:

> Er Grah, or The Stone of the Fairies, sometimes known as Le Grand Menhir Brisé, is now broken in four pieces which when measured show that the total length must have been at least 67 ft. From its cubic content it is estimated to weigh over 340 tons.
>
> Hulle thinks it came from the Cote Sauvage on the west coast of the Quiberon Peninsula. His suggestion that it was brought round by sea takes no account of the fact that the sea level relative to this coast was definitely lower in Megalithic times; neither does he take account of the fact that a raft of solid timber about 100 x 50 x 4 ft. would be necessary—with the menhir submerged. It is not clear how such a raft could be controlled or indeed moved in the tidal waters round the Peninsula.
>
> Assuming that the stone came by land, a prepared track (? of timber) must have been made for the large rollers necessary and a pull of perhaps 50 tons applied (how?) on the level, unless indeed the rollers were rotated by levers. It took perhaps decades of work and yet there it lies, a mute reminder of the skill, energy and determination of the engineers who erected it more than three thousand years ago.
>
> In Britain we find that the tallest stones are usually lunar backsights, but there seems no need to use a stone of this size as a backsight. If, on the other hand, it was a foresight, the reason for its position and height becomes clear, especially if it was intended as a universal foresight to be used from several directions. There are eight main values to consider, corresponding to the rising and setting of the Moon at the

An old print of an inspection of Carnac.

standstills when the declination was plus or minus. …It has now been shown that there is at least one site on each of the eight lines which has the necessary room for side movement.

We must now try to think of how a position was found for Er Grah which would have satisfied the requirements. Increasingly careful observations of the Moon had probably been made for hundreds of years. These would have revealed unexplained anomalies due to variations in parallax and refraction, and so it may have been considered necessary to observe at the major and minor standstills at both rising and setting. At each standstill there were 10 or 12 lunations when the monthly declination maximum and minimum could be used. At each maximum or minimum, parties would be out at all possible places trying to see the Moon rise or set behind high trial poles. At night these poles would have needed torches at the tops because any other marks would not be visible until actually silhouetted on the Moon's disc. Meantime some earlier existing observatory must have been in use so that erectors could be kept informed about the kind of maximum which was being observed; they would need to know the state of the perturbation.

Then there would ensue the nine years of waiting till the next standstill when the other four sites were being sought. The magnitude of the task was enhanced by the decision to make the same foresight serve both standstills. We can understand why this was considered necessary when we think of the decades of work involved in cutting, shaping, transporting and erecting one suitable foresight. It is evident that whereas some of the sites, such as Quiberon, used the top of the foresight of Er Grah, others, such as Kerran, used the lower portion. This probably militated against the use of a mound with a smaller menhir on the top. Much has rightly been written about the labour of putting Er Grah in position, but a full consideration of the labour of finding the site shows that this may have been a comparable task.

We now know that for a stone 60 ft. high the sighting is perfect. We do not know that all the backsights were completed. But the fact that we have not yet found any trace of a sector to the east does not prove that the eastern sites were not used because the stones may have been removed. Perhaps the extrapolation was done by the simpler triangle method or perhaps it was done at a central site like Petit Menec.[17]

Francis Hitching in *Earth Magic*[244] also agrees that this was a central sighting megalith for sighting moonrises and moonsets.

Much of this gigantic astronomical observatory is probably under water. Many of the megaliths along the Brittany coast are apparently submerged. Many famous sites actually lead into the water, and some megaliths can be seen at low tide when they are barely above the surface.

Many of the long lines of standing stones at Carnac and around the Morbihan Gulf were apparently built when the geography of Brittany was quite different.

Near the town of Carnac is the famous alignment of hundreds of standing stones. They too are apparently part of some huge astronomical observatory. In another article by the Thoms entitled "The Carnac Alignments" for the *Journal for the History of Astronomy* (No. 3, pages 11-26, 1972),[17] they conclude that Carnac is also a lunar observatory of vast proportions. Say the Thoms about the Menec alignments at Carnac, "A remarkable feature is the great accuracy of measurement with which the rows were set out. It cannot be too strongly emphasized that the precision was far greater than could have been achieved by using ropes. The only alternative available to the erectors was to use two measuring rods (of oak or whale bone?). These were probably 6.802 ft. long, shaped on the ends to reduce the error produced by malalignment. Each rod would be rigidly supported to be level but we can only surmise how the engineers dealt with the inevitable 'steps' when the ground was not level.

"It may be noted that the value for the Megalithic yard found in Britain is 2.720 plus or minus 0.003 ft. and that found above is 2.721 plus or minus 0.001 ft. Such accuracy is today attained only by trained surveyors using good modern equipment. How then did Megalithic Man not only achieve it in one district but carry the unit to other districts separated by greater distances? How was the unit taken, for example, northwards to the Orkney Islands? Certainly not by making copies of copies of copies. There must have been some apparatus for standardizing the rods which almost certainly were issued from a controlling, or at least advising, centre."[17]

The Thoms see Carnac as part of an ancient and huge system that was used over much of Europe. In their article they conclude, "The organization and administration necessary to build the Breton alignments and erect Er Grah obviously spread over a wide area, but the evidence of the measurements shows that a very much wider area was in close contact with the central control. The geometry of the two egg-shaped cromlechs at Le Menec is identical with that found in British sites. The apices of triangles with integral sides forming the centres for arcs with integral radii are features in common, and on both sides of the Channel the perimeters are multiples of the rod.

"The extensive nature of the sites in Brittany may suggest that this was the main centre, but we must not lose sight of the fact that so far none of the Breton sites examined has a geometry comparable with that found at Avebury in complication of design, or in difficulty of layout.

"It has been shown elsewhere that the divergent stone rows in Caithness could have been used as ancillary equipment for lunar observations, and in our former paper we have seen that the Petit Menec and St. Pierre sites were probably used in the same way." The Thoms confess at the end of their article, "We do not know how the main Carnac alignments were used..."[17]

Carnac likens itself to the important Egyptian Temple of Karnak. The Egyptian Karnak is a huge building which also has long rows of megalithic columns which once supported a huge roof.

Are there other, even larger menhirs under the water near Carnac? One example of a known submerged megalithic structure is the Covered Alleyway of Kernic in the District of Plousescat, Finistére, now submerged at high tide.[250]

The Amazing Megaliths of the Andes

At a leveling-off of a hill overlooking the Cuzco Valley in Peru, is a colossal fortress called Sacsayhuaman, one of the most imposing edifices ever constructed. Sacsayhuaman consists of three or four terraced walls going up the hill and the ruins include doorways, staircases and ramps.

Gigantic blocks of stone, some weighing more than 200 tons (400 thousand pounds) are fitted together perfectly. The enormous stone blocks are cut, faced, and fitted so well that even today one cannot slip the blade of a knife, or even a piece of paper between them. No mortar is used, and no two blocks are alike. Yet they fit perfectly, and it has been said by some engineers that no modern builder with the

An old photo of a wall in Cuzco.

aid of metals and tools of the finest steel, could produce results more accurate.

Each individual stone had to have been planned well in advance; a twenty-ton stone, let alone one weighing 80 to 200 tons, cannot just be dropped casually into position with any hope of attaining that kind of accuracy! The stones are locked and dove-tailed into position, making them earthquake-proof. Indeed, after many devasting earthquakes in the Andes over the last few hundred years, the blocks are still perfectly fitted, while the Spanish Cathedral in Cuzco has been leveled twice.

Even more incredibly, the blocks are not local stone, but by some reports come from quarries in Ecuador, almost 1,500 miles away! Others have located quarries a good deal closer, only five miles or so away. Though this fantastic fortress was supposedly built just a few hundred years ago by the Incas, they leave no record of having built it, nor does it figure in any of their legends. How is it that the Incas, who reportedly had no knowlege of higher mathematics, no written language, no iron tools, and did not even use the wheel, are credited with having built this cyclopean complex of walls and buildings? Frankly, one must literally grope for an explanation, and it is not an easy one.[57]

When the Spaniards first arrived in Cuzco and saw these structures, they thought that they had been built by the devil himself, because of their enormity. Indeed, nowhere else can you see such large blocks placed together so perfectly. I have traveled all over the world searching for ancient mysteries and lost cities, but I had never in my life seen anything like this!

The builders of the stoneworks were not merely good stone masons—they were beyond compare! Similar stoneworks can be seen throughout the Cuzco Valley. These are usually made up of finely-cut, rectangular blocks of stone weighing up to perhaps a ton. A group of strong people could lift a block and put it in place; this is undoubtably how some of the smaller structures were put together. But in Sacsayhuaman, Cuzco, and other ancient Inca cities, one can see gigan-

The strange ruins of the Nekromonteion in northern Greece, which look identical to ruins around Cuzco.

tic blocks cut with 30 or more angles on each one.

At the time of the Spanish conquest, Cuzco was at its peak, with perhaps 100,000 Inca subjects living in the ancient city. The fortress of Sacsayhuaman could hold the entire population within its walls in case of war or natural catastrophe. Some historians have stated that the fortress was built a few years before the Spanish invasion, and that the Incas take credit for the structure. But, the Incas could not recall exactly how or when it was built!

Massive walls at Ollantaytambo.

Only one early account survives of the hauling of the stones, found in Garcilaso de la Vega's *The Incas.*[145] In his commentaries, Garcilaso tells of one monstrous stone brought to Sacsayhuaman from beyond Ollantaytambo, a distance of about 45 miles. "The Indians say that owing to the great labor of being brought on its way, the stone became weary and wept tears of blood because it could not attain to a place in the edifice. The historical reality is reported by the Amautas (philosophers and doctors) of the Incas who used to tell about it. They say that more than twenty-thousand Indians brought the stone to the site, dragging it with huge ropes. The route over which they brought the stone was very rough. There were many high hills to ascend and descend. About half the Indians pulled the stone, by means of ropes placed in front. The other half held the stone from the rear due to fears that the stone might break loose and roll down the mountains into a ravine from which it could not be removed.

"On one of these hills, due to lack of caution and co-ordination of effort, the massive weight of the stone overcame some who sustained it from below. The stone rolled right down the hillside, killing three- or four-thousand Indians who had been guiding it. Despite this misfortune, they succeeded in raising it up again. It was placed on the plain where it now rests."[145]

Even though Garcilaso describes the hauling of one stone, many doubt the truth of this story. This stone was not part of the Sacsayhuaman fortress, and is smaller than most used there, according to some researchers, although the stone has never been positively identified. Even if the story is true, the Incas may have been trying to duplicate what they supposed was the construction technique used by the ancient builders. While there is no denying that the Incas were master craftsmen, if one credits this tale one would have to wonder how they would have transported and placed the 100-ton blocks so perfectly, given the trouble they had with only *one* stone.

That the Incas actually found these megalithic ruins and then built on top of them, claiming them as their own, is not a particularly alarming theory. In fact, it is most probably the truth. It was a common practice in ancient Egypt for rulers to claim previously existing obelisks, pyramids, and other structures as their own, often literally erasing the cartouche of the real builder and subsituting theirs. Indeed, the Great Pyramid itself would seem a victim of such a ruse. The pharoah Kufu, or Cheops as he was known in Greek, had his cartouche chiseled into the Great Pyramid at its base. This is the only writing to be found anywhere on the pyramid, but every indication is that the pyramid was not built by Cheops. It may not have ever been meant to be a tomb, but that is another story.

If the Incas came along and found walls and basic foundations of cities already in existance, why not just move in? Even today, all one needs to do is a little repair work and add a roof on some of the structures to make them habitable. Indeed, there is considerable evidence that the Incas merely found the structures and added to them. There are numerous legends that exist in the Andes that Sacsayhuaman, Machu Picchu, Tiahuanaco, and other megalithic remains were built by a race of giants. Alain Gheerbrant comments in his footnotes to de la Vega's book, "Three kinds of stone were used to build the fortress of Sacsayhuaman. Two of them, including those which provided the gigantic blocks for the outer wall, were found practically on the spot. Only the third kind of stone (black andesite), for the inside buildings, was brought from relatively distant quarries; the nearest quarries of black andesite were at Huaccoto and Rumicolca, nine and twenty-two miles from Cuzco respectively.

"With regard to the giant blocks of the the outer wall, there is nothing to prove that they were not simply hewn from a mass of stone existing on the spot; this would solve the mystery."[57]

Gheerbrant is close in thinking that the Incas never moved those gigantic blocks in place, yet even if they did cut and dress the stones on the spot, fitting them together so perfectly

Ruins near Chavin in Peru. They look as if they had some kind of machinery. From Squire's Peru (1886).

would still require what modern engineers would call superhuman effort. Furthermore, the gigantic city of Tiahuanaco in Bolivia is similarly hewn from 100-ton blocks of stone. The quarries are many miles away, and the site is definitely of pre-Inca origin. Proponents of the theory that the Incas found these cities in the mountains and inhabited them, would then say that the builders of Tiahuanaco, Sacsayhuaman, and other megalithic structures in the Cuzco area were the same people.

Again quoting Garcilaso de la Vega, who wrote about these structures just after the conquest, "...how can we explain the fact that these Peruvian Indians were able to split, carve, lift, carry, hoist, and lower such enormous blocks of stone, which are more like pieces of a mountain than building stones, and that they accomplished this, as I said before, without the help of a single machine or instrument? An enigma such as this one cannot be easily solved without seeking the help of magic, particularly when one recalls the great familiarity of these people with devils."[145]

The Spanish dismantled as much of Sacsayhuaman as they could. When Cuzco was first conquered, Sacsayhuaman had three round towers at the top of the fortress, behind three concentric megalithic walls. These were taken apart stone by stone, and the stones used to build new structures for the Spanish.

One interesting theory about the building of the gigantic and perfectly fitted stones is that they were constructed by using a now-lost technique of softening and shaping the rock. Hiram Bingham, the discoverer of Machu Picchu, wrote in his book *Across South America,* of a plant he had heard of whose juices softened rock so that it could be worked into tightly fitted masonry.[88, 57]

In his book *Exploration Fawcett,*[88] Colonel Fawcett told of how he had heard that the stones were fitted together by means of a liquid that softened stone to the consistency of clay. Brian Fawcett, who edited his father's book, tells the following story in the footnotes: A friend of his who worked at a mining camp at 14,000 feet at Cerro di Pasco in Central Peru, discovered a jar in an Incan or pre-Incan grave. He opened the jar, thinking it was chicha, an alcoholic drink, breaking the still intact ancient wax seal. Later, the jar was accidentally knocked over onto a rock.

Quotes Fawcett, "About ten minutes later I bent over the rock and casually examined the pool of spilled liquid. It was no longer liquid; the whole patch where it had been, and the rock under it, were as soft as wet cement! It was as though the stone had melted, like wax under the influence of heat."[88, 57]

Fawcett seemed to think that the plant might be found on the Pyrene River in the Chuncho country of Peru, and described it as having dark reddish leaves and being about a foot high. Another story is mentioned of a biologist observing an unfamiliar bird in the Amazon. He watched it making a nest on a rock face by rubbing the rock with a twig. The sap of the twig dissolved the the rock, making a hollow in which the bird could make its nest.

All of this speculation may be put to rest by more recent findings, reported in *Scientific American* (February, 1986). In a fascinating article, a French researcher, Jean-Pierre Protzen, relates his experiments in duplicating the construction of Sacsayhuaman and Ollantaytambo. Protzen spent many months around Cuzco experimenting with different methods of shaping and fitting the same kinds of stones used by the Incas (or their megalithic predicessors). He found that quarrying and dressing the stones could be accomplished using the stone hammers found in abundance in the area. The precision fitting of stones was a relatively simple matter, he says. He pounded out the concave depressions into which new stones were fitted by trial and error, until he achieved a snug fit. This meant continually lifting and placing the stones together, and chipping at them a little at a time. This process is very time consuming, but it's simple, and it works.

Yet even for Protzen, some mysteries remain. He was not able to figure out how the builders handled the larger stones. The fitting process necessitated the repeated lowering and raising of the stone being fitted, with trial-and-error pounding in between. He does not know just how 100-ton stones were manipulated at this stage, and some stones are actually far heavier.

According to Protzen, to transport the stones from the quarries, special access roads and ramps were built. Many of the stones were dragged over gravel-covered roads, which in his theory gave the stones their polished surfaces. The largest stone at Ollantaytambo weighs about 150 tons. It could have been pulled up a ramp with a force of about 260,000 pounds, he says. Such a feat would have required a mimimum of some 2,400 men. Getting the men seemed possible, but where did they all stand? Protzen says that the ramps were only eight meters wide at most. Further perplexing Protzen is that the stones of Sacsayhuaman were finely dressed, yet are not polished, showing no signs of dragging. He could not figure out how they were transported the 22 miles from the Rumiqolqa quarry.

Protzen's article reflects good research, and points out that modern science still cannot explain or duplicate the building feats found at both Sacsayhuaman and Ollantaytambo. Continually lifting and chipping away at a 100-ton stone block to make it fit perfectly is just too great of an engineering task to have been practical. Protzen's theory would work well on the smaller, precisely square, later construction, but fails with the older megalithic construction beneath. Perhaps the theories of levitation and softening stones cannot be discarded yet! One last intriguing observation which Protzen makes is that the cutting marks found on some of the stones are very similar to those found on the pyramidion of an unfinished obelisk at Aswan in Egypt. Is this a coincidence, or was there an ancient civilization with links to both sites?

Most "scientists" are bottlewashers and button sorters.
—Robert Heinlein

The World's Largest Computer

The magnificent monument in England called Stonehenge sits alone on the Salisbury Plain, flanked by a parking lot and gift shop for tourists. It is famous for its large stones and curious architecture: a circle of masssive, well-cut stones.

In 1964 the British astronomer Gerald S. Hawkins first published his now-famous treatise on Stonehenge as an astronomical computer. His article, entitled "Stonehenge: A Neolithic Computer," appeared in issue 202 of the prestigious British journal *Nature.* In 1965, Hawkins' famous book *Stonehenge Decoded* was published.[95]

Hawkins upset the archaeological world by claiming that the megalithic site was not just a circular temple erected by some egocentric kings, but rather a sophisticated computer for observing the heavens.

He begins his *Nature* article with a quote from Diodorus on prehistoric Britain from his *History of the Ancient World,* written about 50 BC: "The Moon as viewed from this island appears to be but a little distance from the Earth and to have on it prominences like those of the Earth, which are visible to the eye. The account is also given that the god [Moon?] visits the island every 19 years, the period in which the return of the stars to the

same place in the heavens is accomplished... There is also on the island, both a magnificent sacred precinct of Apollo [Sun] and a notable temple... and the supervisors are called Boreadae, and succession to these positions is always kept in their family."

Hawkins' basic theory was that "Stonehenge was an observatory; the impartial mathematics of probability and the celestial sphere are on my side." Hawkins' first contention was that alignments between pairs of stones and other features, calculated with a computer from small-scale plans, compared their directions with the azimuths of the rising and setting sun and moon, at the solstices and equinoxes, calculated for 1500 BC. Hawkins claimed to have found thirty-two "significant" alignments.

His second contention was that the fifty-six Aubrey holes were used as a "computer" (that is, as tally marks) for predicting movements of the moon and eclipses, for which he claims to have established a "hitherto unrecognized 56-year cycle with 15 percent irregularity; and that the rising of the full moon nearest the winter solstice over the Heel Stone always successfully predicted an eclipse. It is interesting to note that no more than half these eclipses were visible from Stonehenge."

Says Hawkins in *Stonehenge Decoded,* "The number 56 is of great significance for Stonehenge because it is the number of Aubrey holes set around the outer circle. Viewed from the centre these holes are placed at equal spacings of azimuth around the horizon and therefore, they cannot mark the Sun, Moon or any celestial object. This is confirmed by the archaeologist's evidence; the holes have held fires and cremations of bodies, but have never held stones. Now, if the Stonehenge people desired to divide up the circle why did they not make 64 holes simply by bisecting segments of the circle—32, 16, 8, 4 and 2? I believe that the Aubrey holes provided a system for counting the years, one hole for each year, to aid in predicting the movement of the Moon. Perhaps cremations were performed in a particular Aubrey hole during the course of the year, or perhaps the hole was marked by a movable stone.

"Stonehenge can be used as a digital computing machine. ...The stones at hole 56 predict the year when an eclipse of the Sun or Moon will occur within 15 days of midwinter—the month of the winter Moon. It will also predict eclipses for the summer Moon."[95]

The critics of Hawkins, the ruling academic minds of their time, immediately jumped on his discoveries and denounced them. In 1966 an article by the British astronomer R. J. Atkinson appeared in *Nature* (volume 210, 1966), entitled "Decoder Misled?" in which Atkinson criticized Hawkins for many of his statements about Stonehenge being an astronomical computer.

Said Atkinson of Hawkins' book *Stonehenge Decoded*, "It is tendentious, arrogant, slipshod and unconvincing, and does little to advance our understanding of Stonehenge.

"The first five chapters, on the legendary and archaeological background, have been uncritically compiled, and contain a number of bizarre interpretations and errors. The rest of the book is an unsuccessful attempt to substantiate the author's claim that 'Stonehenge was an observatory; the impartial mathematics of probability and the celestial sphere are on my side.' Of his two main contentions, the first concerns alignments between pairs of stones and other features, calculated with a computer from small-scale plans ill-adapted for this purpose."

Atkinson's scathing criticism of Hawkins is revealing because it shows how resistant to new ideas established academics can be. Atkinson's reluctance to believe that Stonehenge was some sort of astronomical computer is probably largely due to the popular belief that ancient man simply didn't have a state of civilization that allowed him to pursue such topics of higher knowledge.

But these critics are heard from no more, and there seems little doubt to even the most conservative archaeologist that Stonehenge is some sort of astronomical temple. There are a number of simple astronomical truths that can be discerned from Stonehenge. For instance, there are 29.53 days between full moons and there are 29 and a half monoliths in the outer Sarsen Circle.

There are 19 of the huge 'Blue Stones' in the inner horseshoe which has several possible explanations and uses. There are nearly 19 years between the extreme rising and setting points of the moon. Also, if a full moon occurs on a particular day of the year, say on the summer solstice, it is 19 years before another full moon occurs on the same day of the year. Finally, there are 19 eclipse years (or 223 full moons) between similar eclipses, such as an eclipse that occurs when the sun, moon and earth return to their same relative positions. Other planets' positions may vary in even larger cycles.

It is also suggested that the five large trilithon archways represent the five planets visible to the naked eye: Mercury, Venus, Mars, Jupiter and Saturn.

The British writer on antiquities, John Ivimy, makes a stirring suggestion at the end of his popular book on Stonehenge, *The Sphinx and the Megaliths.*[96] He spends the bulk of the book trying to prove his thesis that Stonehenge was built by a group of adventurous Egyptians who were sent to the British Isles to establish a series of astronomical sites at higher latitudes in order to accurately predict solar eclipses, which the observatories in Egypt could not do, because they were too close to the equator.

Ivimy gives such evidence as the megalithic construction, keystone cuts in the gigantic blocks of stone, the obvious astronomical purpose, and most of all, the use of a numbering system that is based on the number six, rather than the number ten, as we use today. Ivimy shows that the Egyptians used a numbering system based on the number six, and that Stonehenge was built using the same system. He then suggests that the Mormons used a number system based on six when building their temples, especially the great temple in Salt Lake City.

In the end, Ivimy's thesis is quite controversial: he believes that Brigham Young and the original Mormon settlers in Utah are the reincarnation of the same Egyptian group of settlers who were sent to Britain to build Stonehenge. Says Ivimy, "Reference has already been made to the vast wooden dome, built entirely without metal, that roofs the Mormon Tabernacle. Could its construction have been inspired by a dim recollection of how the same people, in another incarnation some centuries earlier, had built a dome over what had then become the Temple of Hyperborean Apollo?"[96]

It is a fascinating idea that the Egyptians came to Britain to build a megalithic observatory to accurately predict lunar eclipses. It is recorded that about 2000 BC a Chinese emperor put to death his two chief astronomers for failing to predict an eclipse of the sun. Asks ancient astronaut theorist Raymond Drake, "Today, would any king care?"

The Egyptians, Chinese, Mayans and many other ancient cultures were obsessed with eclipses as well as other planetary-solar phenomena. It is believed that they associated catastrophes, including the sinking of Atlantis, with planetary movements and eclipses. Perhaps the ancient Egyptians, Mayans and other civilizations thought they could predict the next cataclysm by monitoring lunar eclipses and the positions of the planets in relation to the earth.

Herodotus writes about ancient Egyptian astronomy and cataclysms in *Book Two,* chapter 142: "...Thus far the Egyptians and their priests told the story. And they showed that there had been three hundred and forty-one generations of men from the first king unto this last, the priest of Hephaestus. ...Now in all this time, 11,340 years, they said that the sun had removed from his proper course four times; and had risen where he now setteth, and set where he now riseth; but nothing in Egypt was altered thereby, neither as touching the river nor as touching the fruits of the earth, nor concerning sicknesses or deaths."

If Herodotus is to be believed, then the earth has shifted around its axis in what is called a pole shift. The sun then appears to rise in a different direction from normal. Pole shifts are accompanied by a wide variety of devastating earth changes and severe weather phenomena. Therefore, if the Egyptians were familiar with this sort of occurrence, and having been thus far unaffected by the cataclysm, they may well have gone to great length to improve their astronomical knowledge, including the colonization of England and the building of Stonehenge.

Indeed, the megalithic masterminds virtually colonized the world, from Egypt to England to the Americas, Easter Island and Tonga. Megaliths exist in such remote places as Manchuria, the Phillipines, Mongolia and the Assam Hills of northeast India. The megalithic masterminds were once everywhere. Yet, what was the technology that these master builders used?

The megalithic computer called Stonehenge.

Top: The Sphinx, still buried, with a tour group, ca. 1912. Bottom: The Sphinx with the Great Pyramid in the background.

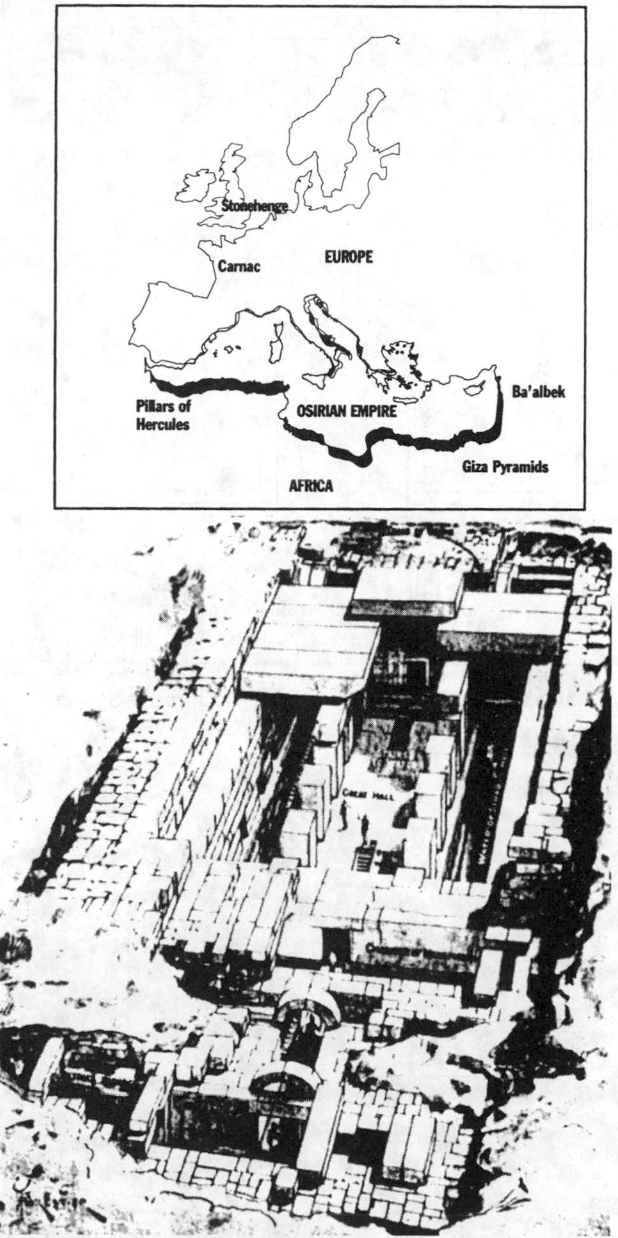

Top: A map of the theoretical Osirian Empire. Bottom: The Osirion at Abydos.

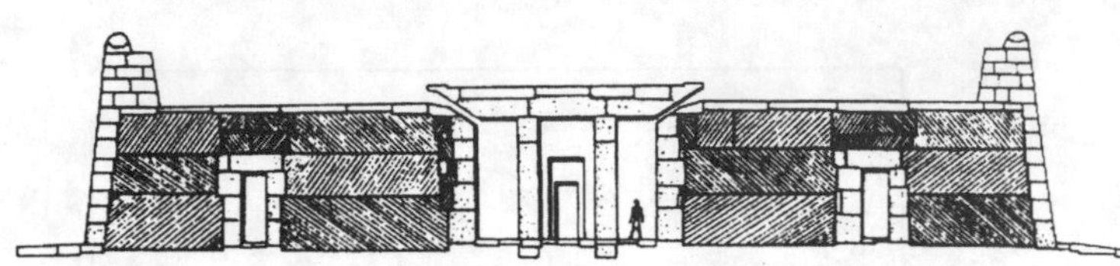

The Valley Temple of the Sphinx. Notice the massive core blocks.

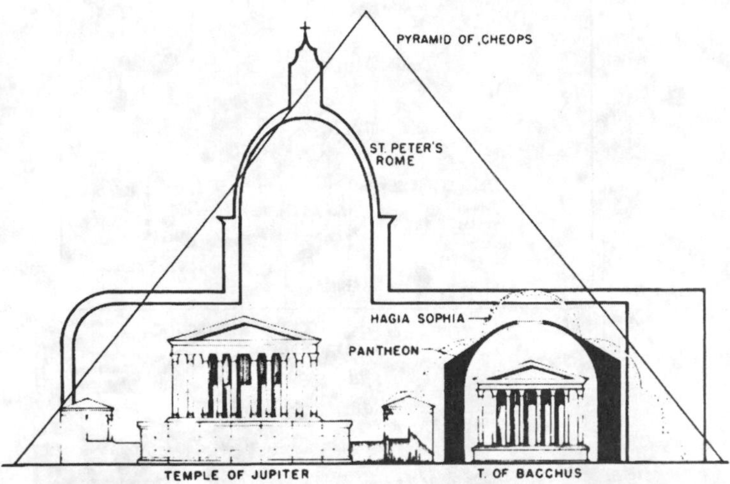

Ba'albek compared with the Great Pyramid and St. Peter's in Rome.

The three largest stones at Ba'albek.

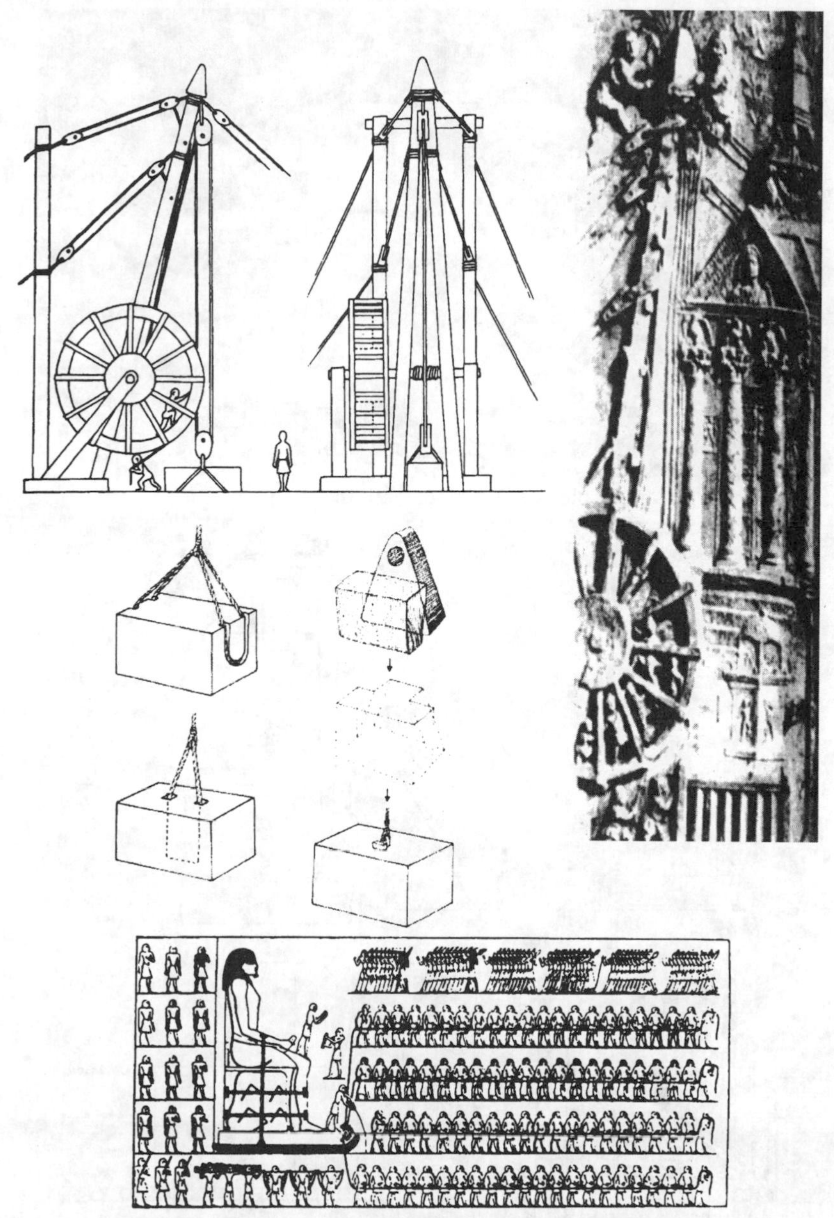

Top and Right: A Roman crane used for building. Such methods could not have been used in building Ba'albek. Center: Methods for lifting blocks of stone. Bottom: Egyptian depiction of a large group of workers dragging a sled with a large stone statue.

The largest block at Ba'albek, still in the quarry.

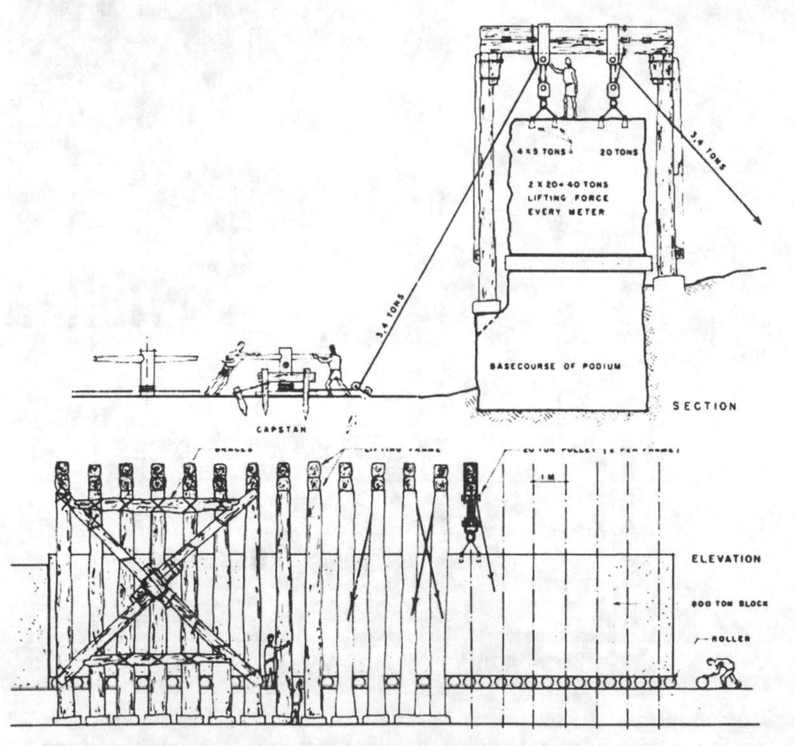

Aside from levitation, the only method that has been contrived to move these huge blocks, even a few inches, involves the use of this frame of pulleys and "Lewis Stones."

The giant door and keystone at Ba'albek.

The "Sun God" at the Gate of the Sun in Tiahuanaco. Legend says that he is crying for the "Red Land," a vanished super-civilization.

Carnac, France

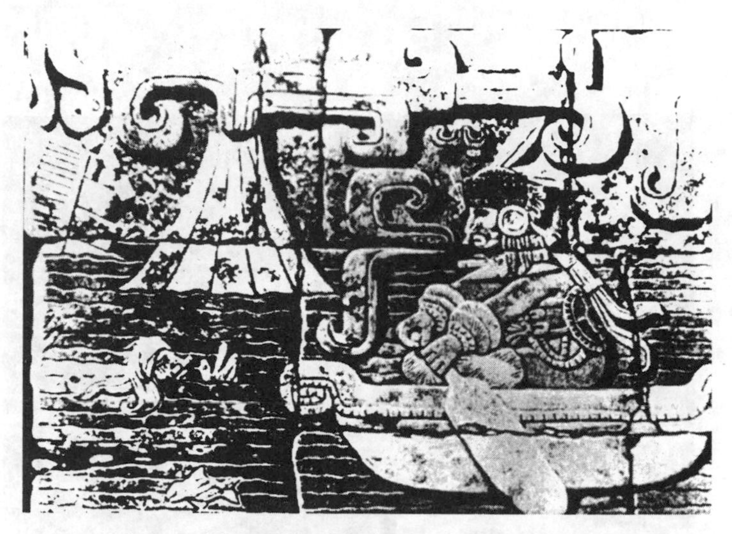

A Mayan frieze showing the destruction of Atlantis.

Ruins near Chavin in Peru. They look as if they had some kind of machinery. From Squire's *Conquest of Peru* (1886).

The massive walls of Sacsayhuaman, above Cuzco.

The massive walls of Sacsayhuaman, above Cuzco.

The giant blocks of Puma Punku, near Tiahuanaco.

The giant blocks of Puma Punku, near Tiahuanaco, restored.

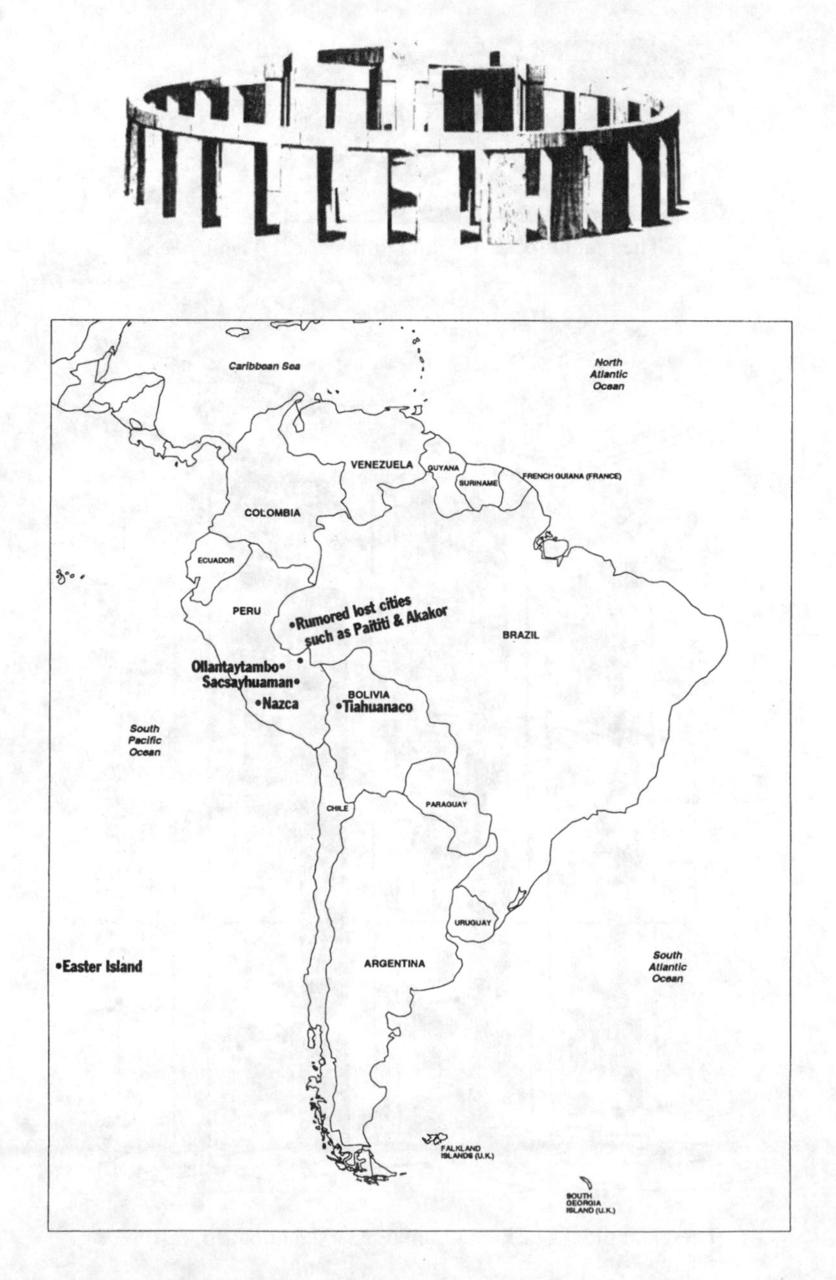

Caribbean Sea
North Atlantic Ocean
VENEZUELA
GUYANA
SURINAME
FRENCH GUIANA (FRANCE)
COLOMBIA
ECUADOR
PERU
•Rumored lost cities such as Paititi & Akakor
BRAZIL
Ollantaytambo•
Sacsayhuaman•
•Nazca
BOLIVIA
•Tiahuanaco
South Pacific Ocean
CHILE
PARAGUAY
URUGUAY
•Easter Island
ARGENTINA
South Atlantic Ocean
FALKLAND ISLANDS (U.K.)
SOUTH GEORGIA ISLAND (U.K.)

3.
Ancient Metallurgy & Machines

Up on the Madison Fork, the Wasiches had found much of the yellow metal that they worship and makes them crazy.
—*Black Elk Speaks*

Any smoothly functioning technology will have the appearance of magic.
—Arthur C. Clarke

Ancient Mining and Smelting

In order to have high technology, a civilization must have strong metals with which to create machines; metals such as iron and steel. Generally, mainstream science says that man's use of smelted iron is a story of slow and sporadic technological development starting about 5,000 years ago. There is evidence, as we shall see, that metallurgy and the manufacture of metal objects, goes back to 50000 BC and before.

The origin of iron, and metallurgy in general, is shrouded in mystery, legend and the mists of time. The Biblical legend of Tubal Cain is about one of the keepers of the secrets of metallurgy. As we have seen, the legend of Osiris tells of how, after his resurrection, he travelled around the world spreading the knowledge of metallurgy and science.

The original discovery of the technique for smelting iron and ultimately making steel, is said to have occurred among the Hittites of central Turkey circa 2700 BC. Knowledge of iron and steel is said not to have come into wide use in the west until about 1200 BC.

Except for anomalous artifacts, the current archaeological record begins finding iron objects from the third millennium BC down to the present period. These accepted specimens, all inferred to be bloomery iron, were found in various places. At Tell Chagar

Bazar, in Northern Syria, a fragment of iron was dated to about 2700 BC; excavations at Tell Asmar, Iraq, yielded an iron knife blade in a bronze hilt dating to the end of the Early Dynastic period of Sumer (ca. 2450-2340 BC); a dagger with an iron blade and a gold-sheathed handle, comes from the Royal Tombs at Alaca Huyuk, Anatolia, and dates from circa 2600-2300 BC.

However, iron objects have been found that are older than 2700 BC, even according to established archaeologists. Their explanation for these older objects is that they may be "meteoric iron" rather than actually smelted objects. According to South African archaeologist Nikolass van der Merwe in his book *The Carbon-14 Dating of Iron:*[97]

> Before the knowledge of iron smelting was acquired, man was able to use meteoric iron. Skills developed in cutting and grinding stone, common since Neolithic times, were sufficient to fashion meteoric iron objects. The knowledge of reducing iron from its ores, however, was not acquired until the third millennium BC. The resultant metal was lacking in quality, and only isolated occurrences in Anatolia, Mesopotamia, and adjoining areas, have been recorded. Bronze, then only in its early stage of cultural significance, proved to be both cheaper and more durable than the early forms of iron for the fashioning of cutting edges. The influence of iron as an important material of manufacture was not felt until the development, by Hittite subjects, of the basic techniques of steel production. After an initial period of development, during the course of some five centuries prior to 1200 BC, iron spread rapidly. By 500 BC, iron was in use in most of Europe, in the Far East, and in Africa as far south as Nubia and Nigeria.
>
> As the technique of iron smelting became more widely known, new metallurgical procedures were added. In the Mediterranean area, techniques for the manufacture and improvement of steel developed rapidly. By the beginning of the Christian era, the techniques of carburization, annealing, quenching, and tempering were widely known and the use of the direct process had become firmly established. In China, a different metallurgical tradition arose; as soon as the utility of iron became known, cast iron was manufactured. The process of steel production through decarburization was quickly developed and became the hallmark of iron metallurgy in the East. In

Europe, the direct process held full sway until the fourteenth century, when the introduction of cast iron and the indirect process laid the foundation of the modern iron industry.[97]

Mining had no doubt been going on for many tens of thousands of years. The metals copper, gold, and silver have been mined since at least 50000 BC. The reason for this is that these metals can be extracted and used straight out of the ground. In other words, pure copper can be taken from the ground and hammered directly into a spearhead, knife or sword. Gold and silver are softer, but usable for a variety of items.

Alloys of metal are a different story, but certain alloys are relatively simple to achieve, such as electrum, a mixture of gold and silver. Other alloys, such as tin and bronze, require a certain refining process, and it is here that high technology becomes involved. Platinum has a high melting point, and is also a difficult process.

The discovery of meteoric iron may have spurred the curiosity of the ancients, but is all early iron really from a meteorite, or is it from a genuine smelter?

Says Nikolass van der Merwe:

> The list of early meteoric iron artifacts in the archaeological record is fragmentary and brief. This is due, in part, to the fact that available sources of meteoric iron are extremely limited and that a correspondingly small number of artifacts have been manufactured from this material. Equally important is the fact that a chemical determination of the nickel content of a piece of iron or a metallographic analysis of its structure is required to identify its meteoric origin; it is significant to note that meteoric iron objects have generally been identified only when large-scale archaeological projects, in which experts from many disciplines participated, were involved, or in cases where iron artifacts occur unexpectedly early in the archaeological record.
>
> From a list of early meteoric iron objects, compiled by Coghlan, a number of examples deserve mention here. The earliest known occurrence comes from Gerzah in Egypt, where Wainwright discovered a number of iron beads. These beads were dated, according to the Petrie system, to S.D. 60-63 (ca. 3500 BC), and have a nickel content of 7.5 percent, clearly in the meteoric iron range. In Mesopotamia, Woolley recovered iron fragments with a 10.9 percent nickel content from the Royal Tombs of Ur (ca. 2500 BC). At the Anatolian site of Alaca Huyuk, two iron specimens with 5.08 percent and 4.3 percent nickel, respectively, have been identified in the Early Bronze Age II levels (ca. 2600—2300 BC).

Some of these early specimens, notably in the case of Alaca Huyuk, were contemporary with smelted iron objects in the same deposits. It is reasonable to assume, therefore that many objects of meteoric iron have gone unnoticed for lack of chemical or metallographic analysis. Knowledge of the use of this material is likely to remain confined largely to times and places where iron objects occur in an unexpected context.[97]

The Origin of Smelting

It is theorized that the origin of smelting came from the simple heating of gold-bearing sand in order to extract the easily melted metal. The extraction of mercury from cinnabar is similar, though this would seem to have occurred much later. The main reason for this is that mercury is not particularly useful as a metal, or a liquid, except in electrified switches and gyros, as we will see later.

This author believes that mining began at least 40,000 years ago on this planet, and smelting began shortly after that, if not at the same time. While mainstream science believes that iron smelting began with the Hittites, there is still a great deal of mystery in the process.

Says van der Merwe:

Some attempts have been made, by inductive reasoning, to reconstruct the procedures by which iron was first smelted. The simplest of these reconstructions involve the production of gold from gold-bearing sand. The ancient Egyptians melted gold from Nubian desert sands, which also contain quantities of magnetite. Under the proper set of conditions, reduced iron would form on top of the melted gold in the crucible, and under a layer of iron slag. This would take place if a reducing atmosphere were used accidentally and if the ratio of magnetite to sand were in the order of 2:1—a situation which could result if a flotation or washing process were used to purify the sand. The iron so produced would, of course, be solid and may well have been discarded. The terms for meteoric iron and smelted iron in Ancient Egyptian clearly indicate, however, that the relationship between the two was known; knowledge of meteoric iron may have enabled the gold smelters to recognize smelted iron.

Another hypothetical reconstruction can be made by considering how iron may have been smelted accidentally in a copper furnace. When copper sulfide ores are roasted before smelting, they are converted to a reddish oxide, not unlike hematite in appearance. If the smelter used hematite in his furnace instead of copper ore, and under reducing conditions, he would obtain useless, molten iron slag in the furnace bottom instead of molten copper. If, however, he paid attention to the reduced lumps of iron immediately above the slag, he would have found these to be malleable above 1,000° C. While this last phase of the argument may be difficult to accept, it seems reasonable to argue that the idea of producing metals from mineral ores would have prompted experimentation with a variety of ores. In fact, it may not be necessary to postulate the accidental charging of a furnace with iron ore; deliberate experiments with different ores may have taken place. Familiarity with the properties of meteoric iron may have facilitated the recognition of iron as a usable material after it had been produced, accidentally or intentionally, a number of times. The fact that the earliest iron was produced in the early phases of the Bronze Age, and in those areas with the most accomplished metallurgical industries, strengthens the viewpoint of deliberate experimentation. It should also be considered that iron was regarded as a precious metal for many centuries after its discovery; the earlier economic success of the production of gold and silver probably provided considerable incentive toward the discovery of metals which could confer similar monetary rewards on the successful smelter. While the exact procedure by which iron was first smelted can thus only be guessed at, we do know what new techniques were required to produce a usable object from these early products of the bloomery process. 'The discovery of man-made iron... had not awaited the evolution of a basically new smelting process; it was almost entirely the outcome of hammering a hot, spongy aggregate of metal, slag and dirt.' Thus the arts of the blacksmith were born, beginning a long period of technological evolution which was eventually to give rise to the Iron Age proper.[97]

There are two general processes in the manufacturing of iron: the "bloomery process," a simpler process, and the "direct process." Says van der Merwe, "A major impetus of the Iron Age proper was the discovery of cementation, the technique by which steel can be produced from bloomery or wrought iron and which is generally associated with the bloomery process. The discovery of this technique is usually attributed to the Chalybes, subjects of the Hittites, and dates to about 1500—1400 BC. The Hittites are thought to have maintained a strict monopoly on

A Hittite relief.

the manufacture of the new alloy, enabling them to keep prices at an artificially high level. This viewpoint is based on interpretations of a letter from the Hittite king, Hattusilis III (1281—1260 BC), to an unknown correspondent and is the subject of some dispute."[97]

Iron was the most expensive metal in ancient times—when it could be obtained at all! Van der Merwe mentions that "the price of iron during the early stages of the Hittite confederacy (early second millennium BC) is known to have been five times that of gold and forty times that of silver, and it must have been even more expensive during the third millennium BC. At such prices, iron objects are likely to have been traded as marks of status among the royalty of ancient Near Eastern kingdoms, thus achieving a distribution which is much wider than the actual areas of manufacture."[97]

Ultimately, the Hittites were destroyed, their capital city of Hattusas vitrified by intense heat, and the modern Iron Age began, according to historians. The secrets of smelting iron were disseminated around the Mediterranean. One question remains: did other nations, such as India and China already possess the secret of iron?

Metallurgy in Ancient India and China

The mystery of the use of iron in India and China is one that largely baffles modern metallurgists. It is assumed that these countries developed iron and other metallurgical skills after the west, but the evidence points otherwise. Nikolass van der Merwe gives the orthodox view: "Spreading east from the Mediterranean, iron was diffused throughout most of Asia before the Christian era. By 1100 BC it was in use in Persia, from where it spread to Pakistan and India. The date of the arrival of iron in India is still a matter of some dispute; until recently, iron was assumed to have reached Northern India around 500 BC, where it appears at the sites of Taxila, Histinapura, and Ahichatra in association with the distinctive 'Northern Black Polished' pottery type. Recent excavations at Atranjikhera, in Uttar Pradesh, however, have uncovered iron artifacts in association with 'Painted Grey' pottery, from an earlier period of the Ganges civilization, and have been dated between 1100 and 1000 BC. Further archaeological work will be necessary to gauge the impact of iron-working knowledge on Northern India, especially regarding the forces which contributed towards the urbanization of the peoples of that area. In the southern part of India, at least, especially in the Deccan, iron seems to have stimulated a veritable 'revolution' of this kind.

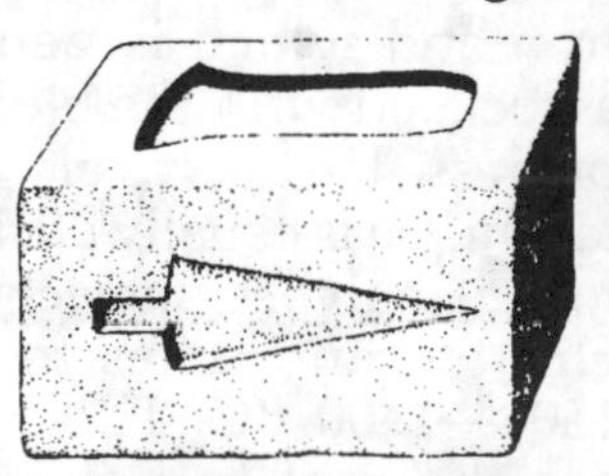
A simple stone mold.

"The transmission of iron-working knowledge to China, if it took place at all, is a problem which

remains unsolved. The possibility exists that iron was carried to China by the nomadic tribes of the Eurasian steppes. During the latter half of the first millennium BC the Sarmatians, a tribe closely related to the Scythians, are known to have occupied the region adjoining Kansu, in Northwest China. The Sarmatians relied primarily on bronze as a source of metal, although using iron to a limited extent. Their penetration to Northwest China is marked by the appearance of their distinctive 'Animal Art Style' in Mongolia and Ordos, where it is dated to circa 500 BC—and possibly earlier. Since iron appears in China during the sixth century BC, and perhaps earlier, it is doubtful whether the Sarmatians did, in fact, carry the knowledge of iron to China. If they did, it was at best a case of stimulus diffusion, since the Chinese did not adopt the direct process, which had been the exclusive method for the production of iron until that time. Instead, cast iron was apparently manufactured in China from this early date on, and the techniques of the indirect process were evolved." [97]

Iron is traditionally said not to have been worked in the Americas. Says van der Merwe, "In the New World, iron cannot be said to have achieved any widespread use before Colonial times. Small quantities of trade iron, however, penetrated to Northern Alaska by way of Siberia. Iron has been found in an early context in a site of the Ipiutak culture at Point Hope, Alaska; on the other side of the Bering Strait, iron occurs in an Old Bering Sea site at Uelen on the Chukchi coast. Both of these cultures have been dated to about AD 300. Iron was not manufactured in the New World, however, until Viking colonists introduced it to Newfoundland around AD 1000." [97]

Archaeologists, however, are ignoring the evidence for iron-smelting furnaces discovered in Ohio. Arlington Mallery in his book, *The Rediscovery of Lost America*, [132] gives details on the discovery of several iron furnaces from southern Ohio that were used in prehistoric times. One furnace that Mallery uncovered in the Allyn Mound near Frankfort, Ohio was of a beehive type with charcoal and iron ore found inside. The mound was about 60 feet in diameter and seven feet high. Mallery compared the furnace to the primitive Agaria iron smelters still used in India.

Mallery's book had an introduction by Matthew W. Sterling, then Director of the Bureau of American Ethnology of the Smithsonian Institution. Sterling said in the introduction, "It will be difficult to convince American archeologists that there was a pre-Columbian iron age in America. This startling item, however, is one that should not long remain in doubt. The detailed studies of metallurgists and the new carbon-14 dating method should be sufficient to give a definite answer on this point." [132]

in doubt. The detailed studies of metallurgists and the new carbon-14 dating method should be sufficient to give a definite answer on this point."[132]

The Iron Pillar of Delhi

In the southern district of New Delhi is the famed Iron Pillar, generally believed to date from the fourth century AD, but said by some scholars to be over four thousand years old. It was built as a memorial to a king named Chandra. It is a solid shaft of iron sixteen inches in diameter and twenty-three feet high. What is most astounding about it is that it has never rusted even though it has been exposed to wind and rain for centuries! The pillar defies explanation, not only for not having rusted, but because it is apparently made of pure iron, which can only be produced today in tiny quantities by electrolysis! The technique used to cast such a gigantic, solid iron pillar is also a mystery, as it would be difficult to construct another of this size even today. The pillar stands as mute testimony to the highly advanced scientific knowledge that was known in antiquity, and not duplicated until recent times. Yet still, there is no satisfactory explanation as to why the pillar has never rusted![43]

The Iron Pillar of Delhi.

To add to the evidence that ancient India had highly advanced smelting works, the monthly *Motilal Banarsidass Newsletter* from New Delhi, India reported in its July 1998 edition that findings by the State Archaeology Department after excavations in Sonebhadra district, Lucknow, India, may revolutionize history as regards to the antiquity of iron. The department has unearthed iron artefacts dated between 1200—1300 BC at the Raja Nal Ka Tila site in the Karmanasa river valley of north Sonebhadra.

Said the newsletter, "Radio carbon dating of one of the samples done by the Birbal Sahani Institute of Palaeobotany has established that it belongs to 1300 BC, taking the antiquity of iron at least 400 years back, even by conservative estimates. This date of iron is one of the earliest in the Indian sub-continent."

And, these are conservative estimates indeed. As we have already seen, there is considerable evidence that mining and iron working have gone on long before 1300 BC. Indeed, if the futuristic (it seems odd to call tales of the past "futuristic") epics of ancient India are any indication, there must have been a great deal of metallurgical activity in ancient India, starting over 20,000 years ago!

The Mysterious Origin of Aluminum

In 1959 Communist Chinese archaeologists claimed that they had discovered ancient Chinese belt buckles in a tomb. They were several thousand years old, the newpaper accounts said, but, incredibly, they were made of aluminum. Aluminum is a curious metal, because the smelting process from bauxite requires electricity! Photos of the belt buckles appeared in the French language magazine *Revue de l'Aluminum,* issue number 283, published in 1961 and reproduced here.

The modern process for extracting aluminum from bauxite was not perfected until 1886. This discovery, as well, is very curious. Most aluminum produced today is made from bauxite. First discovered in 1821 near Les Baux, France (from which its name is derived), bauxite is an ore rich in hydrated aluminum oxides, formed by the weathering of such siliceous aluminous rocks as feldspars, nepheline, and clays. During weathering the silicates are decomposed and leached out, leaving behind a residue of ores rich in alumina, iron oxide, titanium oxide, and some silica. In general, economically attractive ores contain at least 45 percent alumina and no more than five percent to six percent silica.

Most of the large bauxite deposits are found in tropical and subtropical climates, where heavy rainfall, warm temperatures, and good drainage combine to encourage the weathering process. Because bauxite is always found at or near the surface, it is mined by open-pit methods. It is then crushed if necessary, screened, dried, milled, and shipped for processing. Australia, Guinea, Jamaica, Brazil, and India are leading bauxite producers.

Although proof of the existence of aluminum as a metal did not exist until the 1800s, clays containing the metallic element were used in Iraq as long ago as 5300 BC to manufacture high-quality pottery. Certain other aluminum compounds such as the "alums" were used widely by Egyptians and Babylonians as early as 2000 BC. Despite these early uses of the "metal of clay," however, it was almost 4,000 years before the metal was freed from its compounds, which made it a commercially usable metal.

Credit for first separating aluminum metal from its oxide goes to the Danish physicist Hans Christian Oersted. In 1825 he reported to the Royal Danish Academy that he accomplished this by heating anhydrous aluminum chloride with potassium amalgam and distilling off the mercury. His product was so impure, however, that he did not succeed in determining its physical properties beyond observing a metallic luster.

In 1845, after many years of experimentation, Friedrich Wohler succeeded—by substituting potassium for the amalgam—in producing globules of aluminum large enough to allow the determination of some of its properties. In 1854, Henri Sainte-Claire Deville substituted sodium for the relatively expensive potassium and, by using sodium aluminum chloride instead of aluminum chloride, produced the first commercial quantities of aluminum in a small plant near Paris. Bars and various objects made of this metal were shown at the Paris Exposition in 1855, and the ensuing publicity was in large measure responsible for launching the industry.

In 1886, Charles Martin Hall of Oberlin, Ohio, and Paul L. T. Heroult of France, discovered and patented almost simultaneously the process by which alumina is dissolved in molten cryolite and decomposed electrolytically. This reduction process, generally known as the Hall-Heroult process, has survived many attempts to supplant it; it remains the only method by which aluminum is produced in commercial quantities today. The inventors' families made millions, and ultimately billions, of dollars. Aluminum is made all over the world, usually where bauxite can be found and electricity is cheap, such as at hydroelectric plants.

Aluminum is the most abundant metal on the planet, but requires electricity to create metal in a usable form. Indeed, the invention of aluminum extraction is of incalculable benefit to mankind, providing us the advanced metallurgy science that is necessary for inventions such as flight and space travel.

The belt buckles discovered by the Chinese in 1959 make us wonder, were these artifacts made using electricity? The aluminum smelting process from bauxite requires electricity! French scientists studied the buckles and published their studies in 1961. They concluded that the ancient Chinese were making aluminium by an unknown process.

Mining and Metal Anomalies

There are many ancient mines in southern Africa, and many have curious stone ruins to go along with them. The archaeologist J. Theodore Bent, who excavated some of the ruins in 1891 and wrote *The Ruined Cities of Mashonaland* in 1892, said that a Roman coin of the reign of Antoninus Pius (138 AD) was found in a mine shaft at Umtali.[58]

But mines in southern Africa have been dated to much older periods than that, going back 5,000 years or more. Some mines in southern Africa have been dated to 50000 BC. William Corliss quotes from a 1967 article in the British science journal *Nature* on the subject of mines in southern Africa that have been dated to circa 26000 BC! Among the amazingly ancient mines were manganese and iron mines.

Says the article, "The only ancient manganese mine yet recorded is in southern Africa, at Chowa near Broken Hill, Zambia... The Kafufulamadzi Hills, 3 miles away, revealed Later Stone Age Assemblages in quartz, together with manganese tools identical to those found in the working ...[W]orkings at the Ngwenya Iron Mine in western Swaziland... yielded flaked stone mining tools similar to those from Chowa Found in 1934."[5]

Carbon dating of charcoal nodules at the lowest levels of the mines gave the astonishing dates of 22,280 BP and 28,000 BP (Before Present). Samples of the charcoal nodules were given to Yale University and the University of Groningen (Netherlands) Laboratories for carbon-14 dating. Yale came up with the date spread of 22,280±400 BP/ 20,330±400 BP. The Groningen Laboratories came up with the date spread of 28,130 ±260 BP/ 26,180 ±260 BP.[5] Clearly, there is evidence that iron and other metals have been mined for thousands of years in Southern Africa, and probably other areas of the world as well.

Rene Noorbergen in his book, *Secrets of the Lost Races,*[3] tells a bizarre tale. Under the subtitle *Who Shot Rhodesian Man?*, Noorbergen states that someone apparently shot one of these ancient miners. At the Museum of Natural History in London there is an exhibit of a human skull discovered near Broken Hill, Rhodesia, in 1921. "On the left side of the skull is a hole, perfectly round. There are none of the radial cracks that would have resulted had the hole been caused by a weapon such as an arrow or a spear," says Noorbergen. "Only a high-speed projectile such as a *bullet* could have made such a hole. The skull directly opposite the hole is shattered, having been blown out *from the inside*. This same feature is seen in victims of head wounds received from shots from a high-powered rifle. No slower projectile could have produced either the neat hole or the shattering effect. A German forensic authority from Berlin has positively stated that the cranial damage to Rhodesian man's skull could not have been caused by anything but a bullet. If a bullet was indeed fired at Rhodesian man, then we may have to evaluate this in the light of two possible conclusions: Either the Rhodesian remains are not as old as claimed, at most two or three centuries,

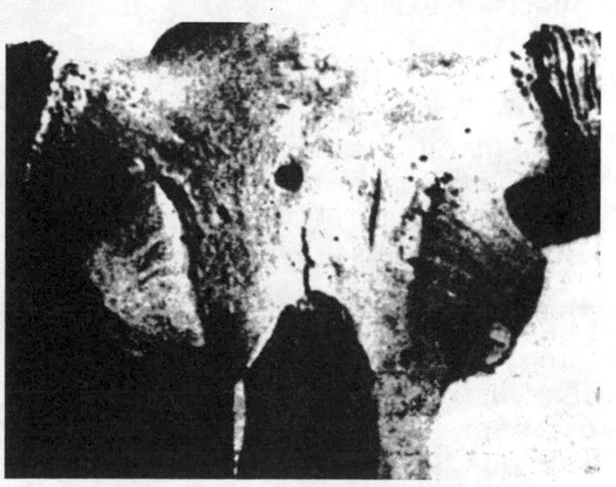

The auroch skull with a bullet hole in it.

and he was shot by a European colonizer or explorer; or the bones are as old they are claimed to be, and he was shot by a hunter or a warrior belonging to a very ancient yet highly advanced culture.

"The second conclusion is the more plausible of the two, especially since the Rhodesian skull was found 60 feet below the surface. Only a period of several thousand years can account for a deposit of that depth. To assume that nature could have accumulated that much debris and soil over only two or three hundred years would be ridiculous."[3]

Noorbergen concludes by mentioning the skull of an auroch, a type of extinct ox, which was discovered west of the Lena River and has been judged as several thousand years old at the Paleontological Museum in Moscow. The curator of the museum, Professor Constantin Flerov, was curious about a small round hole piercing the forehead. The hole has a polished appearance, without radial cracks, indicating the projectile entered the skull at a very high velocity. The auroch survived the shot, as is evidenced by calcification around the hole. The auroch later died of other causes.[3]

One reason that we don't have many iron or other metallic objects that are tens of thousands of years old is that such an object wouldn't last that long. Most metals, such as iron, copper, bronze and tin will corrode and oxidize into nothing. An iron nail, if exposed to water, will rust and disappear in a matter of a few years. This is why gold is particularly valuable—it is indestructable. All gold that ever existed in ancient times still exists today, as jewelry, coins, bullion, or whatever. Gold is too soft, however, to be used for weapons or machines, at least in its pure form. Other metals that also last for exended periods of time are lead and mercury. In order to find artifacts of the rusting metals, it is necessary for them to have been somehow sheltered from the enviroment. The following stories prove that artifacts do exist.

Back of the sun and way deep
under our feet,
at the earth's center, are not a
couple of noble mysteries
but a couple of joke books.
—Tennessee Williams

The Coso Artifact—a spark plug in a geode?

A Spark Plug Found in a Geode

In 1961, Wally Lane, Mike Mikesell, and Mrs. Virginia Maxey, co-owners of the LM&V Rockhounds Gem and Gift Shop in Olancha, California, went into the

Coso Mountains in the Inyo National Forest near Death Valley to look for unusual rocks. Near the top of a 4,300 ft. peak overlooking the dry bed of Owens Lake they found a fossil-encrusted geode. When they opened the geode, which are generally hollow with crystals inside them, they found something that resembled a spark plug.

In the middle of the geode was a metal core, about .08 inch (2 millimeters) in diameter, which responded to a magnet. Enclosing this was what appeared to be a ceramic collar that was itself encased in a hexagonal sleeve carved out of wood that had become petrified, presumably at a later date. A fragment of copper still remaining between the ceramic and petrified wood suggests that the two may once have been separated by a now-decomposed copper sleeve. Around this was the outer layer of the geode, consisting of hardened clay, pebbles, bits of fossil shell, and "two non-magnetic metallic objects resembling a nail and a washer." Based on the fossils contained in the geode, the object was estimated to be at least 500,000 years old![16, 39]

When Ron Calais, Brad Steiger's researcher, did the basic research on the Coso artifact for Vol. 1, No. 4 of Ivan T. Sanderson's *INFO Journal*, editor Paul J. Willis accepted the challenge to come up with an idea of what the object might have been. After examining X-ray photos of the geode and doodling a bit with his pencil, Willis ventured his opinion that the hexagonal part reminded him of a spark plug.[16]

"I was thunderstruck," his brother, Ron Willis, writes, "for suddenly all the parts seemed to fit. The object sliced in two shows a hexagonal part, a porcelain or ceramic insulator with a central metallic shaft—the basic components of any spark plug." The Willis brothers then set about attempting to saw a common spark plug in half near its hexagon. They soon found the porcelain was too hard for their hacksaw, but they did manage to get the plug apart.

"We found all the components similar to the Coso artifact," Ron writes, "but with some differences. The copper ring around the halves displayed in the object seems to correspond to a copper sealer ring in the upper part of the steel casing of any spark plug."

It is their belief that the hexagonal area in the geode is probably composed of rust, the remains of a steel casing. The Willis brothers also noted that the central shaft of the spark plug they had dismembered had a tint reminiscent of brass, and they recalled Virginia Maxey's words that the metal core had a "slightly brassy appearance."

The upper end of the object appears to end in a spring, but Ron and Paul Willis theorized that what is seen in the X-ray photograph might be "the remains of a corroded piece of metal with threads." Although the larger metallic piece in the upper section of the Coso artifact may not seem to correspond exactly with a contemporary, ordinary spark plug, the overall effect is certainly that of some kind of electrical apparatus. If it is some bizarre trick of Nature, it is indeed a good one.

The Willis brothers asked an INFO member to call upon Wallace A. Lane, who at that time (circa 1969) was residing in Vista, California, and who was in possession of the Coso artifact. Virginia Maxey had told Ron Calais that the object had been displayed at the Southeastern California Museum in Independence for approximately three months during 1963, but when INFO investigated, Lane had the artifact in his home. The Coso artifact, Lane said, could be purchased for $25,000. If a buyer should be interested, Lane went on, he had better hurry, because several museums were after it.

"There is no indication that any professional scientist has ever carefully examined the object, so what it may be is still questionable," Ron Willis concluded his article. "The Coso artifact now seems to join the club with the Casper, Wyoming, mummy, the Voynich manuscript and other Fortean objects whose owners refuse to allow anyone to examine the object in question without an exorbitant payment."[8, 16]

Oddities Found in Solid Stone

A book by Frank Edwards entitled *Strangest of All*[53] recounts the discovery of several similar out-of-place objects: "Somewhere in the dusty storage room of a museum there lies a chunk of feldspar which was taken from the Abbey mine near Treasure City, Nevada, in November of 1869. This fist-sized piece of stone was unusual because firmly embedded in it was a metal screw about two inches long. Its taper was clearly visible as was the regular pitch of the threads. Having originally been of iron, it had oxidized, but the hard stone which held its crumbling remains had faithfully preserved its delicate contours. The trouble with this exhibit was that the feldspar in which the screw was embedded was millions of years older than man himself (as estimated by science), so the annoying exhibit was sent off to a San Francisco academy and quietly forgotten."

It was reported in *Scientific American* in 1852 (No. 7, page 298) that during blasting work at Dorchester, Massachusetts, in 1851, the broken halves of a bell-shaped vessel were thrown by the force of an explosion from a bed of formerly solid rock. The vase, just under five inches high, was made of an unknown metal and embellished with floral inlays of silver—the "art of some cunning workman," according to the local newspaper report.

The editor of *Scientific American* gave as his opinion that the vase had been made by Tubal Cain, the biblical father of metallurgy. In response, Charles Fort, who collected such stories of oddities in his four books, said, "Though I fear that this is a little arbitrary, I am not disposed to fly rabidly at every scientific opinion."[39]

The metallic sphere from the Ottosdal Mines in South Africa believed to be 2.8 billion years old.

In 1891, Mrs. S. W. Culp of Morrisonville, Illinois, was breaking a lump of coal for her stove and noticed a gold chain firmly embedded in the now-split chunk. In 1851 Hiram de Witt, of Springfield, Massachusetts, accidentally dropped a fist-sized piece of gold-bearing quartz that he had previously brought back from California. The rock was broken apart in the fall and, inside it, de Witt found a two-inch cut-iron nail, slightly corroded. "It was entirely straight and had a perfect head," reported the *Times of London*.[39]

Similarly, Frank Edwards remarks, "In 1851, in Whiteside County, Illinois, the twisting bit of a well driller brought up two artifacts from sand at one hundred twenty feet. One was a copper device shaped like a boat hook; the other a copper ring of unknown purpose. And in 1971, near Chillicothe, Illinois, drillers brought up a bronze coin from one hundred fourteen feet—another bit of evidence that man had been there. But when, no man can say."[53]

There are probably hundreds of reports of anomalistic items like these, reports of artifacts that are unquestionably man-made, yet, according to uniformitarian geology, must be hundreds of thousands, if not millions of years old! Geological dating of coal, fossils, geodes, etc., is done on the basis of geological strata. Lower strata are deemed older than the strata above them. On the assumption that geological change is slow and uniform, then strata can be said to coincide with certain time periods during which the components were deposited (5 million years, etc.).

Given the distinct possibility that uniformitarian geology and dating are completely erroneous, objects that would initially appear to have a startlingly ancient date, say hundreds of thousands or millions of years, might actually be of much more recent manufacture. I suggest that this is the case with most of these artifacts. While it seems that most of them are authentic, they are probably closer to tens of thousands of years old, rather than millions of years old.

Another interesting thing to note here is the mechanism for burying artifacts in coal, stone and geodes. This is the same mechanism that creates fossils—not slow geological change, but sudden geological cataclysms, like the one that supposedly sank the ancient continents. It appears that such cataclysms are not isolated, rare events, but occur with alarming regularity!

A curious discovery along these lines was first reported in 1982. According to various reports, including one in the book *Forbidden Archeology*,[113] over the past several decades, South African miners have found hundreds of metallic spheres, several of which have three parallel grooves running their equator.

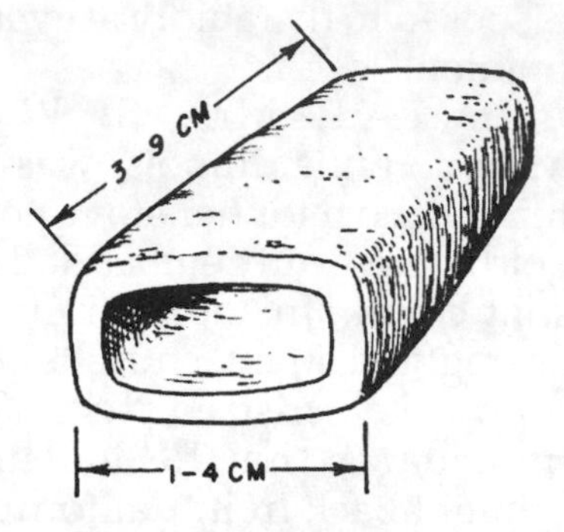

French metallic tube.

The spheres are of two types—"one of solid bluish metal with white flecks, and another which is a hollow ball filled with a white spongy center." Roelf Marx, curator of the museum in Klerksorp, South Africa, where some of the spheres are currently being housed said in a 1984 letter, "There is nothing scientific published about the globes, but the facts are: They are found pyrophyllite, which is mined the little town of Ottosdal in the Western Transvaal. This pyrophyllite ($Al_2Si_4O_{10}(OH)_2$) is a quite soft secondary mineral with a count of only 3 on the Mohs' scale and was formed by sedimentation about 2.8 billion years ago. On the other hand the globes, which have a fibrous structure on the inside with a shell around it, are very hard and cannot be scratched, even by hard steel." The Moh scale is a hardness scale using ten minerals as reference points, with diamond the hardest (10) and talc the softest (1).[113]

The grooved metallic spheres from the Ottosdal mines are thought, by uniformitarian geology, to come from a stratum that is termed Precambrian, a mineral deposit that is "believed" to be 2.8 billion years old. 2.8 billion! It seems unlikely that such a gap would occur in the history of metallurgy, and I would theorize that these metal spheres are tens of thousands, even hundreds of thousands of years old. Much of uniformitarian dating is overly conservative and it has been proven that large deposits of strata, up to several meters thick, can be deposited in a manner of days, rather than over millions of years, as uniformitarians are fond of guessing. It is sometimes said that "strata are dated by the fossils, and the fossils are dated by the strata." Such circular reasoning is used with the grooved metallic spheres; they are undoubtedly old, but are they billions of years old?

Another similar find is cataloged by William Corliss in *Ancient Man: A Handbook of Puzzling Artifacts:*[5] the discovery of molded metallic objects found in a chalk bed in France. The discovery was made in Caen, France on September 30, 1968. Some metallic nodules were formed in a hollow in an "Aptian" chalk bed in a quarry being worked in Saint-Jean de Livet. These metallic nodules have a reddish brown color, and a semi-ovoid, identical form (of different sizes). The chalk bed was thought to be 65 million years old and the metallic nodules were deemed artificial, created by "intelligent beings" that had lived in remote antiquity.[5, 113]

More Ancient Artifacts

The files of history are full of strange reports of unaccountable objects. I mused over a report from the *The American Antiquarian* published in 1883 which said that in about 1880, a Colorado rancher went on a journey to fetch coal from a seam driven into a hillside. The particular load that he collected was mined about 150 feet (45 meters) from the mouth of the seam, and about 300 feet (90 meters) below the surface.

When he returned home, the rancher found the coal lumps were too big to burn on his stove. He split some of them—and out of one of the lumps fell an iron thimble!

At least, it looked like a thimble—and "Eve's thimble" was the name given to the object in the locality, where it became well known. It had the indentations that modern thimbles have, and a slight raised "shoulder" at the base. The metal crumbled easily, and flaked away with repeated handling by curious neighbors. Eventually it was lost.

In 1883 it was not thought that tribes of American Indians had ever used thimbles, nor metallic objects at all. Besides, this seam of coal was dated as between the Cretaceous and Tertiary periods, which are generally dated at about 70 million years ago.

It was an impossible artifact, yet it fit snugly into a cavity in the coal. Like similar out-of-place-artifacts (Ivan T. Sanderson called them *ooparts*) it appears to be quite genuine, yet totally impossible by today's geological dating and accepted history of the planet.[8, 16]

Field Museum of Natural History (Neg. #92967), Chicago

In 1967, human bones were reported to have been discovered in a vein of silver in a Colorado mine. A copper arrowhead four inches (ten centimeters) long accompanied them. The silver deposit was, of course, several million years old and much more ancient than humanity, according to generally accepted ideas.[18]

Although the following story has nothing to do with ancient metals, per sé, it is fascinating and bears repeating here. It is absolutely true and still mystifies researchers to this day. In October of 1932, a couple of gold prospectors were working a gulch at the base of the Pedro Mountains about 60 miles west of Casper, Wyoming when they spotted some "color" in the rock wall of the gulch and used an extra heavy charge of dynamite to rip a section of the rock out in their search for mineral wealth.

The powerful blast exposed a small natural cave in solid granite, a cave *not more than four* feet wide, four feet high and about 15 feet deep. When the smoke had cleared, the miners got down and peered into the opening. What they saw was shocking, for peering back at them was a tiny mummy of a man-like creature!

He was on a tiny ledge, legs crossed, sitting on his feet, arms folded in his lap. He was dark brown, deeply wrinkled with a face that was almost monkey-like in some respects. One eye had a definite droop as though this strange little fellow might be winking at his discoverers. The ancient mummy was astonishingly small, only about 14 inches high!

The prospectors carefully picked him up, wrapped him in a blanket and headed back for Casper, where the news of their discovery attracted considerable attention. Scientists were skeptical, but interested; for according to conventional archaeology it would be impossible for a living being to be entombed in solid granite. Yet, the creature was real!

The mummy was examined and X-rayed by scientists. It was only 14 inches tall, weighing only about 12 ounces. The X-rays showed unmistakably that the tiny mummy had been an adult. Biologists who examined it declared it to be about 65 years old at the time of death. The X-rays showed a full set of teeth, a tiny skull, a full backbone and ribs and completely formed arms and legs. The mummy was not a clever hoax but a genuine biological entity with normal, though miniature, features.

The features had had an overall bronze-like hue. The forehead was very low, the nose flat with widespread nostrils, the mouth very wide with thin lips twisted in a sardonic smile.

According to the popular science writer Frank Edwards, the Anthropology Department of Harvard said that there was no doubt about the the genuineness of the mummy. Dr. Henry Shapiro, head of the Anthropology Department of the American Museum of Natural History, said that the X-rays revealed a very small skeletal structure covered by dried skin, obviously of extremely great age, historically speaking, and of unknown type and origin. The mystery mummy, said Dr. Shapiro, is much smaller than any human types now known to man.

Common speculation was that the mummy was a deformed, diseased infant, though anthropologists who examined the mummy were of the opinion that, whatever it was, it was fully grown at the time of death. Edwards says that the curator of the Boston Museum Egyptian department examined the creature and declared that it had the appearance of Egyptian mummies which had not been wrapped to prevent exposure to the

air. Still another expert, Dr. Henry Fairfield, ventured the supposition that the mystery mummy of the Pedro Mountains might be a form of anthropoid which roamed the North American continent about the middle of the Pliocene Age.[19]

The cave was examined as well, but no traces of human residence, no artifacts, no carvings, or writings—nothing but the tiny stone ledge on which this mummy had been sitting for countless ages. How had it come to be entombed in the solid granite wall anyway? To my knowledge, no carbon dating was ever done on the mummy.

While the mummy was on display in Casper for many years, it has since disappeared, and its current whereabouts are unknown.

After all, what is reality anyway?
Nothin' but a collective hunch.
—Jane Wagner

Robots & Automatons of the Ancients

Ancient man made a large number of machines, many of them virtually identical to machines that we use today. Ancient man had water pumps, cranes, hoists, catapults, water wheels, and even amusing toys and "gadgets." They had coin operated machines, automatons, and even computers, radio and television incredible as it may seem.

Some of the automatons are actual inventions that we know existed, others are only inferred from texts and "legends." Says historian Andrew Tomas in *We Are Not the First*,[24] "According to Greek legends, Haephestus, the "blacksmith of Olympus," made two golden statues which resembled living young women. They could move of their own accord and hastened to the side of the lame god to aid him as he walked. It cannot be denied that the concept of automation was present in ancient Greece.

"The engineers of Alexandria had over one hundred different automatons over 2,000 years ago. The legendary Daedalus, the father of Icarus, is reported to have constructed humanlike figures which moved of their own accord. Plato says that his robots were so active that they had to be prevented from running away! By what energy were they operated?"

Similarly, in the temples of ancient Egypt, such as Thebes, there were images of gods which could make gestures and speak. It is not at all improbable that some were manipulated by the priests hiding inside them, others may have had mechanical movements. The flashing of lights, such as in the case of the famous flashing eyes on the statue of Isis at Karnak,

were probably done with simple electric lights of some kind.

The legends of the Greeks, Romans, Persians, Hindus and Chinese all have references to what we would call robots or automatons: machines that looked and acted like people. The ancient Chinese, for instance, were fond of bronze dragons whose tails wagged in an automated fashion of one sort or another.

In the old Greek story of the quest for the golden fleece, Jason and the Argonauts came to Crete in the course of their voyages and legendary adventures. Medea told them that Talus, the last man left of the ancient bronze race, lived there. Then a metallic creature appeared, threatening to crush the ship Argo with rocks, if they drew nearer. A robot?

Says Tomas in *We Are Not the First,* "The know-how of robot construction was recorded in ciphered books on magic and thus preserved for long centuries. The monk Gerbert d'Aurillac (920-1003), professor at the University of Rheims who later became Pope Sylvester II, was reported to have possessed a bronze automaton which answered questions. It was constructed by the pope 'under certain stellar and planetary aspects.' This early computer said yes or no to questions on important matters concerning politics or religion. Records of this 'programming and processing' may still be extant in the Vatican Library. The 'magic head' was disposed of after the pope had died.[24]

Again quoting from Tomas, "Albertus Magnus (1206-1280), the Bishop of Regensburg, was a very learned man. He wrote extensively on chemistry, medicine, mathematics, and astronomy. It took him over twenty years to construct his famous android. His biography says that the automaton was composed of "metals and unknown substances chosen according to the stars." The mechanical man walked, spoke, and performed domestic chores. Albertus and his disciple Thomas Aquinas lived together and the android looked after them. The story goes that one day the talkative robot drove Thomas Aquinas mad with his chatter and gossip. Albertus' pupil grabbed a hammer and smashed the machine.

"This account should not be dismissed as mere fiction. Albertus Magnus was a true scholar—in the thirteenth century he explained the Milky Way as a conglomeration of very distant stars. Albertus Magnus and Thomas Aquinas were later canonized by the Catholic Church. The word android has even been adopted by science to signify an automaton or robot."[24]

Celestial globes, of various sizes, were cast metal machines with automatically moving parts. The earth was in the middle and remained stationary while the heavens revolved about it. The globe was constantly revolving by a mechanical device, and the whole thing agreed with the actual motion of the heavens.

Says Tomas, "According to Cicero (first century BC), Marcus Marcellus possessed just such a globe from Syracuse, Sicily, which demonstrated the motion of the sun, moon, and planets. Cicero assures us that the machine was a very ancient invention, and that a similar astronomical model was displayed in the Temple of Virtue at Rome. Thales of Miletus (sixth century BC) and Archimedes (third century BC) were considered to be the constructors of these mechanical devices.

"The memory of planetariums has persisted for many a century. The historian Cedrenus writes about Emperor Heraclitus of Byzantium, who, upon entry into the city of Bazalum, was shown an immense machine. It represented the night sky with the planets and their orbits. The planetarium had been fabricated for King Chosroes II of Persia (seventh century AD)."[24]

Ancient Technology and the Antikythera Device

In 1900, an amazing discovery took place on the small island of Antikythera, 25 miles northwest of Crete. A sunken Greek galley had been discovered just offshore from the tiny island and some fishermen and sponge divers managed to salvage its cargo of marble, pottery and other objects.

Among the items was an encrusted bronze object of undetermined use. It languished in the reserve section of a museum until 1955, when a curious scientist decided to clean it. He found that it was a complex instrument with cogwheels fitting one into another. Finely graduated circles and inscriptions marked on it in ancient Greek were obviously concerned with its function. The object seems to have been a sort of astronomical clock without a pendulum.

The cargo has enabled the shipwreck to be dated to around the 1st century BC. No Greek or Roman writer has ever described the workings of such an ancient computer, though other marvels of antiquity are mentioned that seem incomprehensible to us.

In 1958, a British scientist named Derek de Solla Price was researching the

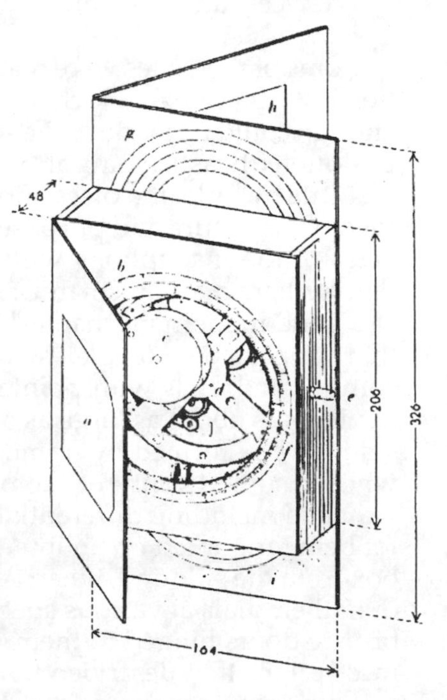

The Antikythera Device reconstructed.

history of scientific instruments when he came across the Antikythera device in the Athens Museum. He was astonished at the complexity of the device and later wrote, "Nothing like this instrument is preserved elsewhere. Nothing comparable to it is known from any ancient scientific text or literary allusion. On the contrary, from all that we know of science and technology in the Hellenistic Age we should have felt that such a device could not exist."[4]

Price was later quoted as saying, "Finding a thing like this is like finding a jet plane in the tomb of King Tut."

Price had previously believed that the first time that such complicated gear-work had appeared was in a clock made in 1575. For more than a decade Price studied the fragments of the machine and, in 1971, had X-ray photographs taken of it by the Greek Atomic Energy Commission. This finally revealed the astonishing array of intercogged wheels.[4,5]

Price described the computer in an article that appeared in the March, 1962, issue of *Natural History* (71:8-17) entitled "Unworldly Mechanics." (It was entitled "Unworldly Mechanics" because Price and others never imagined that the ancient Greeks, Egyptians or other cultures of 100 BC could have had the astronomical or mechanical knowledge to construct such a device—an idea which is just plain wrong.) As Price explains:

> Some of the plates were marked with barely recognizable inscriptions, written in Greek characters of the first century BC, and just enough could be made of the sense to tell that the subject matter was undoubtedly astronomical.
>
> Little by little, the pieces fitted together until there resulted a fair idea of the nature and purpose of the machine and of the main character of the inscriptions with which it was covered. The original Antikythera mechanism must have borne a remarkable resemblance to a good modern mechanical clock. It consisted of a wooden frame that supported metal plates, front and back, each plate having quite complicated dials with pointers moving around them. The whole device was about as large as a thick folio encyclopedia volume. Inside the box formed by frame and plates was a mechanism of gear wheels, some twenty of them at least, arranged in a non-obvious way and including differential gears and a crown wheel, the whole lot being mounted on an internal bronze plate. A shaft ran into the box from the side and, when this was turned, all the pointers moved over their dials at various speeds. The dial plates were protected by bronze doors hinged to them, and dials and doors carried the long inscriptions that described how to operate the machine.
>
> It appears that this was, indeed, a computing machine that could work out and exhibit the motions of the sun and moon and prob-

ably also the planets. Exactly how it did this is not clear, but the evidence thus far suggests that it was quite different from all other planetary models. It was not like the more familiar planetarium or orrery, which shows the planets moving at their various speeds, but much more like a mechanization of the purely arithmetical Babylonian methods. One just read the dials in accordance with the instructions, and legends on the dials indicated which phenomena would occur at any given time.[5]

The British-Greek historian Victor J. Kean maintains in his book *The Ancient Greek Computer from Rhodes*[4] that the Antikythera device was made on the island of Rhodes around 71 BC. Kean theorizes that the machine was made at the ancient metallurgical science city known as Kamiros and was destined for Rome when the transport ship sank.

What the Antikythera device has shown historians is that the ancient world did in fact have a higher science than we had previously given it credit for. As in tales of the Rama empire, Osiris and Atlantis, the ancient past was a world in which isolated areas had complex machinery, electricity, and metallurgical sciences. History has been destroyed, just as Solon the Greek told Plato.

Zoomorphic pendant from Panama.

Zoomorphic Glyphs of Ancient Heavy Machinery

It has been suggested as well that the ancients must have had heavy machinery for construction purposes. Today, bulldozers, mechanical diggers or pneumatic power tools for such things as quarry work, are commonplace. Many individuals, especially farmers, even have heavy machinery to dig around their ditches. But did the ancients have John Deere tractors and Caterpillar backhoes?

Ivan T. Sanderson says in *Investigating the Unexplained*[8] that he investigated small gold models of airplanes from Columbia, as well as a gold model of a "dozer." The dozer model was found in Panama by archaeologists in the 1920s, says Sanderson, who was apparently at the site at some point.

Sanderson says that the site was on the

land "of a family named Conté, in the Province of Colclé, which is on the southern coast of Panama to the west of the Canal Zone. This site was near the town of Penonomé... Here were found hundreds of graves containing masses of pottery, some furnerary urns of children, and masses of gold ornaments, body shields, and jewelry. The Peabody Museum of Harvard University carried out somewhat extensive diggings on this site in 1930, 1931, and 1933." The object is currently in the University Museum of Philadelphia.[8]

The dozer is described as being made by a consummate artist out of gold and containing a huge green gemstone (probably an emerald). It is apparently intended to be a pendant and is four and a half inches long. It was first described as a crocodile, but later as a jaguar by others. The object is clearly covered with mechanical devices, however, including two cogged wheels.

Sanderson mentions that jewelry substituted for coinage in the ancient world of the Americas. Therefore, jewelry often travelled large distances over many hundreds, or even thousands of years. Though the graves in Panama contained material from only about 1000 AD, the pendant was very possibly much older. And, it does look like a dozer, with mud guards and a digging hoe. Indeed the whole thing looks like the model of some sort of digging machine, but it is a zoomorphic depiction.

There is a very strange and interesting rock sketch that was discovered at Merowe, a city of ancient Kush, a country just to the south of Egypt, in present-day Sudan. It appears in the German archaeologist Philipp Vandenberg's book *Curse of the Pharaohs*[25] and is reproduced here. The sketch shows two men operating a device that is said to be identical to a radiation condenser or a laser gun. Others believe it is a rocket of some sort, or a telescope or some sophisticated ray gun. Readers of this book may decide for themselves what it looks like. Academic "experts" have nothing to say on the subject, except that it can't be a laser, rocket or radiation gun because they didn't have such devices at the time. Perhaps it was an excavating device or a rock quarrying device—the possibilities are endless if we assume that ancient societies had access to high technology.

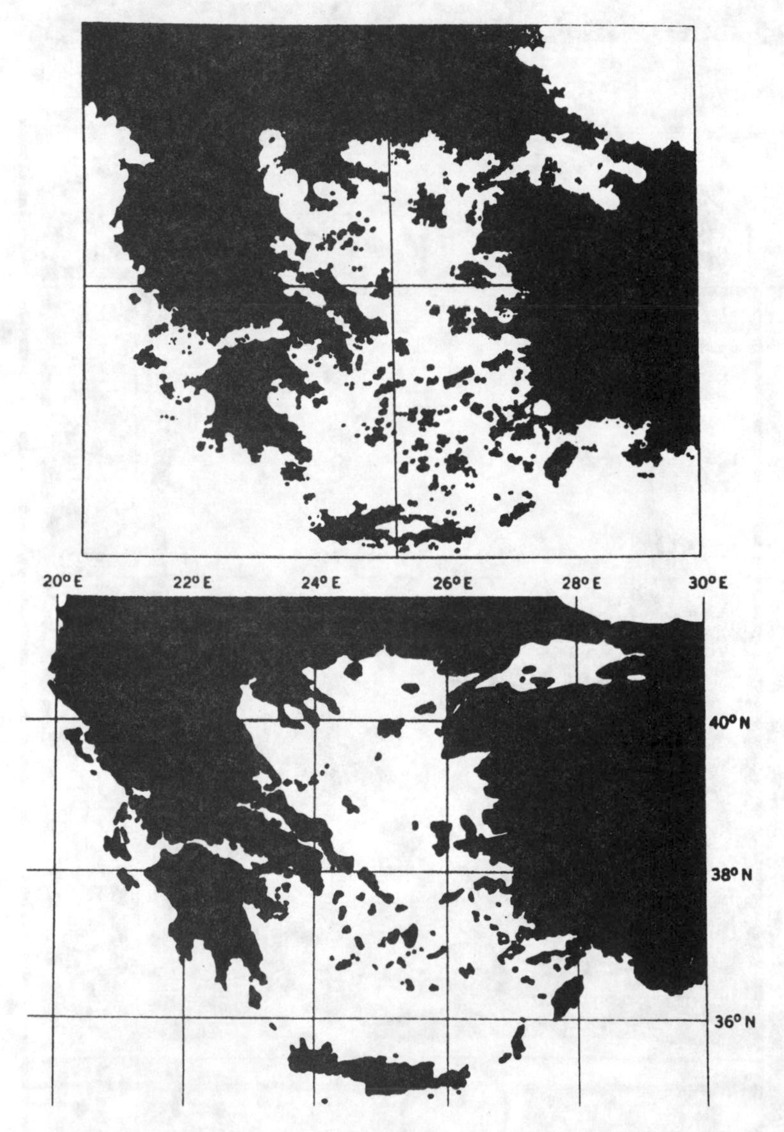

Top: A detail of the Aegean from the Ibn Ben Zara map compared with a modern map of the Aegean. Below: A modern map of the Aegean showing fewer islands than were recorded on earlier maps.

A Hittite stele from central Turkey. The Hittites controlled iron making in Asia Minor for hundreds, if not thousands of years.

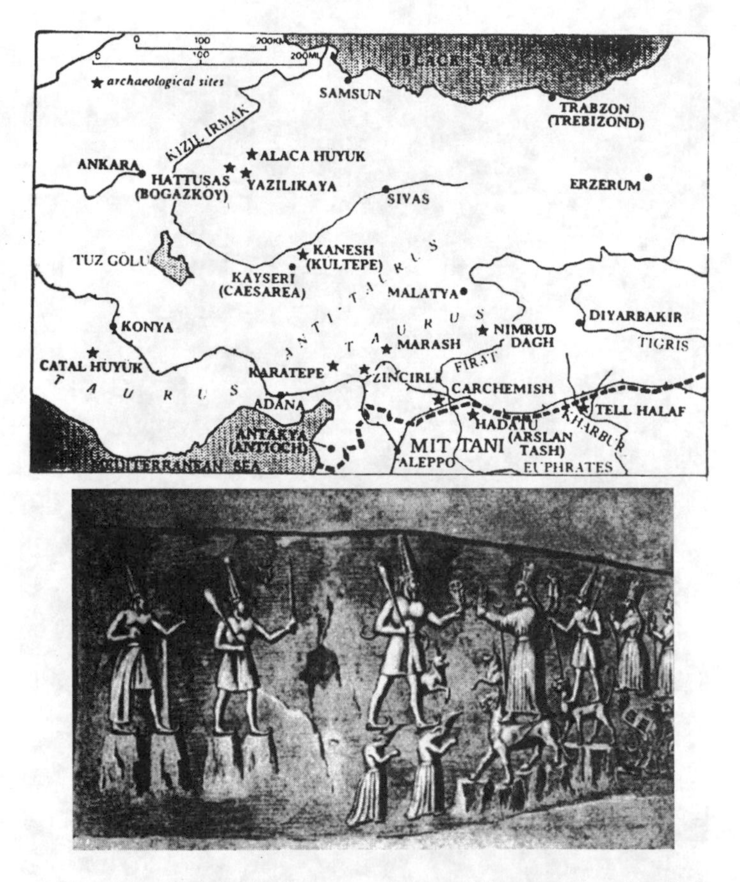

Top: A map of the Hittite areas of control. Bottom: A rock relief at Haṭtusas. The Hittites wore pointy shoes and hats, but were fierce warriors and terrorized Asia Minor with their iron weapons. Ultimately, they were completely destroyed and their cities literally vitrified by intense heat.

The metallic sphere from the Ottosdal Mines in South Africa. The mineral strata was believed to be Precambrian, or 2.8 billion years old. The sphere has three parallel grooves around its equator.

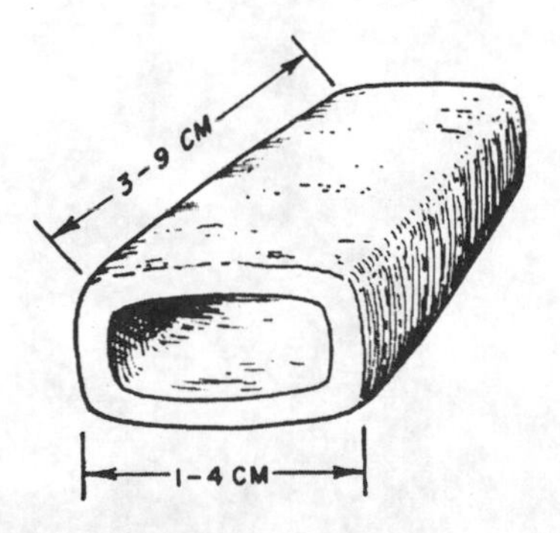

One of many metallic tubes found at Saint-Jean de Livet, France, in a chalk bed that was thought to be 65 million years old.

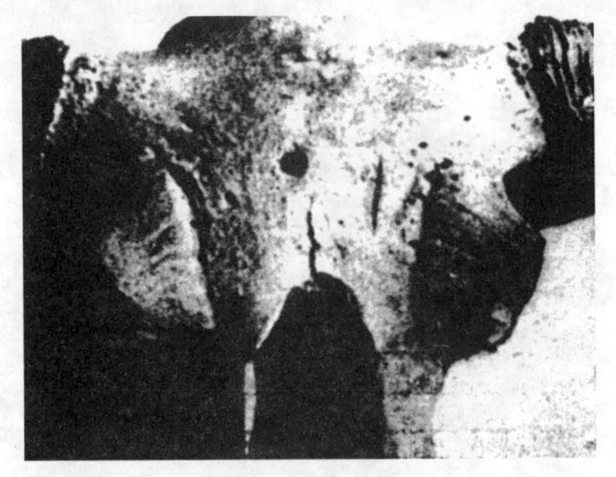

An auroch skull from Russia with a bullet hole in it.

This metalic vessel was blown out of solid rock in Dorchester, Massachusetts in 1851.

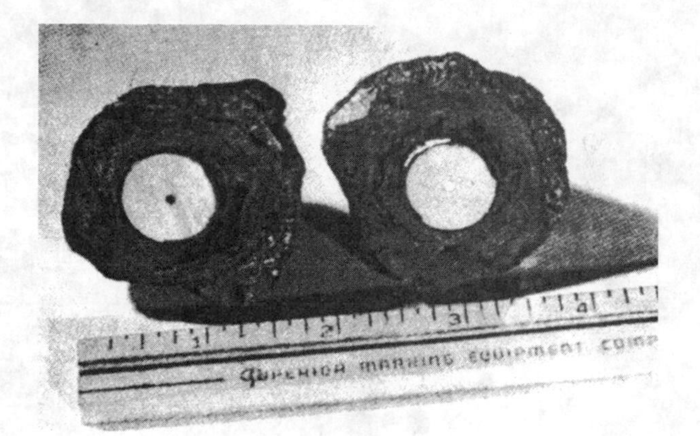

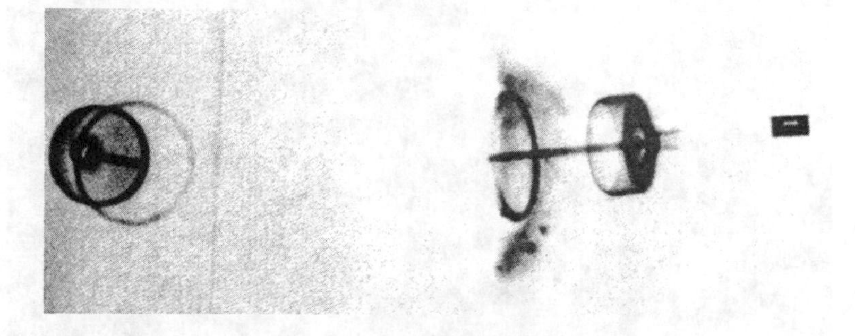

Top: The Coso Artifact—a spark plug in a geode? Bottom: An x-ray of the metal disks and rods inside the geode.

The iron pillar of Delhi.

Foto: *Revue de l'Aluminium*, n.º 283, 1961

Aluminum belt buckles found in an ancient Chinese grave.

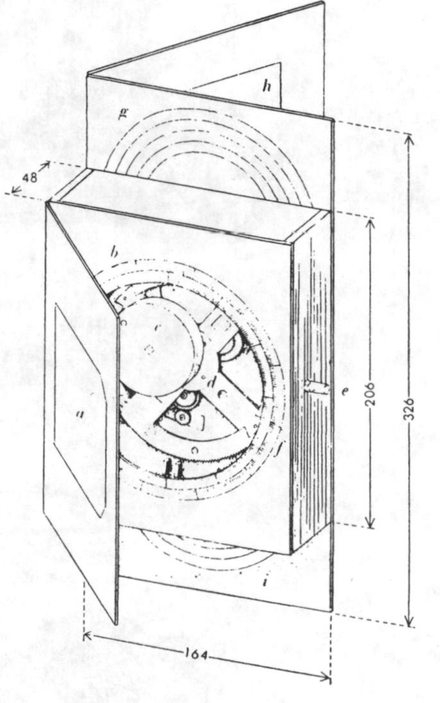

The parts of the mechanism of the 'computer' or astronomical clock were in a sorry state when discovered (*above*), but through the careful work of the museum technicians, and with the help of George Stamires who deciphered the inscriptions, Mr. Price was able to reconstruct the position of the pieces.

The labelled parts in the reconstruction are: *a*: front door inscription; *b*: front dial; *c*: eccentric drum; *d*: front mechanism; *e*: input shaft; *f*: fiducial mark; *g*: four slip rings of upper back dial; *h*: back door inscription; *i*: three slip rings of lower back dial. The dimensions are given in millimetres.

Left. Segment of the lower back dial. At right is a fixed scale; within it were three slip rings and within them a subsidiary dial.

Right. Segment of front dial. The upper scale pertains to the months, the lower to the zodiac. The inscribed area is a parapegma (astronomical calendar) plate. The various dials show the annual motion of the sun in the zodiac and also the main risings and settings of bright stars and constellations throughout the year.

Photograph *Derek Price*
Diagram *Scientific American*

ANTIKYTHERA

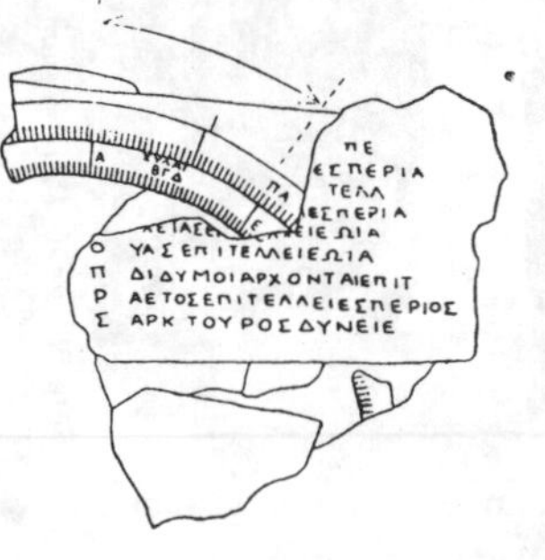

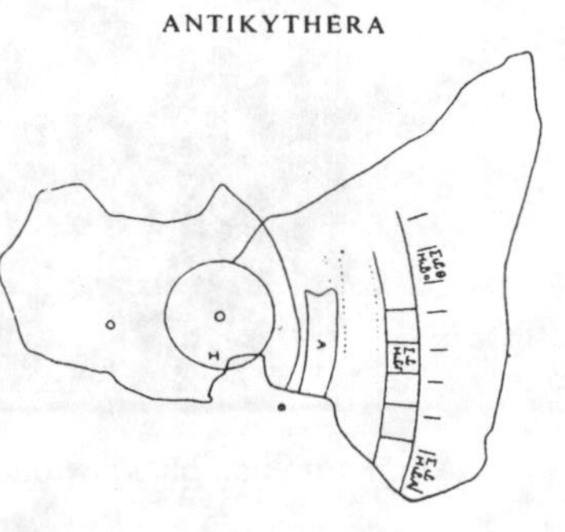

The Antikythera Device.

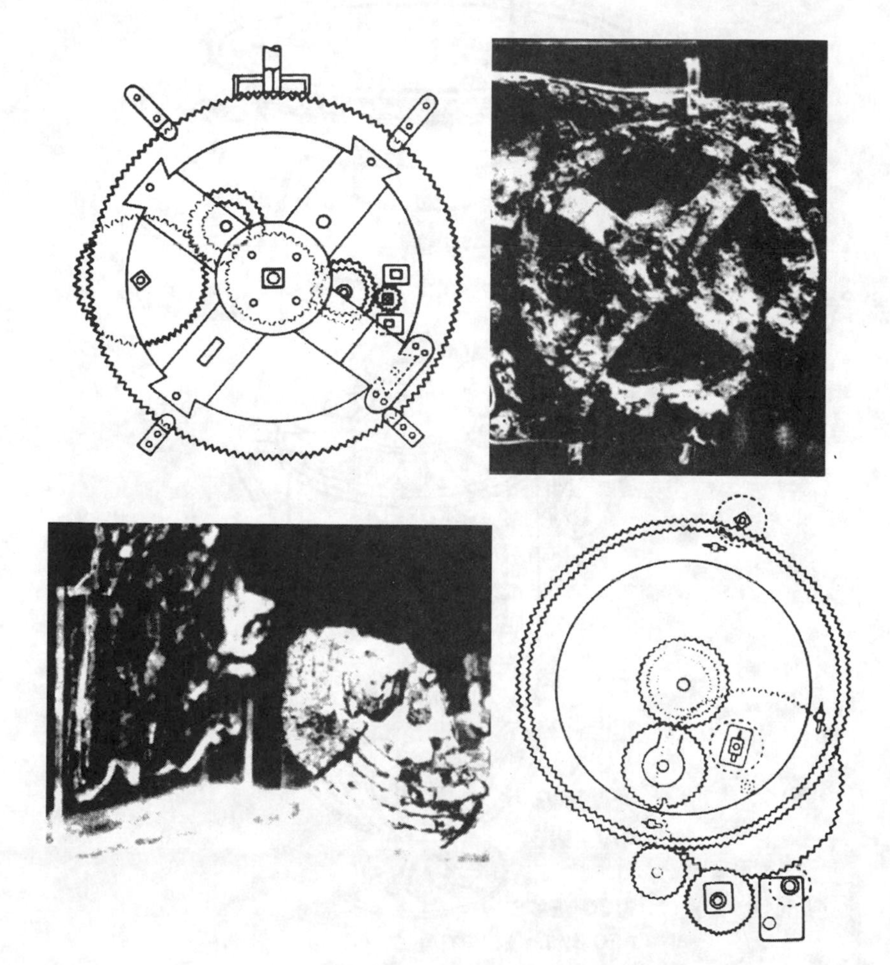

The Antikythera Device.

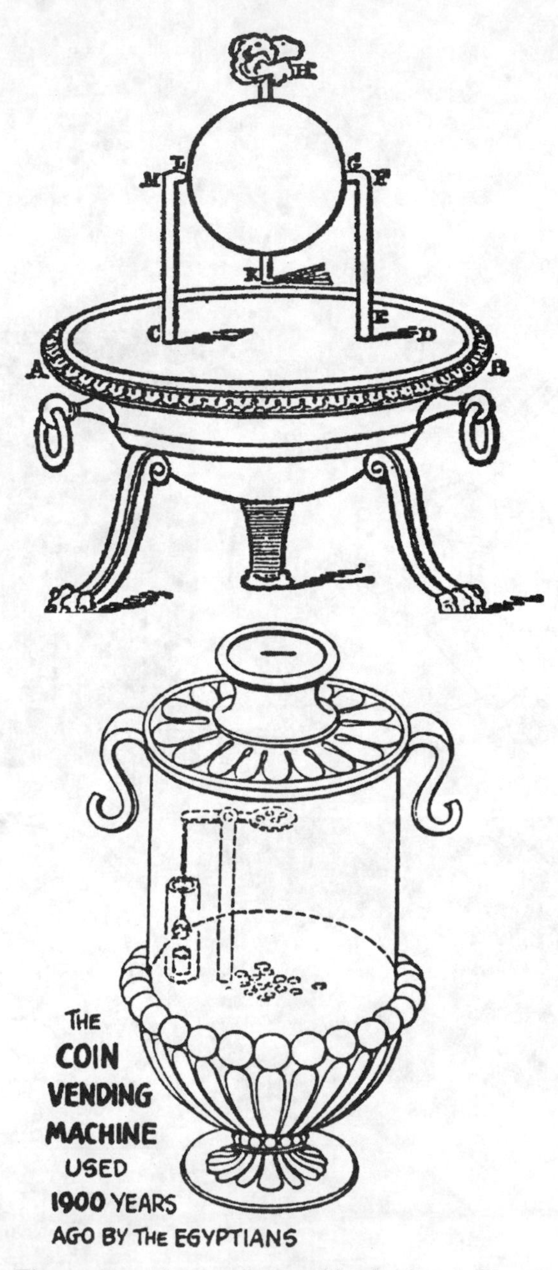

To: The steam engine-toy invented by the Greek-Egyptian inventor Hero of Alexandria. Bottom: One of Hero's in ventions, a coin operated holy water dispenser.

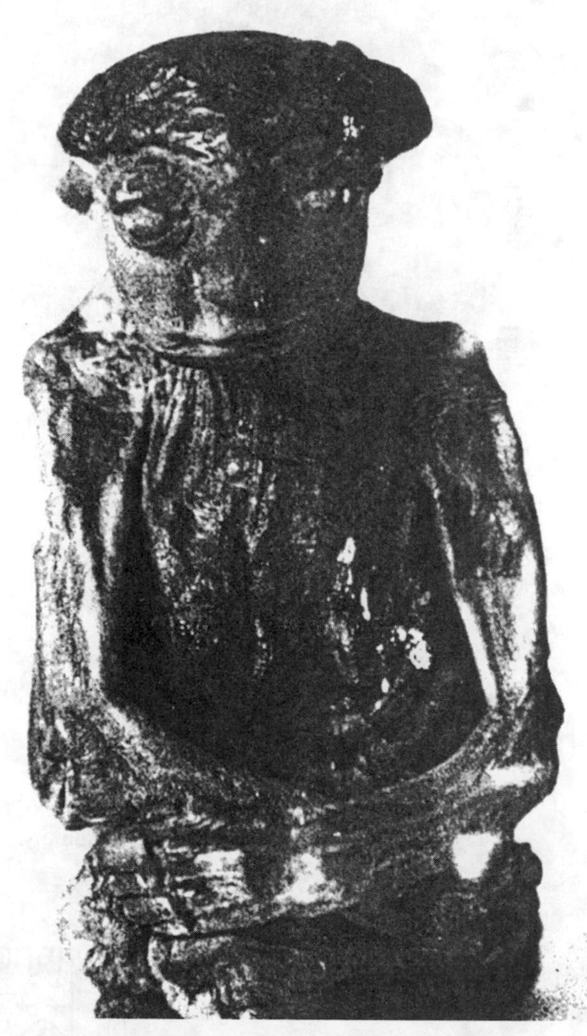

One of the few photographs of the Pedro Mountain, Wyoming, mummy. It was blasted out of solid rock in 1932. Today, its whereabouts are unknown.

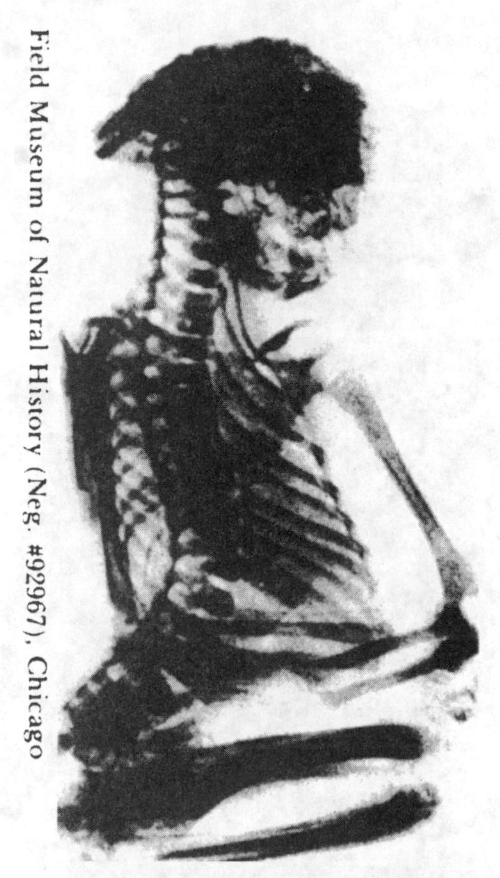

Field Museum of Natural History (Neg. #92967), Chicago

An x-ray of the Pedro Mountain, Wyoming, mummy.

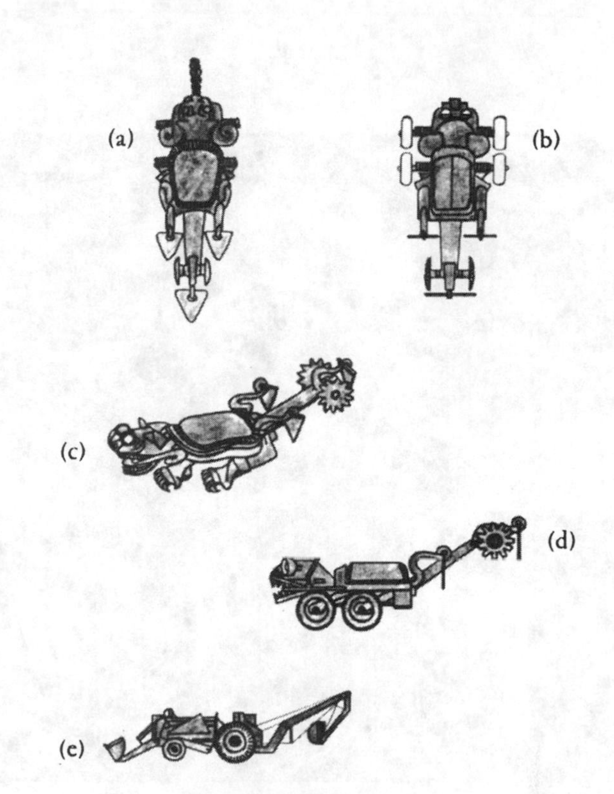

The zoomorphic gold and "emerald" pendant from Coclé on the south coast of Panama: (a) hung as a pendant; (b) as seen from above, "squared-off," with "mudguards" hanging down, and possible riding wheels indicated; (c) object as from a photograph taken in the University Museum of Philadelphia; (d) the same, rectified for lateral view, also "squared-off" and with wheels added; (e) a modern backhoe with dozer-bucket scoop as front attachment.

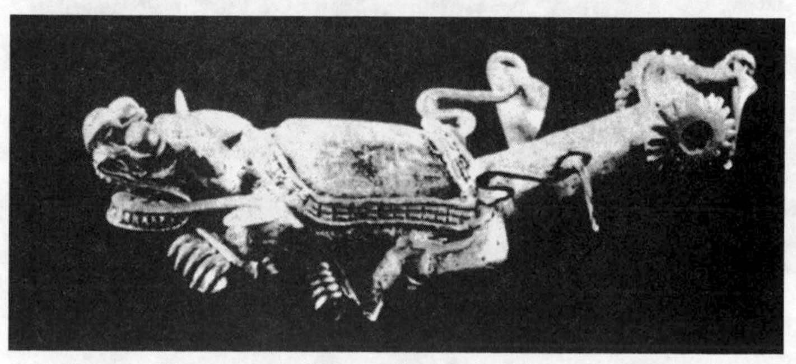

The zoomorphic pendant from Panama. Side view.

The zoomorphic pendant from Panama. Front view.

The zoomorphic pendant from Panama. Side view.

A modern zoomorphic depicition of heavy machinery from a 1940s science fiction magazine.

Ivan T. Sanderson

Hat and Har-hat

sphinx

Top: Horus, Osiris, Toth, and other gods bring science and civilization to the world. Bottom: The Sphinx appears in ancient art from North Africa to Central Asia and China.

This strange tablet is attributed to Narmer, the legendary first Pharoah of a united Egypt.

The Tombs of the Genii in Siberia, possibly the largest megaliths ever discovered. Now seemingly lost, these monstrous megaliths (note the horse) are located on the Kora River in what was Soviet Turkestan and were depicted in the 1876 book *The Early Dawn of Civilization* (Victoria Institute Journal of Transactions).

An old photo of a megalithic structure in Madagascar.

4. Ancient Electricity & Sacred Fire

I never pass by a wooden fetish,
a gilded Buddha, a Mexican idol without reflecting:
perhaps this is the true God.
—Charles Baudelaire

... that is what learning is.
You suddenly understand something
you've understood all your life,
but in a new way.
—Doris Lessing

2,000-Year-Old Electrical Batteries

Electric batteries 2,000 years ago? Shocking but true! True technology of the advanced kind requires some kind of power, usually electricity, or at least a control panel that uses electricity. Think of the amazing array of devices that we use today, from automobiles to airplanes, from toaster ovens and refrigerators to power tools and computers—all of them use electricity in one form or another.

That the ancients could harness electricity is absolutely essential to the belief that a high technology once existed in the remote past. We are all familiar with story of the great American statesman, writer and inventor Benjamin Franklin flying his kite in a thunderstorm; but the study of electricity has no doubt gone on for thousands and thousands of years. Benjamin Franklin is credited with the invention of the lightning rod, though, like most everything else, this too was no doubt used by the ancients. Andrew Tomas, in his 1971 book *We Are Not the First* cites this example: "In 1966 the author visited Kulu Valley in the

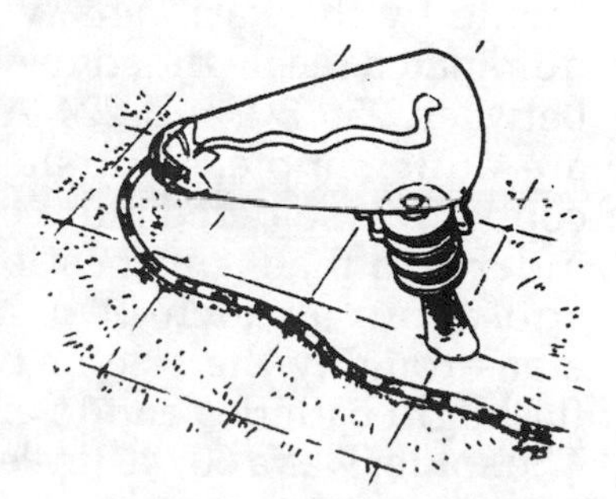

Himalayas. In the town of Kulu there is a remarkable old temple on a hill, dedicated to the god Shiva. Its special feature is an eighteen-meter iron mast erected near the temple. In an electric storm the pole attracts the 'blessing of Heaven' in the form of lightning which flashes down the mast and hits a statuette of Shiva at its base. The pieces of shattered Shiva are then pasted together by the priest and used for the next 'blessing.' The custom has existed since time immemorial, which would mean that the presence of electric conductors in India has been a reality from the most ancient times."[24]

Electric batteries were in use more than 2,000 years ago says Dr. Wilhelm Koenig, a German archaeologist employed by the Iraq Museum in Baghdad, who discovered one in 1938 while conducting a dig at Khujut Rabu'a, not far to the southeast of Baghdad. The Museum had begun scientific excavations, and in the digging turned up a peculiar object that—to Koenig—looked very much like a present-day dry-cell.[5] Other similar finds followed.

A *Popular Electronics* article in the July, 1964 issue said that the ancient electrochemical batteries had central cell elements that included "...a copper cylinder containing an iron rod that had been corroded as if by chemical action. The cylinder was soldered with a 60/40 lead-tin alloy, the same solder alloy we use today." Two thousand years ago they not only had electricity, they had also come up with exactly the same tin-alloy solder that we use today!

An earlier article on this amazing ancient technology appeared in the April 1957 issue of *Science Digest* entitled "Electric Batteries of 2,000 Years Ago." (Harry M. Schwalb, *Science Digest*, 41:17-19). Says the article, "...in Cleopatra's day, up-and-coming Baghdad silversmiths were goldplating jewelry—using electric batteries. It's no myth; young scientist Willard F. M. Gray, of General Electric's High Voltage Laboratory in Pittsfield, Mass., has proved it. He made an exact replica of one of the 2,000-year-old wet cells and connected it to a galvanometer. When he closed the switch—current flowed!

"These BC-vintage batteries (made by the Parthians, who dominated the Baghdad region between 250 BC and 224 AD) are quite simple. Thin sheet-copper was soldered into a cylinder less than 4 inches long and about an inch in diameter—roughly the size of two flashlight batteries end to end. The solder was a 60/40 tin-lead

Parts of the Baghdad Battery.

alloy—'one of the best in use today,' Gray points out.

"The bottom of the cylinder was a crimped-in copper disc insulated with a layer of asphaltum (the 'bitumen' that the Bible tells us Noah used to caulk the Ark). The top was closed with an asphalt stopper, through which projected the end of an iron rod. To stand upright, it was cemented into a small vase.

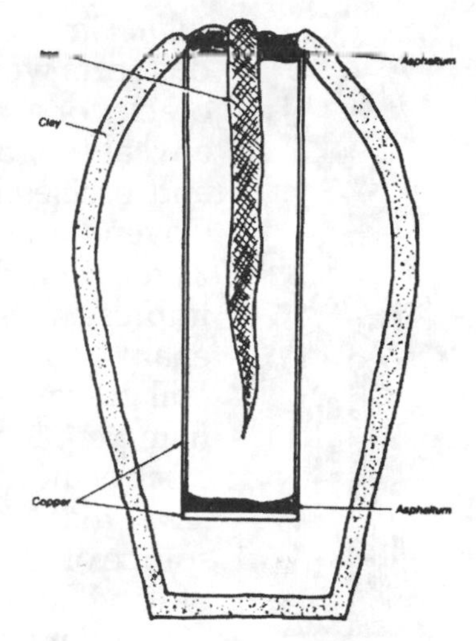

"What electrolyte the Parthian jewelers used is a mystery, but Gray's model works well with copper sulfate. Acetic or citric acid, which the ancient chemists had in plenty, should be even better."

This is conclusive proof that the Babylonians did indeed use electricity. As similar jars were found in a magician's hut, it can be surmised that both priests and craftsmen kept the knowledge as a trade secret. It should be noted here that electroplating and galvanization were again introduced only in the first part of the nineteenth century.[24]

Andrew Tomas (whose book is quoted above) was an Australian who travelled widely. He mentions that during his stay in India he was told about an old document, preserved in the Indian Princes' Library at Ujjain and listed as *Agastya Samhita*, which contains instructions for making electrical batteries:

> Place a well-cleaned copper plate in an earthenware vessel. Cover it first by copper sulphate and then by moist sawdust. After that put a mercury-amalgamated-zinc sheet on top of the sawdust to avoid polarization. The contact will produce an energy known by the twin name of Mitra-Varuna. Water will be split by this current into Pranavayu and Udanavayu. A chain of one hundred jars is said to give a very active and effective force.

Says Tomas, "The Mitra-Varuna is now called cathode-anode, and Pranavayu and Udanavayu are to us oxygen and hydrogen. This document again demonstrates the presence of electricity in the East, long, long ago."[24]

Electricity and Religion

Yet, the knowledge of electrical devices was not limited to batteries for electroplating. Authors such as Jerry Ziegler in his books *YHWH*[22] and

Asar (Osiris)

Indra Girt by Maruts[89] claim that electrical devices of various sorts were used in temples and were often utilized as oracles or awesome manifestations of deities. Ziegler cites a wealth of ancient sources on ancient lights, sacred fires and oracles in his books. He argues that the Ark of the Covenant as well as the sacred flames of Mithraic and Zoroastrian oracles were ancient electrical devices used for impressing the congregation.[22] The famous Ark of the Covenant is often described as an electrical device, and several passages in the Old Testament describe how unfortunate people who touch the relic are killed, seemingly by electrocution. Ancient Hebrew legends tell of a glowing jewel that Noah hung up in the Ark to provide a constant source of illumination and of a similar object in the palace of King Solomon about 1000 BC.[9]

Similar devices were apparently used by Native Americans in special ceremonies held in underground chambers known as "kivas." For instance, David Chandler mentions in his book *100 Tons of Gold*[91] that the Hopi of Northern Arizona had a fascinating generator for making light which was made out of luminescent quartz. It consisted of a rectangular base of pure white-vein quartz with a groove in it and a bolster-shaped upper piece of the same material. Friction produced by rapid rubbing made a strong glow in the dark, which was used to light the sacred kivas.

Says Chandler, "The machine still worked perfectly when it was discovered by archaeologist Alfred Kidder in the Pecos ruins, as he reported in 1932. Archaeologist S.H. Ball remarked upon it, 'Here we have a perfected machine perhaps seven hundred years old; the first Indian to observe luminescence of quartz must have done so centuries earlier.'"[91]

Chandler goes on to say that similar "lightning machines" or "glow stones" have been found at several other localities in north-central New Mexico. Chandler is quoting from Stuart A. Northrop's *Minerals of New Mexico* (1959, University of New Mexico Press, Albuquerque) on the existence of the quartz light machine that the ancient Indians used. These machines may still be being used by Hopi or other tribes to generate light in secret ceremonies in their kivas.

Ancient electrics in many cases were apparently only used by special priesthoods and not by the masses. In Ziegler's *Indra Girt by Maruts*[89] it is maintained that many of the ancient Vedas also describe electrical devices, and these were typically used in religious ceremonies.

A similar book to Ziegler's is the difficult to find *Ka: A Handbook of Mythology, Sacred Practices, Electrical Phenomena, and their Linguistic Connections in the Ancient World,* by Hugh Crosthwaite.[90] Crosthwaite's fascinating 1992 book maintains that the ancients built simple—and more complicated—electrical devices that were used in religious ceremonies. These sacred "fires" ranged from amber disks that created sparks of static electricity when rubbed together (easily seen in a darkened room) to static electrical condensers such as the famous Ark of the Covenant.

What is important about Crosthwaite's book is that he shows how much of early religion was built around electrical phenomena. The many famous temples may have had as their center of attraction an electrical light of some sort that amazed the pilgrim and gave him something to genuinely wonder about.

Tomas mentions that Lucian (AD 120-180), the Greek satirist, gave a detailed account of his travels. In Hierapolis, Syria, he saw a shining jewel in the forehead of the goddess Hera which brilliantly illuminated the whole temple at night. Nearby, the Roman temple of Jupiter at Ba'albek was said to be lit by "glowing stones."[24]

Crosthwaite says the Ka of the ancient Egyptians is related to electrical phenomena and that much of the teachings of the so-called Mystery Religions, such as at Delphi in Greece, were related to various electrical devices as well. Over time, civilization slipped into the Dark Ages and the old religions were swept away by Christianity and Islam.

Electric Eternal Flames

Australian author and researcher Andrew Tomas, who was well versed on classical texts of both east and west, has an entire chapter entitled "Electricity in the Remote Past" in *We Are Not the First.*[24] This chapter has a long list of classical authors who have made many statements in their works testifying to the reality of ever-burning lamps in antiquity. Some of these ever-burning lamps may have used ancient electrical devices of various design.

A beautiful golden lamp in the temple of Minerva, said to burn for a year at a time, was described by the second-century historian Pausanias. Saint Augustine (AD 354-430) wrote of an ever-burning lamp which neither wind nor rain could extinguish.

Tomas relates that when the sepulchre of Pallas, son of Evander, immortalized by Virgil in his *Aeneid,* was opened near Rome in 1401, the tomb was found to be illuminated by a perpetual lantern which had apparently been alight for hundreds of years.[24]

Tomas also says that Numa Pompilius, the second king of Rome, had a perpetual light shining in the dome of his temple. Plutarch wrote of a lamp which burned at the entrance of a temple to Jupiter-Ammon, and its priests claimed that it had remained alight for centuries.

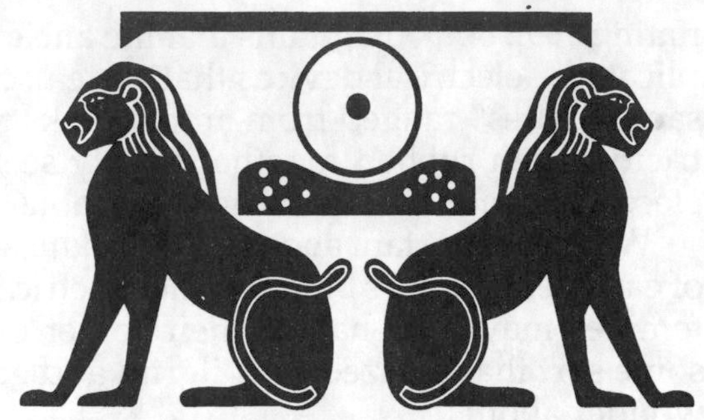

He claims that an ever-burning lamp was found at Antioch during the reign of Justinian of Byzantium (sixth century AD). An inscription indicated that it must have been burning for more than five hundred years. During the early Middle Ages a third-century perpetual lamp was found in England, and it had burned for several centuries.

Tomas also mentions a sarcophagus containing the body of a young woman of patrician stock that was found in a mausoleum on the Via Appia near Rome in April, 1485. When the sealed mausoleum which had housed the sarcophagus was opened, a lighted lamp amazed the men who broke in. It must have been burning for 1,500 years! When the dark ointment preserving the body from decomposition had been removed, the girl looked lifelike with her red lips, dark hair, and shapely figure. It was exhibited in Rome and seen by 20,000 people.

Quoting Tomas on other examples of ancient lighting:

> In the temple of Trevandrum, Travancore, the Reverend S. Mateer of the London Protestant Mission saw 'a great golden lamp which was lit over one hundred and twenty years ago,' in a deep well inside the temple.
>
> Discoveries of ever-burning lamps in the temples of India and the age-old tradition of the magic lamps of the Nagas—the serpent gods and goddesses who live in underground abodes in the Himalayas—raises the possibility of the use of electric light in a forgotten era. On the background of the *Agastya Samhita* text's giving precise directions for constructing electrical batteries, this speculation is not extravagant.
>
> In Australia the author learned of a village in the jungle near Mount Wilhelmina, in the western half of New Guinea, or Irian. Cut off from civilization, this village has "a system of artificial illumination equal, if not superior, to the 20th century," as C. S. Downey stated at a conference on street lighting and traffic in Pretoria, South Africa, in 1963.
>
> Traders who penetrated this small hamlet lost amid high mountains said that they "were terrified to see many moons suspended in the air and shining with great brightness all night long. These artifi-

cial moons were huge stone balls mounted on pillars. After sunset they began to glow with a strange neon-like light, illuminating all the streets.

Ion Idriess is a well-known Australian writer who has lived amongst the Torres Strait islanders. In his *Drums of Mer* he tells of a story about the booyas which he received from the old aborigines. A booya is a round stone set in a large bamboo socket. There were only three of these stone scepters known in the islands. When a chief pointed the round stone toward the sky, a thunderbolt of greenish-blue light flashed. This "cold light" was so brilliant that the spectators seemed to be enveloped in it. Since Torres Strait washes the shores of New Guinea, one can perceive some connection between these booyas and the "moons" of Mount Wilhelmina."[24]

Other mysterious lights and "glowing stones" have been reported in lost cities around the world. Tibet is said to have such glowing stones and lanterns mounted on pillars in towers. Tomas relates that Father Evariste-Regis Huc (1813-1860), who travelled extensively in Asia in the 19th century, left a description of ever-burning lamps he had seen, while the Russian Central Asian explorer Nicholas Roerich reported that the legendary Buddhist secret city of Shambala was lit by a glowing jewel in a tower.

Atlantis and ever-lasting stone lamps were featured in the beliefs of the famous British explorer Colonel Percy Fawcett, who vanished in the Brazilian jungles while searching for a lost city which he believed was lit by glowing stones on pillars. Tomas quotes a letter sent by Fawcett to British Atlantis authority Lewis Spence about the lost city in the jungle and what the natives had told him about the glowing stones. "These people have a source of illumination which is strange to us—in fact, they are the remnant of a civilization which has gone and which has retained old knowledge."[24, 57] Colonel Fawcett disappeared with his oldest son in 1925 but his youngest son published a book of his father's material in 1953 entitled *Expedition Fawcett*[88] (called *Lost Trails, Lost Cities* in the US edition).

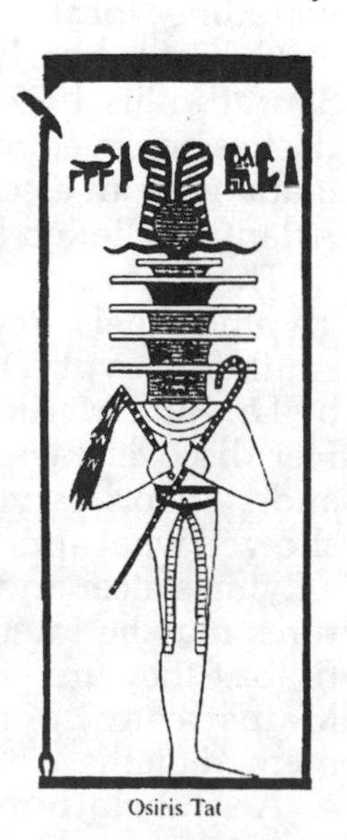
Osiris Tat

Colonel Fawcett never reported finding his city, but Tomas (probably quoting from Harold Wilkins' books on South America),[124,125] relates that in 1601 the Spanish author Barco Centenera recorded the discovery of a similar-sounding place. Centenera wrote of the discovery of the lost city of Gran Moxo, located near the source of the Paraguay River in the Mato Grosso. In the center of the island city he says "on the summit of a 20 foot pillar was a great moon which illuminated all the lake, dispelling darkness."[22]

As Tomas avers:

> History shows that the priests of India, Sumer, Babylon, and Egypt, as well as their confreres on the other side of the Atlantic—in Mexico and Peru—were custodians of science. It appears likely that in a remote epoch these learned men were forced to withdraw into inaccessible parts of the world to save their accumulated knowledge from the ravages of war or geological upheavals. We still are not certain as to what happened to Crete, Angkor, or Yucatan and why these sophisticated civilizations suddenly came to an end. If their priests had foresight, they must have anticipated these calamities.
>
> In that case they would have taken their heritage to secret centers as the Russian poet Valery Briusov depicted in verse:
>
> *The poets and sages,*
> *Guardians of the Secret Faith,*
> *Hid their Lighted Torches*
> *In deserts, catacombs and caves.*[24]

Electric Lights in Ancient Egypt?

Tomas says that the Jesuit Kircher, in his *Oedipus Aegyptiacus* (Rome, 1652), tells of lighted lamps found in the subterranean vaults of Memphis. Here we have reference to electric lights in Egypt, these still functioning, incredibly, for thousands of years.[24]

One of the early proponents of electricity in ancient Egypt was Denis Saurat in his 1957 book *Atlantis & the Giants*.[81] Saurat suggests that the flashes observed in the eyes of Isis in her temples throughout Egypt were made with an electrical apparatus. Like many other authors, Saurat saw Atlantis as linked to the sciences of the ancient world.

There have also been high-tech devices depicted in Egyptian hieroglyphic panels. Recently in the news was one found at Abydos Temple in southern Egypt. This had been discovered in 1987 by Dr. Ruth McKinley-Hover of Sedona, Arizona. Her discovery was of a lintel with hieroglyphics and symbols carved into the granite rock which showed what appeared to be a helicopter, a rocket, a flying saucer and a jet plane. These unusual pictures may be interpreted by the reader as seems fit, but they are genuine, and not a clever hoax. Mainstream Egyptologists have not yet commented on these hieroglyphs.

An Egyptian temple flame.

At the Hathor Temple at Dendera, near to

Abydos, is found an unusual depiction of what appears to be an ancient Egyptian electrical device. Like the Temple of Osiris, Dendera is a beautiful and massive edifice with huge columns that tower over one's head like redwoods. The temple is of quite recent origin, built in the 1st century BC, but it encloses earlier temples. An inscription in one of the subterranean vaults says that the temple was built "according to a plan written in ancient writing upon a goatskin scroll from the time of the Companions of Horus." This is a curious inscription, essentially stating that the Ptolemaic (Greek) architects of the 1st century BC were claiming that the actual plan of the temple dated to the legendary prehistoric era when the "companions of Horus" ruled Egypt. This long era extended for many thousands of years, and in a sense takes us back, once again, to the legendary civilization of Osiris.

The relief at the Temple of Hathor at Dendera.

The temple is richly decorated with inscriptions and hieroglyphics. Probably most interesting to me was an incised petroglyph in the room designated No. XVII that depicts a strikingly unusual scene with what appear to be electrical objects. The famous British scientist Ivan T. Sanderson discusses this petroglyph and ancient Egyptian electricity in his book *Investigating the Unexplained.*[8] In the petroglyph, attendants are holding two "electric lamps" supported by "Djed" pillars and connected via cables to a box. Djed columns are interesting, as they are usually associated with Osiris. They are said to represent the column in which he was found at Byblos in Lebanon by Isis. The Djed columns are explained as insulators, though they are probably electrical generating devices themselves due to the odd "condenser" design at the top of the columns. An electrical engineer named Alfred Bielek explained the petroglyph to Sanderson as depicting some sort of projector with the cables being a bundle of many multi-purpose conductors, rather than a single high-voltage cable.

Another depiction from an 18th dynasty papyrus scroll shows "sacred baboons" and priests worshipping a Djed column with an ankh with hands holding up an orb. Sanderson likens the object to static generators such as a Van de Graaff generator or Wimshurst generator. Sanderson had Michael R. Freedman, an electrical engineer, draw up plans for his version of a Djed-column static-electricity generator. They indeed looked very similar to the modern Van de Graaff generator found in many high school science laboratories.

In such a device, static electricity builds up in the orb, and, says Freedman, "...what better 'toy' for an Egyptian priest of ancient times? ...such an instrument could be used to control both the Pharaoh and fellahin (peasant), simply by illustrating, most graphically, the powers of the gods; of which, of course, only the priests knew the real secrets. Merely by placing a metal rod or metal-coated stave in the general vicinity of the sphere, said priest could produce a most wondrous display, with electric arcs and loud crashes. Even with nothing more elaborate than a ring on his finger, a priest could point to the 'life-symbol,' be struck by a great bolt of lightning, but remain alive and no worse for wear, thus illustrating the omnipotent powers of the gods—not to mention himself—in preserving life for the faithful."[8]

Though the device may have been some exotic, but simple, static generator, it might also have been a self-generating electric light tower and bulb. A glowing electric ball in the center of an ornate temple would have been an impressive sight. Did the Egyptians use electric lighting? It would seem so!

Part of the evidence for ancient Egyptian electrics is the mystery of why tombs and underground passages are highly painted and decorated, yet there is no smoke residue or evidence of torches on the ceilings! It is usually assumed that the artists and workers would have to have worked by torchlight, just as early Egyptologists did in the 1800s. However, no smoke is found on the tombs. One ingenious theory was that the passageways and chambers were lit by series of mirrors, bringing sunlight from the entrance. However, many tombs are far too elaborate, with deep and twisting turns, for this to work.

Djed column with orb.

Ark of the Covenant—Electrified?

It is my belief that the famous Biblical Ark of the Covenant was in part an ancient electrical device that was Egyptian in origin. Furthermore, it may have come out of the Great Pyramid, or the underground tunnels that have recently been discovered beneath the Giza Plateau. Graham Hancock in his bestselling 1992 book *The Sign and the Seal*[162] says that the nested sarcophagi of the young Pharaoh Tutankhamen was apparently a similar type of box as that described as the Ark of the Covenant. According to Hancock, this sort of special construction for a box was relatively common in ancient Egypt. He too is a believer in Egyptian electrics and other special knowledge left over from ancient civilizations.

Just what was the Ark of the Covenant, anyway? The Ark of the Covenant first appears in the story of the Exodus and approximately 200 other times in the Old Testament. Moses is said to have symbolically placed a copy of the Ten Commandments inside the Ark, which was a nesting of three boxes one inside the other. Descriptions of the Ark in the Bible are brief and scanty, but it seems that the box, or "Ark" was something between four and five feet long and two to three feet in both breadth and width. The three boxes were a sandwiching of gold, a conducting metal, and acacia wood, a non-conductor. There were dangers in handling the ark, which was generally done by the Levites who were said to have worn protective clothing. The Bible reports one tragedy that happened when the Ark was touched incorrectly.

In II Samuel, Chapter 6, the ark is being transported by ox cart. Apparently this gave the ark an unsteady ride and as the Bible says, "And when they came to Nachon's threshingfloor, Uzzah put forth his hand to the ark of God, and took hold of it, for the oxen shook it.

And the anger of the Lord was kindled against Uzzah: and God Smote him there for his error; and there he died by the ark of God."

Uzzah was immediately stunned into death by the force that was part of the ark! This is probably quite true, because such a sandwiching of a conductor and a non-conductor creates what is known as an electrical condenser. A condenser such as the Ark would accumulate static electricity over a period of days (or years) until it was suddenly discharged onto a person, or grounded by means of a conductor, like a wire or metal rod touching the ground. If the Ark had not been grounded for some long period of time, the electrical charge built up in it could give a very nasty and fatal shock to someone who touched it. If the shock itself was not fatal, then the surprise of the shock could well be. After the Ark had been discharged, however, it would be quite safe to touch as many of the Temple Priests would demonstrate.

Another part of the Ark of the Covenant was a golden statue, whose importance is often missed. Indeed, in esoteric literature, it is the most significant part of the Ark. It is described in the Bible as the "Holy of Holies." It was a solid gold statue of two cherubim (angels) facing each other, their wing tips touching above them. They hold between them a shallow dish with their outstretched arms. This was known as the "Mercy Seat."

It is upon this Mercy Seat that an esoteric flame called in Hebrew the "Shekinah Glory" rests. The Shekinah Glory is supposedly a kind of "spirit fire" which was maintained from a distance, originally by Moses and later by an Adept of the Temple. If the person viewing the Holy of Holies was able to detect the Shekinah Glory, that showed psychic talent, as it could only be perceived by a person with clairvoyant ability.

This statue, along with the Ark, is suddenly referred to in the Biblical book of *Exodus*, and it is usually believed that the Hebrews manufactured them while they were out in the desert. This seems unlikely, especially the manufacturing of the solid gold statue of the angels. Rather, it is more likely that the Holy of Holies, and the Ark, were relics from an earlier time, and were being taken out of Egypt by the fleeing Israelites. Indeed, it is quite possible that for this very reason the Egyptian army decided to pursue the fleeing Israelites, even after they had given them permission to depart.

According to an obscure esoteric order known as The Lemurian Fellowship, the Holy of Holies was a statue created many tens of thousands of years ago on the lost continent of the Pacific generally known as Mu or sometimes Lemuria (a term coined by geologists in the late 19th century). The statue was created to test a person's clairvoyant ability, which was shown by whether they could see the Shekinah Glory on the Mercy Seat. Persons of sufficient psychic ability were then offered the chance to take citizenship training and join the Commonwealth of Mukulia, the name which The Lemurian Fellowship attributes to this civilization which reportedly covered the entire Pacific Basin, including Australia (for more information on the belief that the Holy of Holies is from Mu/Lemuria, see my book *Lost Cities of Ancient Lemuria & the Pacific*[11]). This lost continent, a controversial subject among geologists and mys-

The Ark of the Covenant at the fall of Jericho.

tics, then reportedly sank in a cataclysmic pole shift circa 22000 BC.[1,2]

According to The Lemurian Fellowship, with the downfall of the Pacific civilization, the Holy of Holies and plans for rebuilding the Tabernacle were removed to Atlantis where they were kept in a gigantic pyramidal building called the Incalathon, which was sort of the government headquarters and museum at the same time. Just prior to the supposed destruction of Atlantis circa 10000 BC, the Holy of Holies was taken to Egypt, which was part of the Osirian Empire at that time.[2] According to *The Ultimate Frontier*,[12] the relic was first kept in the Temple of Isis and then secreted in the large stone crypt which occupies the King's Chamber of the Great Pyramid at Giza. For 3,400 years it remained there, until the birth of Moses.

The box, or Ark, within which the Holy of Holies was kept, was probably constructed in Egypt. Electricity was used by the Egyptians, as evidenced by electroplated gold objects, electrical lighting reportedly being used in the temples, and the use of the Djed column as an electrical generator. Because many persons still knew the significance of the gold statue, it was important that the Holy of Holies and the Ark be kept away from the evil Amon priests who fostered mummification in Egypt and controlled the country for thousands of years. Therefore, secret Mystery Schools operated in Egypt that kept the ancient traditions of Atlantis and Mu alive. The Holy of Holies and possibly the Ark were sealed in the so-called King's Chamber of the Great Pyramid, and the entrance to the inside was a carefully guarded secret, known only to a select few.

Is it possible that the Biblical Ark of the Covenant was at one time kept in the Isis Temple at Hathor, and could the underground crypt depict a portion of the electrical system that was used in the ancient temples?

Crystal Lenses, Solar Mirrors and Luminous Disks

Though there may be some doubt amongst the more conservative archaeologists that ancient societies like the Egyptians' had electricity, they all agree that ancient societies had relatively sophisticated glass technology as well as glass and crystal lenses. As we have seen,

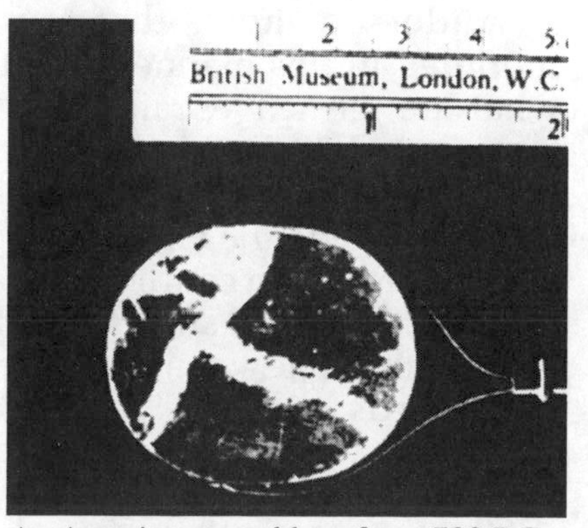

An Assyrian crystal lens from 700 BC.

the ancient arts of glass smelting and metallurgy go back to the very dim mists of human civilization.

The British researcher Harold T. Wilkins mentions luminous disks in his 1952 book *Secret Cities of Old South America*.[19] Says Wilkins, "The Moslem Qu-ran, or Koran, says that old Noah planted an ebony tree and cut planks from it to make his great deluge ship, which is a not unlikely thing. We have a glimpse of some knowledge of physics and electro- or chemi-luminescence possessed by this old Atlantean Noah. The Qu-ran says he placed on the walls of the Ark *two luminous discs to make (or mark) day and night*."

A fascinating book on the use of ancient magnifying lenses is the 1953 book *The Ancient Secret: Fire from the Sun* by Flavia Anderson.[21] This is one of my favorite books on ancient technology, and Ms. Anderson is to be commended for having written this wonderful piece. Anderson says that the Grail legends are based on the existence of ancient lenses made of ground rock crystal that were once used in ancient ceremonies in the great temples of Egypt and the Eastern Mediterranean.

The lenses were set in elaborate supportive stands made of precious metals, and usually they had other precious gems set around the central lens. This central lens was an important sacred relic, yet it was no more than any common magnifying glass used today. These lenses were suspended in a device known as a monstrance. A monstrance (Anderson depicts a Spanish monstrance from the 16th century in her book) used screws to hold a rock crystal or glass lens into place on the silver/copper/gold stand. Anderson suggests that candles were lit with these lenses and they were used in religious ceremonies. Later they were used to develop telescopes, something that the Egyptians and others had known about earlier.

The Thummim or Urim was a crystal set in precious metals.

Anderson shows that crystal lenses were mounted in this manner by the Babylonians in what are termed "Grail Trees." The Grail Tree appears to be a lens held in the center of metal stand looking like a combination of a tree and the sun. Beside the Grail Tree, in her depiction, is a "Solar-hero in conflict with an Eagle-headed monster." She also shows how the Thummin or Urim of the Bible was a crystal set in a metal stand and words such as "Tetragrammaton" and "Elohim" were engraved in the stand, in either Latin or Hebrew letters.

Anderson says that these crystal lenses were

extremely valuable and often the symbol of nobility or authority. She gives several examples of crystal lenses being set in wonderful jewelry. Charlemagne, for instance, had a special crystal talisman. Says Anderson, "At Denderah in Egypt there is a carving of Pharaoh presenting a wonderful necklace to the goddess Hathor on the face of the temple 'chapel,' known as the 'Birth Chamber' (where it was probable that the rebirth of the sun was celebrated yearly). The mysterious Arthurian queen in the Prose Percival whose hand points to her necklace and its pendant 'star,' which it is claimed concerns the mystery of the Grail, could therefore be pointing to a crystal talisman such as that of Charlemagne. ...That the culture of Egypt and the Near East spread in some unknown fashion to Mexico and Peru has long been a supposition... The Spaniards recorded on their arrival in Peru that the heathen priests were accustomed to light their sacred fires from the sun's rays by means of a concave cup set in a metal bracelet."[21]

Anderson says that the legend of the phoenix bird, rising from the ashes of the fire that consumes it, may be based on certain rituals that used a magnifying crystal. The lens was used to focus the sun on some dry straw or other tinder and a trained bird then played in the fire. Anderson demonstrates in her book that a trained bird, a rook in this case, can play with fire in this manner and not be burnt or harmed in any way.

While crystal and glass lenses were apparently used by the ancients to focus the sun and light fires (often in religious ceremonies), this was probably a secondary technology to actual electric lights or other electrical devices such as Van de Graaff generators.

There are several famous stories from ancient literature that speak of giant lenses or mirrors that were used in battle. The most interesting of these stories is that of the Greeks using a fearsome 'solar mirror,' which Archimedes had cooked up in Syracuse in 212-215 BC, to incinerate the invading Roman fleet. He allegedly focused this giant solar mirror on the ships of the Roman fleet and set them on fire! Archimedes was credited with naval the victory, though the Romans eventually got the better of the Greeks in the long run.

To reenact and prove the Syracuse event, Tonnis Sakkas, an Athenian engineer, solar-focused seventy copper-backed mirrors, each measuring 90 cm. x 150 cm., and successfully set fire to a canoe in Skaramanga port at a distance of sixty meters.[40]

Robin Collins, in his book *Laser Beams from Star Cities*, says that old legends from China refer to the terrible 'yin-yang' mirror used by warring supermen to burn the enemy. Another instrument of war possibly utilized by the ancients may have been immense electromagnets. Collins mentions that the *Arabian Nights* stories refer to giant magnets which withdrew the nails of ships as a means of conquering the enemy.[40]

Perseus possessed a magical helmet which, when placed upon his head, instantly rendered him invisible. Robin Collins asks, "Was the 'helmet' an electronic device to diffract or deflect light rays, thereby acting as a protective agent? The 'magic mist' produced by the Druids to render themselves invisible, may have been linked with light diffraction devices."[40]

Says Collins, "It is not technically impossible for the solar mirrors to have reflected light and heat (and electromagnetic?) radiation from a central radiant core, e.g., a plasma radiation energy source positioned in the center of a crystalline/metallic alloyed mirror, and held by a magnetic field. Plastic jelly plasma photo-energy street lights are now experimental in the USSR, while in 1964 Columbia University scientists developed a 'free floating' plasma (ionized gas) only a few centimeters long which emitted heat radiation of +20,0000C., and a brightness three times more intense than the previous brightest artificial light source known to Man. The plasma was as bright, or brighter than the Sun! So, perhaps there is more than a grain of truth in the antiquated legends of the solar mirror engines of destruction?"[40]

Many of the ancient tales of magic mirrors and "fire from heaven" may be stories of exceptionally advanced technology. For instance, crystals could be grown with phosphorescent or luminescent chemicals that would allow them to absorb solar energy during the day and be a glowing light of hard stone at night. Perhaps a remote jungle village in New Guinea still has the ancient street lights, made of "plastic jelly plasma photo-energy" that just come on, night after night, as they have for thousands of years. This should give the modern battery and electric companies a run for their money!

A Bablylonian "Grail Tree."

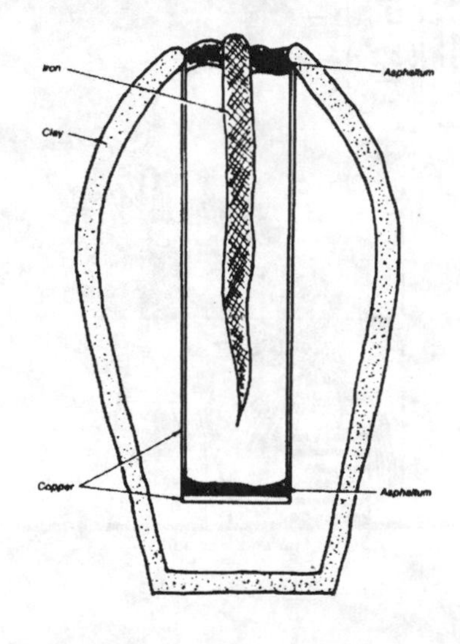

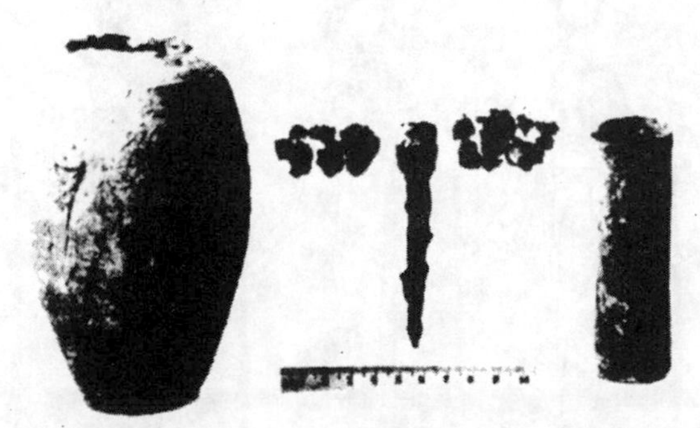

Components of the Baghdad Battery.

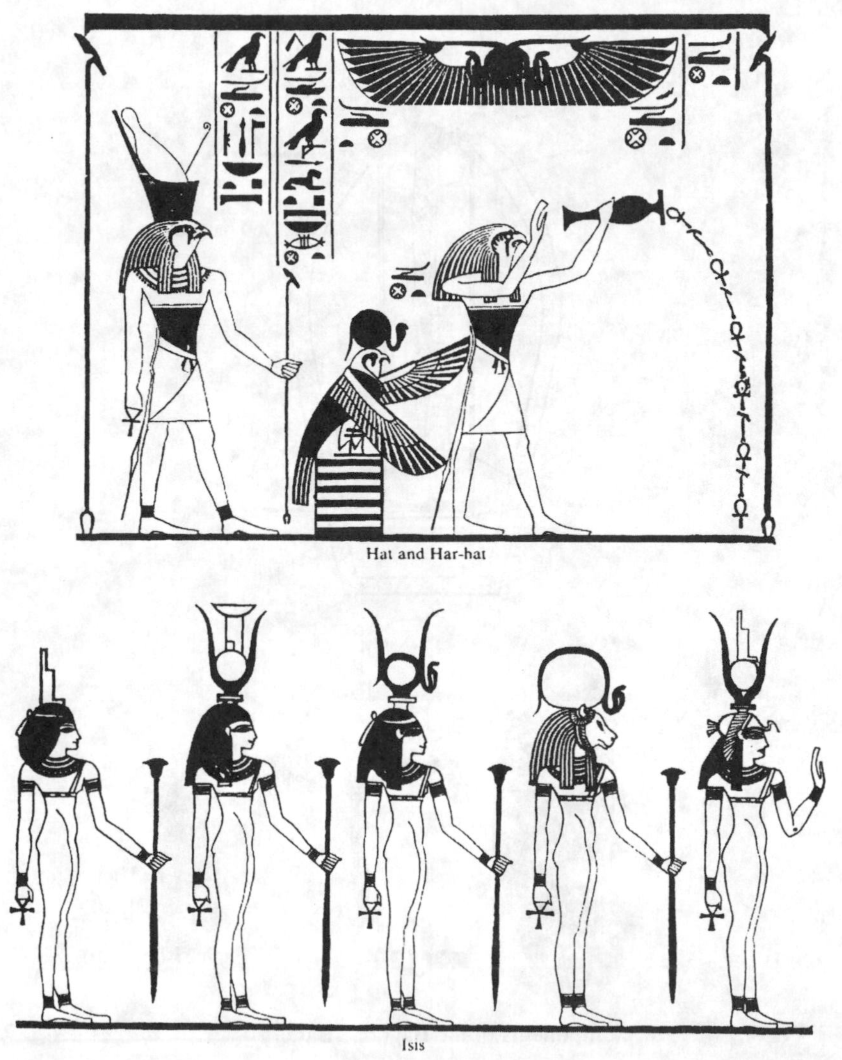

Top: Various aspects of Horus with a winged disk. Bottom: Priestesses with orbs over their heads—an electric light or crystal lens, rather than the sun?

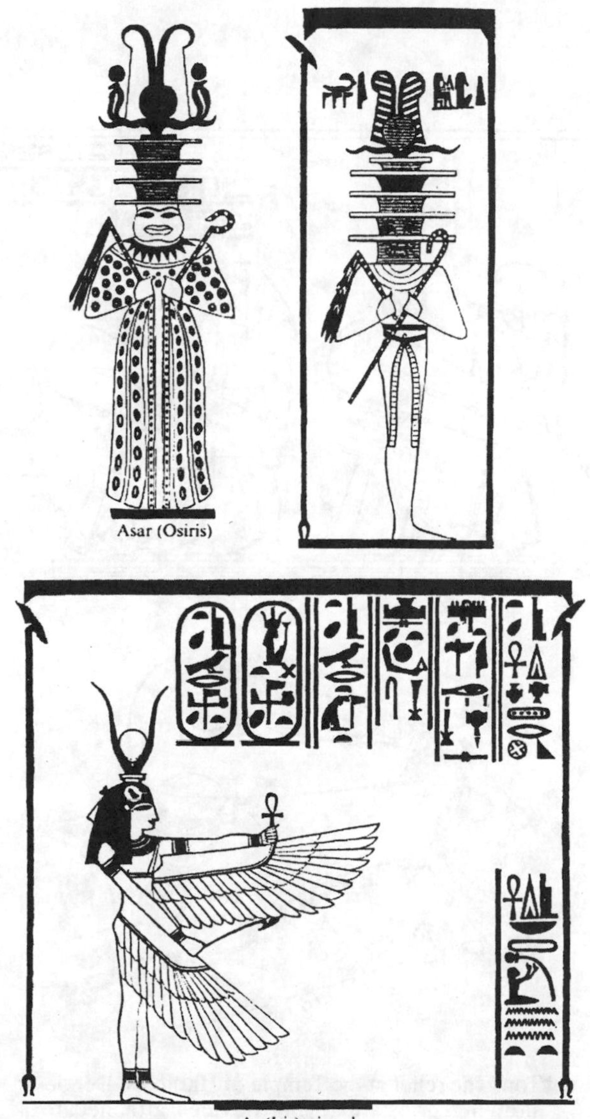

Top: Depiction of Djed columns with orbs at the top. Electrical devices?
Bottom: Depicitions of a winged Isis with an orb over her head.

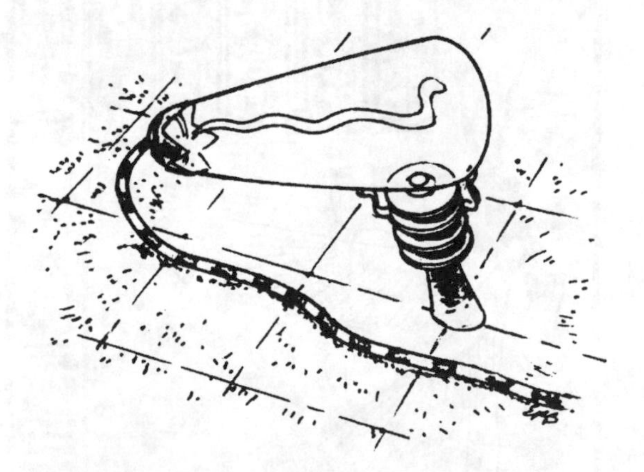

Top: The relief at the Temple of Hathor at Dendera, showing priests carrying devices attached to a braided cable to an altar. Bottom: An interpretive drawing of the ancient relief.

Top: Part of an 18th Dynasty papyrus scroll allegedly depicting sacred baboons worshipping the sun. Note that the device bearing this "sun" and the object itself occlude the "mountains" so that they represent a solid opaque structure. The sun would not move in front of a mountain, as in this drawing. Bottom: A small, stone sphinx holds an orb in each hand.

Top: A Van de Graaff Generator. Bottom: A Wimshurst static electrical device.

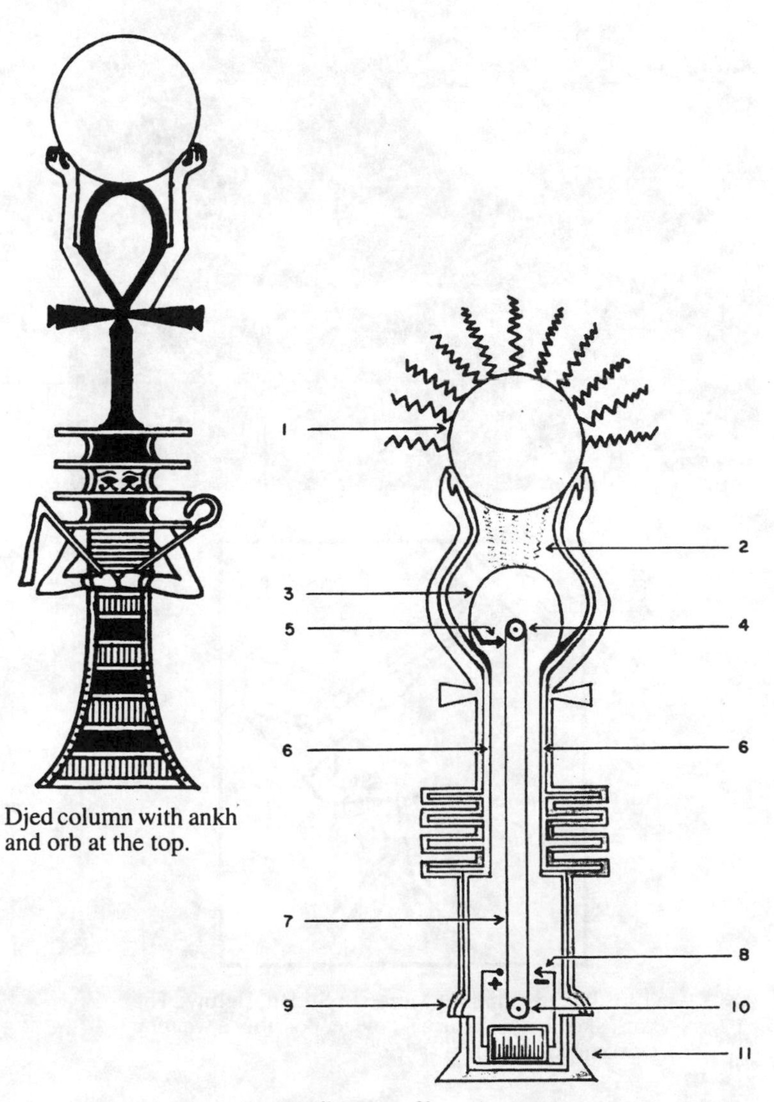

Djed column with ankh and orb at the top.

A diagram of how the static generator might have worked.

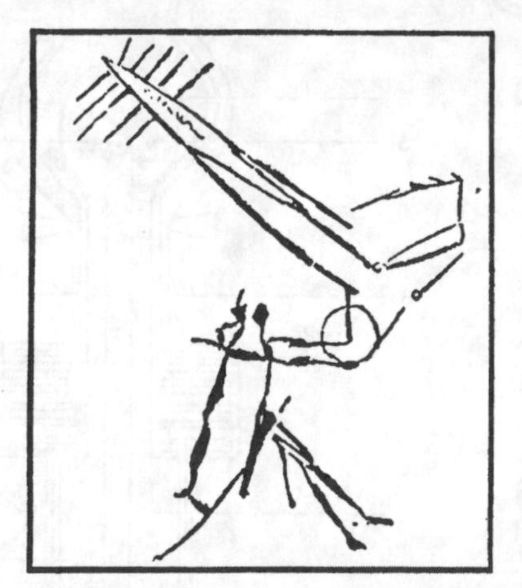

Top: The pyramids at Meroe in Sudan. Below: The rock sketch of an unusual device, possibly a weapon of some kind, or a missile?

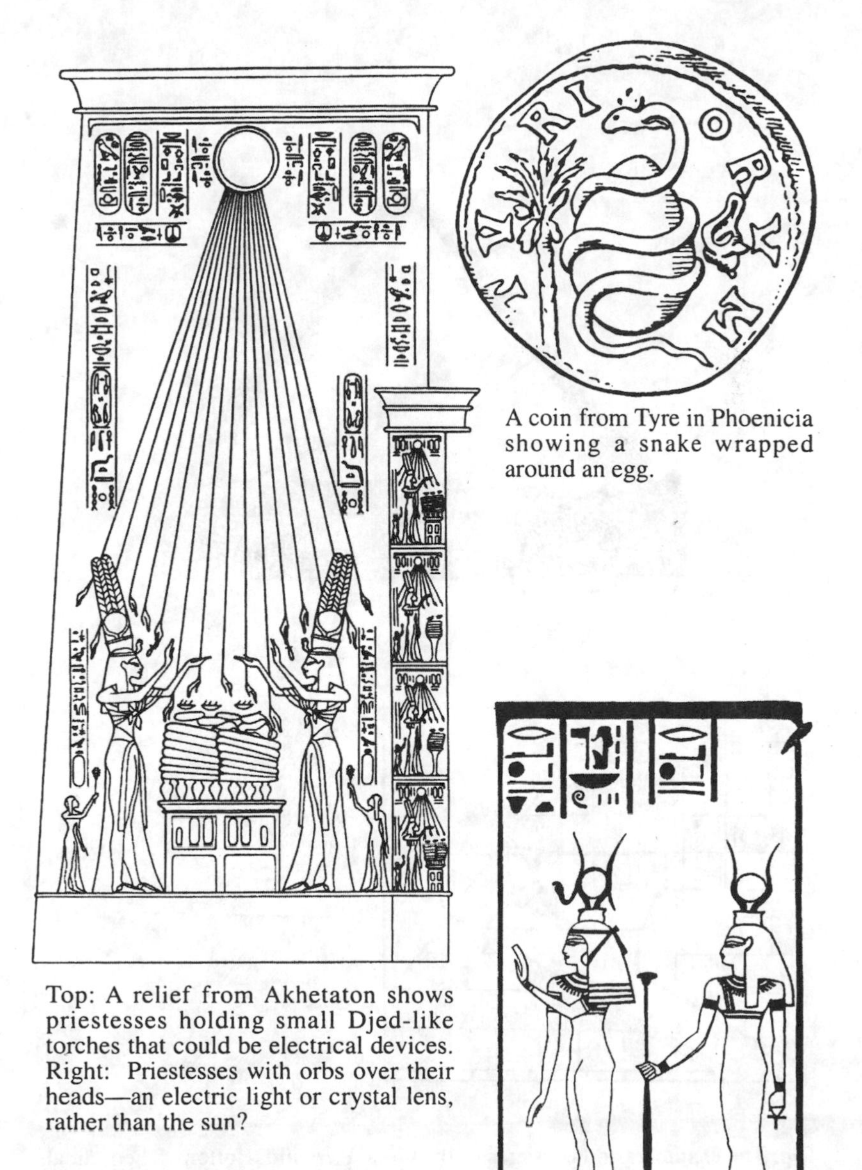

A coin from Tyre in Phoenicia showing a snake wrapped around an egg.

Top: A relief from Akhetaton shows priestesses holding small Djed-like torches that could be electrical devices. Right: Priestesses with orbs over their heads—an electric light or crystal lens, rather than the sun?

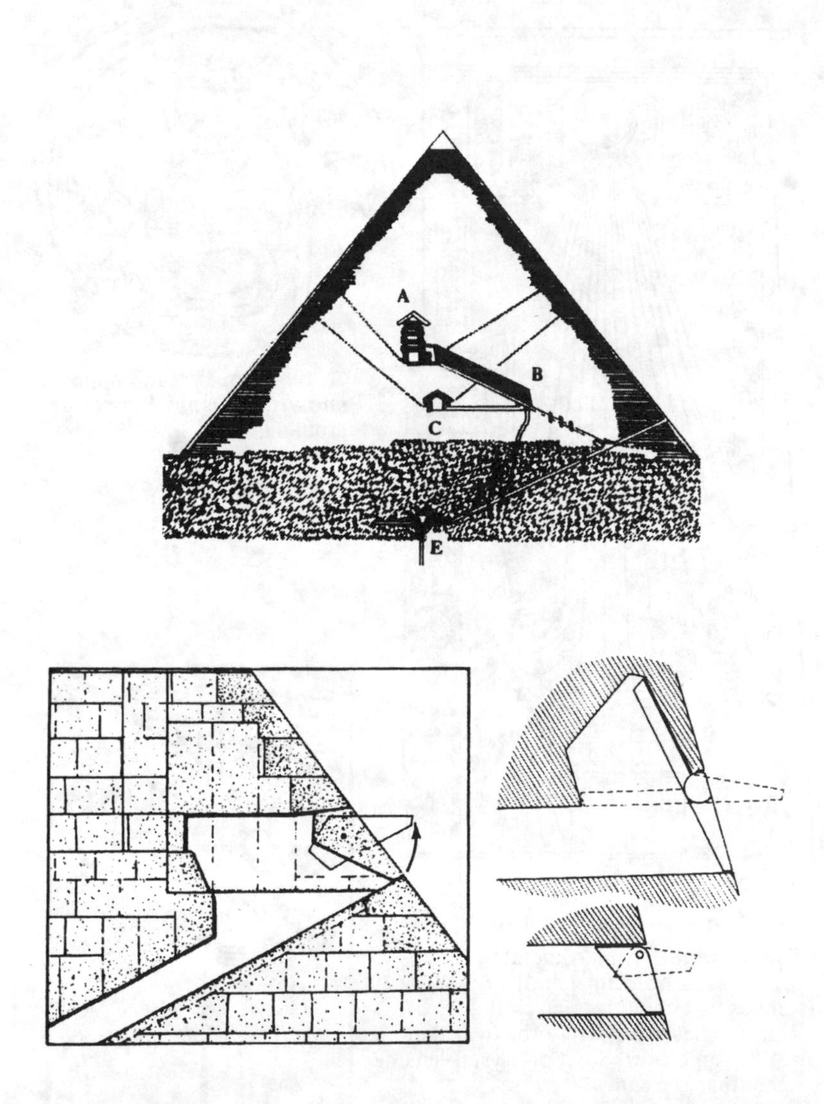

Top: The chambers and passages of the Great Pyramid. Bottom: Theoretical configurations for the massive limestone door that hid the entrance to the pyramid. Moses was said to have entered the pyramid by pushing on the door; he then removed the Ark of the Covenant.

Moses is blinded by the Ark of the Covenant.

The Ark of the Covenant and the Tabernacle in the Wilderness.

The Ark of the Covenant inside Solomon's Temple.

An Assyrian crystal lens from 700 BC.

A Bablylonian "Grail Tree."

The Thummim or Urim was a crystal set in precious metals.

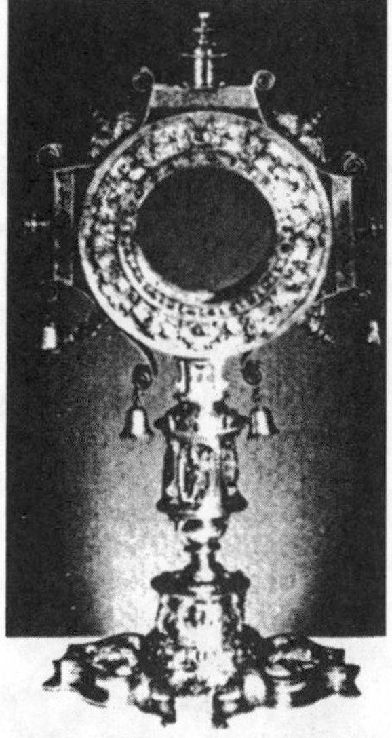

A Spanish monstrance, 16th c.

Platinum artifacts found at the lost city of Coaque in Ecuador by Robin Moore and Howard Jennings. Photo from their 1974 book *The Treasure Hunter.*

5.
Ancient Flight & Aerial War

Man was born free,
and everywhere he is in chains.
—Jacques Rousseau (1712-1778)

No experiment is ever a complete failure.
It can always be used as a bad example.
—Johnny Carson

Ancient Rockets to Ancient Flight

Throughout history there have been tales of flight—from flying carpets to Ezekial's fiery wheels within wheels. Within the myths and legends of ancient history there are countless stories of flying people, flying chariots, flying carpets, and other tales usually dismissed as fantasy and legend.

In his book *Wonders of Ancient Chinese Science,*[135] Robert Silverburg says that Chinese myths tell of a legendary people, the Chi-Kung, who traveled in "aerial carriages." In the ancient Chinese chronicle *Records of the Scholars* it is recorded that the great Han Dynasty astronomer and engineer, Chang Heng, made a "wooden bird" with a mechanism in its belly that allowed it to fly nearly a mile. Propellers seem to be described in a book written about 320 by Ko Hung, an alchemist and mystic: "Some have made flying cars with wood from the inner part of the jujube tree, using ox leather straps fastened to revolving blades so as to set the machine in motion..."[135]

The development of modern spaceflight can be traced to the early use of gunpowder in China, including experimentation with manned rockets. Charcoal and sulfur had long been known as ingredients for incendiary mixtures. As early as 1044 the Chinese learned that saltpeter, added to such a mixture, made it fizz even more alarmingly. We do not know who first learned that if you grind charcoal, sulfur, and saltpeter up very fine, mix them very thoroughly in the proportion of 1:1:3.5 or 1:1:4, and pack the mixture into a dosed container, it will, when ignited explode with a delightful bang. It has been suggested that experimenters, believing that salt made a fire hotter because it made it brighter, tried various salts until they stumbled on potassium nitrate or saltpeter.

The rocket probably evolved in a simple way from an incendiary arrow. If one wanted to make a fire arrow burn fiercely for several seconds using the new powder, one would have to pack the powder in a long thin tube to keep it from going off all at once. It would also be necessary to let the flame and smoke escape from one end of the tube. But, if the tube were open at the front end, the reaction of the discharge would be in the direction opposite to the flight of the arrow and would make the missile tumble wildly. If the tube were open to the rear, on the other hand, the explosion would help the arrow on its way.

Early on it was discovered that with a discharge to the rear the arrow did not even have to be shot from a bow. The forward pressure of the explosion inside the tube would move the device fast enough.

The Chinese created all manner of rocket-powered arrows, grenades and even iron bombs, very similar to those in use today. The first two-stage rocket is credited to the Chinese in the 11th century AD with their "Fire Dragon" rocket. While on its way to its target, the "Fire Dragon" ignited fire arrows that flew from the dragon's mouth. An early two-stage cluster bomb rocket.

When the Mongol army attacked Kaifeng—once the capital of the Sung but now that of the Gin Dynasty—in 1232, the armies of the Gin checked the invincible Mongols for a while by using secret weapons. One called "heaven-shaking thunder" was an iron bomb lowered by a chain from the city's walls to explode among the foe. The other, an early rocket called an "arrow of flying fire," whistled among the Mongols with much noise and smoke and stampeded their ponies. [27]

Celtic design of a flying horse.

The stampeding of horses, or worse yet, war elephants, was one of the primary uses of the early war rockets. It is known that war rockets were used not only in ancient China, but also in ancient India and Southeast Asia. These countries traditionally fought with heavily armored war elephants. A few exploding rockets into the middle of troop of mounted soldiers could send an entire army into chaos.

A curious incident along this line is told by Frank Edwards in *Stranger Than Science.*[19] He says that Alexander the Great's invasion was stopped at the Indus River by an odd historical event: "Flying Shields" or discoid aircraft were buzzing the groups of war elephants that were part of Alexander's invasion army, and made them stampede. Alexander's generals refused to continue with the invasion of the Indian subcontinent, probably the richest, and most civilized group of states in the world at the time. Alexander turned back to Asia Minor and was soon poisoned in Baghdad.

Meanwhile, gunpowder was used in the making of rockets, Roman candles, bombs, and even manned craft.

Russell Freedman in his book *2000 Years of Space Travel*[42] tells the tale of the daring Chinese inventor named Wan Hoo who is credited with launching the first rocket-powered vehicle. Around 1500 AD he built a sturdy wooden framework around a comfortable chair. To the framework he attached 47 skyrockets, and atop it he fastened two large kites. The he strapped himself to the chair.

When he raised his hand, servants carrying blazing torches advanced toward the vehicle and ignited the skyrockets. "A moment later there was a mighty blast, followed by an impressive cloud of black smoke. Wan Hoo vanished, leaving behind but a legend."

There is evidence that bombs and gunpowder were used at the time of Christ and before. However, strictly speaking, it was not yet "gunpowder" because the gun had not been invented. For instance, according to L. Sprague de Camp in *The Ancient Engineers,*[27] at some date in the third century AD an otherwise unknown Marchus or "Mark the Greek," wrote *Liber ignium,* or *The Book of Fire.* Marchus told how to make explosive powder by a mixture of "one pound of live sulfur, two of charcoal and six of saltpeter." This would give a weak explosion. In the 13th century, Albertus Magnus gave the same formula as Marchus, while Albertus' contemporary Roger Bacon recommended "seven parts of saltpeter, five of young hazelnut wood and five of sulfur." This would also produce something of a bang.[27]

About 1280 AD the Syrian al-Hasan ar-Rammah wrote *The Book of Fighting on Horseback and with War Engines.* Ar-Rammah told of the importance of saltpeter for incendiary compounds and gave careful directions for purifying it. He also told of rockets, which he called "Chinese arrows." The Chinese also created the first Roman candles, flame throwers and mortars say modern scholars. Early Roman candles had alternate packings of loose and compressed powder, along with a few nails or small stones, so that as the powder burned down from the muzzle, the solid lumps were thrown out and burned as they flew.[27]

The Roman candle was as close as the Chinese came to the invention of the gun. The invention of the real gun is an obscure and disputed event, generally thought to have taken place in Germany. The "Chronicle of the City of Ghent" for 1313 states that "in this year the use of guns (*bussen*) was found for the first time by a monk in Germany." A manuscript published in 1326, Walter de Milemete's *De officiis regun,* shows a primitive gun called a *vasa* or vase. This is a bottle-shaped device shooting massive darts. An Italian manuscript of the same year mentions guns. In the 1340s, Edward III of England and the cities of Aachen and Cambrai all paid bills for guns and powder.[27]

Some of the earliest guns were thin "barrels" of wood strengthened by iron hoops, or of copper and leather. Guns soon evolved into cannon, rifles and hand guns. The latter at first were small cannon lashed to poles, which the gunners held under their arms like lances at rest. Cannon evolved into long guns for direct fire and very short guns, called mortars from their shape, for high-angle fire. For a time, balls of iron or lead were used in hand guns and balls of stone in cannon.

Iron cannon balls soon replaced those of stone, because iron is much denser than stone and iron balls therefore carried more kinetic energy for a given bore. Now cannon had to be made stronger and of smaller bore, because if a cannon designed for stone balls were fired with an iron ball of the same size, the gun would burst. Grenades were already in use in the Middle East by the time of the Crusades, and the Knights Templars (and other crusaders) were said to have brought this technology back to Medieval Europe.

Meanwhile the hand gun in its turn improved until it outshone the cannon. The flintlock musket, became cheap enough for any citizen to own, simple enough for him to use, and deadly enough to enable him to face professional soldiers with a nearly equal footing. Thus the stage was set for the fall of kings and the setting up of republics. The common man, with this new technology, need not be afraid of thugs or thieves anymore, nor drunken soldiers or others who might threaten him and his family because he was of larger

Icarus

stature and carried a long, heavy sword. The pistol was the great equalizer, a deadly weapon that women could use effectively as well. As it was said popularly at the turn of the 20th century, "God created man, but Sam Colt made all men equal."

Prehistoric Aircraft: From Airplane Models to Flying Chariots

The development of modern weapons was immediately followed by the development of aviation. This was initially quite successful and sparked the imagination of the entire world. By the mid-1800s balloons were a common sight in most major cities of the world. Powered flight, designed after the shape of bird's wings, came shortly afterwards.

But what of ancient flight? Were the Wright brothers really the first to fly through the air on a powered vehicle? Clearly Wan Hoo would argue the point if he could.

When American scientists expressed surprise at the sophistication of the Antikythera Device by saying that it was "like finding a jet plane in the Tomb of King Tut," they weren't far off the mark. Models of what appear to be jet planes have been found in tombs in Colombia, and in Egypt as well.

Several small delta-winged gold "jets" can be found in the Colombian government's Gold Museum in Bogota. The small models are thought to be at least 1,000 years old, if not more. They are variously said to be models of bees, flying fish or other animals, however, unlike any known animal, they have vertical and horizontal tail fins.

When these zoomorphic objects were photographed in a V-formation with nine original artifacts, they looked amazingly like a squadron of delta-wing jets! Sanderson says in *Investigating the Unexplained*[8] that a similar object was on display at the Field Museum of Natural History in Chicago. The label said it was "probably meant to represent a flying fish."

Like the bulldozer in Panama, these gold zoomorphic models were dated as probably from 800 to 1,000 years old. Gold is indestructible, however, and all the gold jewelry and coinage from ancient times still exists today in one form or another. In many cases it has been melted down and reformed into gold bars or new jewelry. Other metals will eventually corrode and oxidize, but, as pointed out previously, gold jewelry or trinkets could be traded for hundreds, even thousands, of years.

Flying Fish or Ancient Airplane?

In 1898 a model was found in an

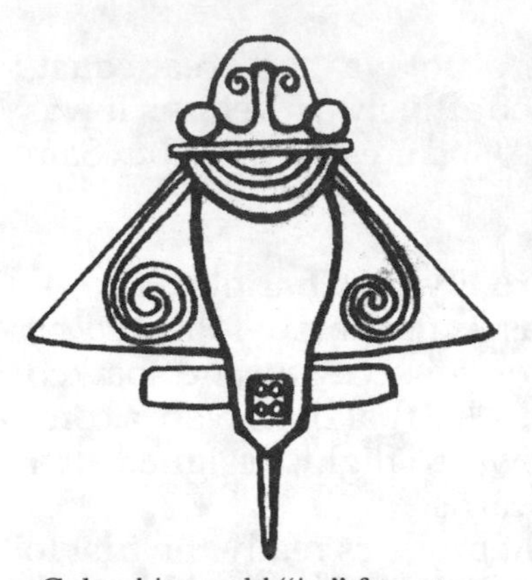

Colombian gold "jet" from top.

Egyptian tomb near Sakhara. It was labeled a "bird" and cataloged object 6347 at the Egyptian Museum in Cairo. Then in 1969 Dr. Khalil Messiha was startled to notice that the bird not only had straight wings, but also an upright tailfin. To Dr. Messiha, the object appeared to be a model airplane.

It is made of wood, weighs 39.12 grams and remains in good condition. The wingspan is 18 cm, the aircraft's nose is 3.2 cm long and the overall length is 14 cm. The extremities of the aircraft and the wingtips are aerodynamically shaped. Apart from a symbolic eye and two short lines under the wings, it has no decorations nor has it any landing legs. Experts have tested the model and found it airworthy.

After this sensational discovery, Egypt's Minister for Culture, Mohammed Gamal El Din Moukhtar, commissioned a technical research group to put other "birds" under the microscope. The team, nominated on Dec. 23, 1971, consisted of Dr. Henry Riad, Director of the Museum of Egyptian Antiquity; Dr. Abdul Quader Selim, Deputy Director of the Egyptian Museum for Archaeological Research; Dr. Hismat Nessiha, Director of the Department of Antiquities; and Kamal Naguib, President of the Egyptian Aviation Union. On January 12, 1972, the first exhibition of Ancient Egyptian model aircraft was opened in the Hall of the Egyptian Museum for Antiquities. Dr. Abdul Quader Hatem, Representative of the Prime Minister, and the Air Minister Ahmed Moh presented fourteen ancient Egyptian model "aircraft" to the public.

Another curious exhibit at the Egyptian Museum in Cairo is a large display of boomerangs found in the tomb of King Tutankhamen. While boomerangs may not be models of ancient aircraft, they demonstrate that the Egyptians were highly interested in the mechanics of flight, as few devices operate like a boomerang after being thrown. A number of Egyptian reliefs show Egyptians hunting with boomerangs, and the curved flying sticks have been found in Florida, Poland, Texas and of course, Australia. Boomerangs may well have been distributed around the world by the ancient Egyptians or some previous culture.

Says Tomas in *We Are Not the First:*[24]

> One of the first aeronautical designers in the world was Daedalus. He constructed wings for his son Icarus and himself but in piloting his glider the boy flew too high and fell into the sea which is now called the Icarian Sea. The Wright brothers were more fortunate 4,500

years later because the basis for aviation technology had already been developed before them.

It is erroneous to think that Daedalus belongs to mythology. His colleagues—the engineers of Knossos—constructed water chutes in parabolic curves to conform exactly to the natural flow of water. Only long centuries of science could have produced such streamlining. And the streamline is also an essential part of aerodynamics, which Daedalus might have mastered.

Friar Roger Bacon left a mysterious sentence in one of his works: "Flying machines as these were of old, and are made even in our days." A statement like that, written in the thirteenth century, is enigmatic, indeed. First of all, Bacon affirmed that engines flying in the air had been a reality in a bygone era, and secondly, that they existed in his day. Both possibilities seem to be farfetched and yet history is replete with legends as well as chronicles of airships in the remote past.

Perhaps more striking are the Chinese annals which relate that Emperor Shun (c. 2258-2208 BC) constructed not only a flying apparatus but even made a parachute about the same time as Daedalus built his gliders.[24]

There was also Emperor Cheng Tang (1766 BC) who ordered a famous inventor named Ki-Kung-Shi to design a flying chariot. The primeval aviation constructor completed the assignment and tested the aircraft in flight, supposedly reaching the province of Honan in his flying machine, possibly a glider. Subsequently, the vessel was destroyed by imperial edict as Cheng Tang was afraid that the secret of its mechanism might fall into the wrong hands.

Circa 300 BC, the Chinese poet Chu Yuan wrote of his flight in a jade chariot at a high altitude over the Gobi Desert toward the snow-capped Kun Lun Mountains in the west. Says Tomas, "He accurately described how the aircraft was unaffected by the winds and dust of the Gobi, and how he conducted an aerial survey."[24]

Chinese folklore is replete with tales about flying chariots and other tales of flight. Tomas mentions that a stone carving on a grave in the province of Shantung, dated AD 147, depicts a dragon chariot flying high above the clouds. And, as mentioned earlier, the fourth century AD Chinese historian Ko-Hung wrote "flying cars with wood from the inner part of the jujube tree, using ox leather straps fastened to rotating blades to set the machine in motion." Leonardo da Vinci had also designed a functional helicopter, possibly from Chinese designs. Helicopters, unlike gliders, do not need long landing areas, but they are much more difficult to control. However, a combination of a balloon with propellers to help move the craft would be a technical feat well within the capability of the dynastic Chinese.

Jim Woodman and his pals experimented with similar technology when they built a reed basket in Peru and then floated it above the Nazca Plain with a crude hot-air balloon made of native fibers and woven cloth. The craft was named the Condor I, and Woodman wrote their story in the 1977 book *Nazca: Journey to the Sun.*[129] They rose to over 1,200 feet into the air and landed successfully, with no one hurt. Woodman believed that the Nazca lines, which can only be fully viewed from the air, were seen by

One of the solid gold mini-models of "Aircraft" kept at the Bogata Gold Museum.

ancient Nazca priests who flew over the desert plain in primitive, but effective, hot air balloons.

The Airships of King Solomon

A number of historical characters have been said to have had airships or flying chariots in modern religious texts. One such famous person was the Northern Indian prince Rama of Ayodha about whom the Ramayana is written. He will be discussed shortly. Another famous owner of an ancient airship was the Hebrew King Solomon the Wise, the son of David.

Solomon is said to have built the famous Temple in Jerusalem to hold the Ark of the Covenant, which we have seen was apparently some sort of electrical device. He had a romantic affair with the Ethiopian Queen of Sheba who had come to visit him in about one thousand BC. According to ancient Ethiopian tradition, recorded in the *Kebra Negast*[101] ("Glory of Kings"—a sort of Ethiopian Old Testament that is the most important document to all Ethiopians) the reigning Queen, Makeda, left Axum, then the capital of Saba, and journeyed across the Red Sea to present day Yemen and up the Hijaz to Jerusalem to visit the court of King Solomon. Seeing the important Ark of the Covenant was a key goal of her visit.

After living with Solomon for some months, she had to return to her own kingdom, where she bore King Solomon's son. He was named Menelik I, and it was with this child, later to become king, that the Solomonic line of rulership over Ethiopia was begun. This line was unbroken for three thousand years until the death of Haile Selassie (born Ras Tafari, 225th Solomonic ruler) in August of 1975.

According to the *Kebra Negast,* King Solomon would visit Makeda and his son Menelik by flying in a "heavenly car." "The king...and all who obeyed his word, flew on the wagon without pain and suffering, and without sweat or exhaustion, and traveled in one day a distance which took three months to traverse (on foot)."[101]

Throughout the Middle East, as far as Kashmir, are mountains known as the "Thrones of Solomon," including one in northwestern Iran, a flat-topped mountain called Takht-i-Suleiman (Throne of Solomon). It has been conjectured that these may have been landing bases for Solomon's airship.

Nicholas Roerich testifies that throughout Central Asia it is widely believed that Solomon flew about in an airship. "Up to now, in the people's conception, King Solomon soars on his miraculous flying device over the vast spaces of Asia. Many mountains in Asia are either with ruins or stones bearing the imprint of his foot or of his knees, as evidence of his long-enduring prayers. These are the so-called thrones of Solomon. The Great King flew to these mountains, he reached all heights, he left behind him the cares of rulership and here refreshed his spirit."[102]

Did King Solomon have some flying vehicle with which he flew to Persia, India, and Tibet? With whom did he meet there? Given the many stories of flying vehicles from the ancient Indian epics, this is not so unusual. Mountains with ruins on their summits do indeed exist all over the world. One amazing city along these lines is Machu Picchu, the mountaintop city in Peru. Were the large grassy areas in each of these cities the landing pads for flying vehicles similar to zeppelins? It is a strange world, full of strange stories, legends and ancient mysteries. Sometimes, indeed, "truth is stranger than fiction!"

The early concept of flight was primitive.

The First Space Programs

Some ancient texts recount not just ancient craft, such as in Ezekial's Biblical vision, but actual eye-witness accounts of going into outer space. The 4,700-year-old Babylonian *Epic of Etana* contains the poem of the *Flight of Etana:*

> "I will take you to the throne of Anu," said the eagle. They had soared for an hour and then the eagle said: "Look down, what has become of earth!" Etana looked down and saw that the earth had become like a hill and the sea like a well. And so they flew for another hour, and once again Etana looked down: the earth was now like a grinding stone and the sea like a pot. After the third hour the earth was only a speck of dust, and the sea no longer seen.

Anu, the Zeus of Babylonian Olympus, was the god of the Heavenly Great Depths—which we now call space. The description of this space flight depicts exactly what happens when man leaves the earth. It is important that we have the presence of a concept of the round earth which becomes small because of perspective as distance increases, indicating an actual eye-witness account.

The *Book of Enoch*, part of the *Apocrypha*, contains a passage that also seems to describe spaceflight:

And they lifted me up into heaven... (14:9)
And it was hot as fire and cold as ice... (14:13)
I saw the places of the luminaries... (17:3)
And I came to a great darkness... (17:6)
I saw a deep abyss. (17:11)

Does this not sound like a graphic account of a trip into space? It is a dark abyss, where objects get hot on the side illuminated by the sun and icy cold on the shaded side. And it is the abode of the sun, moon, planets, and the stars, as Enoch said.

In the second century AD, Lucian, the Greek author who visited Asia Minor, Syria, and Egypt, wrote his novel *Vet-a Historia.* He drew a picture of a voyage to the moon which anticipated the American space program: "Having thus continued our course through the sky for the space of seven days and as many nights, on the eighth day we descried a sort of earth in the air, resembling a large, shining circular island, spreading a remarkably brilliant light around it."[43]

Author Andrew Tomas tells the story of how Chinese historical tradition mentions Hou Yih (or ChihChiang Tzu-Yu), the engineer of Emperor Yao who was acquainted with astronautics. In the year 2309 BC he decided to go to the moon on a celestial bird. This bird advised him of the exact times of the rising, culmination, and setting of the sun. Was it the equipment of a spaceship which provided this information to the prehistoric astronaut? Hou Yih explored space by "mounting the current of luminous air." The exhaust of a fiery rocket?

Hou Yih flew into space, where "he did not perceive the rotary movement of the sun." This statement is of paramount importance in corroborating the story because it is only in space that man cannot see the diurnal movement of the sun.

On the moon the Chinese astronaut saw the "frozen-looking horizon" and built there the "Palace of Great Cold." His wife Chang Ngo also dabbled in space travel. According to the ancient writings of China, she flew to the moon, which she found a "luminous sphere, shining like glass, of enormous size and very cold; the light of the moon has its birth in the sun," declared Chang Ngo.

It is this message from the moon which makes the 4,300-year-old tale so provoking. Chang Ngo's moon exploration report was correct. Apollo 11 astronauts found the moon desolate with a glasslike soil. It is cold in the shade, colder than at our poles. And, of course, moonlight comes from the sun.

Thomas goes on to mention another old Chinese book, *The Collection of Old Tales,* compiled in the fourth century

AD. The book includes an interesting story from the times of Emperor Yao when Hou Yih and Chang Ngo went to the moon. An enormous ship appeared on the sea at night with brilliant lights which were extinguished during the day. It could also sail to the moon and the stars, hence its name "a ship hanging among the stars" or "the boat to the moon." This giant ship which could travel in the sky or sail the seas was seen for 12 years.[24]

Author Tomas says that one of the world's oldest books on astronomy is the Hindu *Sut-ya Siddhanta.* It speaks of Siddhas and Vidyahat-as, or philosophers and scientists, who were able to orbit the earth in a former epoch "below the moon but above the clouds."

Says Tomas:

> Another book from India—the *Samaranagana Sutradhara*—contains a fantastic paragraph about the distant past when men flew in the air in skyships and heavenly beings came down from the sky. Was there a sort of two-way space traffic in a forgotten era?
>
> In his essay on the *Rig Veda* Professor H. L. Hariyappa of Mysore University writes that in a distant epoch "gods came to the earth often times," and that it was "the privilege of some men to visit the immortals in heaven." The tradition of India is insistent upon the reality of this communication with other worlds during the Golden Age.
>
> Old Sanskrit texts speak of the Nagas, or Serpent Gods, who live in underground palaces lighted by luminous gems in the vastness of the Himalayas. The Nagas are flying creatures who go on long voyages in the sky. The belief in the Nagas is so firmly imprinted in the national consciousness of India that even today motion pictures and stage plays exploit this theme to the delight of Indian audiences. The subterranean city of the Nagas—Bhogawati—brilliantly illuminated by diamonds, may perhaps be a folklore image of a space base, lighted and air-conditioned. We wonder if those cosmonauts are still there.
>
> The god Garuda is thought by Brahmins to be a combination of man and bird who travels through space. He is believed to have reached the moon and even the Pole Star, which is fifty light-years away from us.
>
> The fifth volume of the *Mahabharata* contains a passage which has but one meaning—that of life on other planets: "Infinite is the space populated by the perfect ones and gods; there is no limit to their delightful abodes."[24]

Tales of the descent of skygods upon earth can be found all over the globe. The New Testament contains a mear.ingful passage: "Remember to

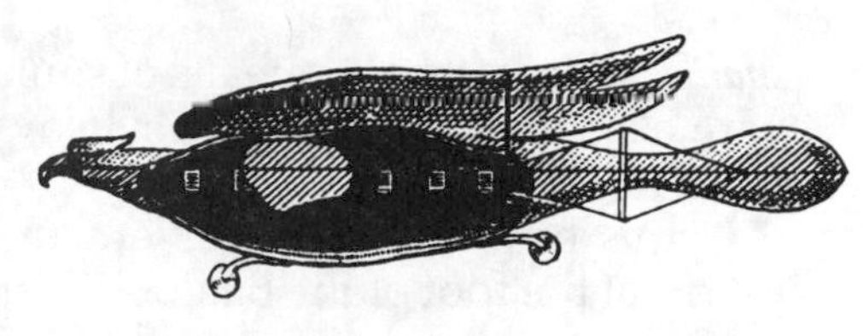

show hospitality. There are some who by so doing, have entertained angels without knowing it." (*Heb.* 13:2). One wonders if angels really need entertaining, but pilots and astronauts can always use a good square meal!

Myths have developed even in this century. The Cargo Cult of Melanesia held the strange belief that "cargo," or manufactured articles such as knives, tinned foods, soap, or toothbrushes, would be brought to their Stone Age tribes by "big canoes" or "big birds." When American planes dropped loads of foodstuffs in the jungle for the advancing Australian and American troops in 1943, the natives took this as a fulfillment of the myth. After the war they continued to build mock airstrips for the big birds to deliver "cargo." They even constructed immense warehouses for the expected goods. Having seen radio installations, they erected masts with aerials and "radio sets" of bamboo by means of which they expected to contact the "gods." Part of their belief was that their dead ancestors were sending all this free food in the form of "cargo."

Influenced by Christianity, some thought they could talk to Jesus Christ or "John Fromme" on these bamboo radio transmitters. But throughout all of these naive beliefs there was some basis in reality: the "big birds" (airplanes), the "big canoes" (steamships), and the "cargo" (industrial society's manufactured articles) were all quite real.

In like manner the ancient legends of the "gods descending upon earth" and an era when "men and gods" mixed could be a folk memory of an epoch when skyships cruised the planet, landing at certain airports in larger cities. Indeed, many ancient cities, such as those in South and Central America, have huge open spaces in front of them, or in the middle of the city—flat fields capable of landing large airships.

Legends and Histories of Levitation

Physicists tell us that there are several "forces" acting on us at any given time. These forces are atomic force, electrical force, magnetic force, and finally, gravitational force. Gravity is the weakest, and least understood of all the forces. Paradoxically, the weakest force is the most difficult to master because we know so little about it. However, levitation, a sort of cancellation of gravitational force, has been known to occur—at least in the historical record!

Says Tomas:

> Some of the most incredible tales of antiquity concern levitation or the power to neutralize gravity. Francois Lenormant writes in

> *Chaldean Magic* that by means of sounds the priests of ancient Babylon were able to raise into the air heavy rocks which a thousand men could not have lifted.
>
> Is this how Baalbek was erected? The gigantic slab left in the quarry at the foot of the Baalbek Terrace by the Titans who had built it is 21 meters long, 4.8 wide, and 4.2 deep. Forty thousand workers would be needed in order to move this huge mass. The question is, how could such a multitude have had access to the slab in order to lift it? Moreover, even in this brilliant era of technology there is not a crane in the world today that could raise this monolith from the quarry!
>
> Certain Arab sources contain curious tales about the manner in which the pyramids of Egypt were erected. According to one, the stones were wrapped in papyrus and then struck with a rod by a priest. Thus they became completely weightless and moved through the air for about 50 meters. Then the hierophant repeated the procedure until the stone reached the pyramid and was put in place. This would explain the absence of chips on edges of the stone blocks for which the author searched in vain and the joints into which it is impossible to insert a sheet of paper. Even though the Khufu pyramid is no longer the tallest edifice in the world, it is still the biggest megalithic structure on earth.
>
> Babylonian tablets affirm that sound could lift stones. The Bible speaks of Jericho and what sound waves did to its walls. Coptic writings relate the process by which blocks for the pyramids were elevated by the sound of chanting. However, at the present level of our knowledge we can establish no connection between sound and weightlessness.[24]

Also mentioned are that Lucian (second century AD) testified to the reality of antigravity feats in ancient history. Speaking about the god Apollo in a temple in Hierapolis, Syria, Lucian related a wonder which he witnessed himself: "Apollo left the priests on the floor and was born aloft."

Tomas, who had travelled widely in China and India in the sixties, mentions that the biography of Liu An in the *Shen Hsien Chuan* (fourth century AD) contains an anecdotal case of levitation. When Liu An swallowed his Taoist elixir, he became airborne. But he had left the container in the courtyard and it was not long before the dogs and poultry licked and drank whatever was left in the vessel. As the historical record says: "They too sailed up to heaven; thus cocks were heard crowing in the sky, and the barking of dogs resounded amidst the clouds." Similarly, he says that a Buddhist *Jataka* narrative speaks of a magic gem capable of raising a man into the air if he holds it in his mouth.[24]

There is a story about Simon the Magus, a first-century AD Gnostic philosopher, addressing thousands in Rome on the subject of his philosophy of gnosis, or knowledge. Tradition says that the "spirits of the air" helped him to raise himself high in the air, for Simon was "a man well versed in magic arts." Although Christian historians were not sure of the source of Simon's powers, the power of levitation was nevertheless attributed to him. The magician was also said to have made statues lose their weight and glide in the air. Iamblichus, a fourth-century AD Neoplatonic philosopher, was also known to have floated in the air to the height of half a meter.[24]

The Catholic Church lists some two hundred saints who were alleged to have conquered the force of gravity. Saint Christianna, a Christian missionary in Spain in the third century AD, is reported by Rufinus to have performed a feat of anti-gravitation. The King and Queen of Iberia were having a church built and it happened that one column was so heavy that it could not be put in place. The story goes that the saint came to the building site at midnight and asked for divine help in prayer. Suddenly, the pillar went up into the air and remained hovering until morning. The astonished workers had no trouble in moving the weightless column in the air to the right spot, upon which it regained its weight and was easily installed on the pedestal.

At Mount Cassino in Italy there is a large and heavy stone which was traditionally lifted by Saint Benedict (AD 448-548) by the neutralization of gravity. The stone was intended for the wall of a monastery being built at the time, and the stonemasons could not budge it. Saint Benedict made the sign of the cross on the block, and while the seven men who could not lift it looked in amazement, he raised it alone without any effort.

Ezekiel's Vision.

Tomas mentions that King Ferdinand I was a host to Saint Francis of Paula (1416-1507) in Naples. Through a half-opened door he saw the monk in meditation, floating high above the floor of his room. Also, Saint Teresa of Avila (1515-1582) used to rise in the air frequently and sometimes at the most inconvenient moments, such as during the visit of an abbess or a bishop to her monastery when she would suddenly rise up to the ceiling.[24]

Probably the most famous of the flying saints was the Italian monk Joseph of Copertino (1603-1663). To help ten men struggling to lift an 11-meter cross, Saint Joseph flew for 60 meters, picked up the cross in his arms, and installed it in its place. In 1645, in the presence of the Spanish ambassador to the papal court, he raised himself and then floated in the church over the heads of those present to the foot of a religious statue. The ambassador, his wife, and the people in the church were all spellbound with astonishment.

The early British press in India gave many accounts of yogis sitting in a Buddha-like posture either upon air or on water. Fakirs would climb a levitated rope or levitate while holding a staff in one hand.

Tomas gives a comparatively recent account (1951) of a case of levitation in Nepal by E. A. Smythies, advisor to the government of Nepal, concerning his young native servant, who was in a trance: "His head and body were shaking and quivering, his face appeared wet with sweat, and he was making the most extraordinary noises. He seemed to me obviously unconscious of what he was doing or that a circle of rather frightened servants—and myself, were looking at him through the open door at about eight or ten feet distance. This went on for about ten minutes or a quarter of an hour, when suddenly (with his legs crossed and his hands clasped) he rose about two feet in the air, and after about a second bumped down hard on the floor. This happened again twice, exactly the same except that his hands and legs became separated."

Furthermore, he says that "According to the 2,000-year-old *Surya Siddhanta,* the Siddhas, Adepts of High Science, could become extremely heavy or as light as a feather. This ancient concept of gravity as a variable force rather than a constant is in itself very remarkable, for there was nothing in the physical experience of the ancient Brahmins that we know of to indicate that objects could possibly become heavier or lighter."[24]

In 1939 a Swedish aircraft designer named Henry Kjellson claimed he witnessed Tibetan monks levitating stones with the beating of large drums. Kjellson claimed in a book published in Swedish that 14 large or medium sized drums, suspended from a frame, and accompanied by trumpeters and a crowd of 200 monks, were beaten in a special rhythm until a large granite block was levitated onto a cliff. The heavy block of stone allegedly flew through the air in an arc and landed on the ledge of cliff on a steep mountain side 250 meters above the crowd.

Kjellson allegedly filmed the entire episode on 16mm film, but this film has never been released. The use of horns and drums to acoustically levitate something has been studied by NASA, and it is interesting to compare a modern stereo speaker cone with photos and diagrams of flying saucers. They look very similar! Horns were used in the biblical battle of

Jericho to bring down the walls of the city. Ultrasonic weapons using soundwaves for destruction are a reality today. Were they in ancient times as well?[127]

The famous French explorer Madame Alexandra David-Neel, who died in 1969 at the age of 101, wrote in *With Mystics and Magicians in Tibet*, about her strange experiences with levitation in that country, where she had lived for fourteen years: "Setting aside exaggeration, I am convinced from my limited experiences and what I have heard from trustworthy lamas, that one reaches a condition in which one does not feel the weight of one's body."[104]

Madame David-Neel was fortunate enough to see a sleepwalking lama, or *lung-gom-pa.* These lamas become almost weightless and glide in the air after a long period of training. The lama she saw on her journey in north Tibet leaped with "the elasticity of a ball and rebounded each time his feet touched the ground."

The Tibetans warned Madame David-Neel not to stop or accost the lama as that could have caused his death from shock. As this lama passed by with extraordinary rapidity on his undulating run, the French explorer and her companions decided to follow him on horseback. In spite of their superior means of transport, they could not catch up with the sleepwalking lama! In this trancelike state the lung-gom-pa is said to be quite aware of the terrain and obstacles on the way.

Madame David-Neel was given some very significant information about this levitation. Morning, evening, and night were said to be more favorable for these sleepwalking marches than noon or afternoon. Therefore, there might be some correlation between the position of the sun and gravity.

The power is said to be developed by deep rhythmic breathing and mental concentration. After long years of practice, the feet of the lama no longer touch the earth and he becomes airborne, gliding with great swiftness, writes David-Neel. She adds that some lamas create artificial gravity by wearing heavy chains in order not to float away into space!

While personal levitation may be convenient for some, and is certainly interesting in that it flouts known physical "laws," it is nuts-and-bolts flying machines that we are really concerned with here.

The Rama Empire of India

In an archaeological sense, the idea that civilization began in Sumeria is fairly recent, beginning with the British and German excavations in the mid-1800s. At this time it was established that Sumeria

was the oldest civilization in the world, and that all others must be more recent. Science essentially held that man lived in unorganized chaos for tens or hundreds of thousands of years until the Sumerians, circa 9000 BC. It is now thought that Sumeria is not the oldest culture in the world. It is theorized now that the cultures of ancient India and Southeast Asia are older.

India's own records of its history claim that its culture has existed for literally tens of thousands of years. Yet, until 1920, all the "experts" agreed that the origins of the Indian civilization should be placed within a few hundred years of Alexander the Great's expedition to the subcontinent in 327 BC However, that was before the cities of Harappa and Mohenjo Daro were discovered in what is today Pakistan. Later, other cities with the same plan were found and excavated, including Kot Diji, near Mohenjo Daro, Kalibangan and Lothal, a port in Gujerat, India. Lothal is a port city that is now miles from the ocean.

The discoveries of these cities forced archaeologists to push the dates for the origin of Indian civilization back thousands of years, just as Indians themselves insisted. A wonder to modern-day researchers, the cities are highly developed and advanced. The way that each city is laid out in regular blocks, with streets crossing each other at right angles, and with each city laid out in sections, causes archaeologists to believe that the cities were conceived as a whole before they were built—a remarkable early example of city planning. Even more remarkable is that the plumbing-sewage system throughout the large cities is so sophisticated that it is superior to that found in many Pakistani (and other) towns today. Sewers were covered, and most homes had toilets and running water. Furthermore, the water and sewage systems were kept well separated.

This advanced culture had its own writing, never deciphered, and used personalized clay seals, much as the Chinese still do today, to officialize documents and letters. Some of the seals found contain figures of animals that are unknown to us today!

Unlike other ancient nations such as Egypt, China, Brittany or Peru, the ancient Hindus did not have their history books all ordered destroyed, and therefore we have one of the few true links to an extremely ancient and scientifically advanced past. Modern scholars value ancient Hindu texts, as they are one of the last tenuous connections to the ancient libraries of the past. The super-civilization known as the Rama Empire is described in the *Ramayana*, which holds many keys to the truth of the past.

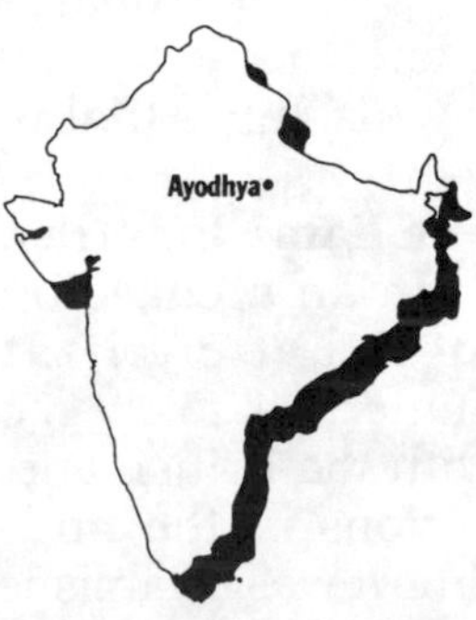

The *Ramayana* describes the adventures of a young prince named Rama who marries a beautiful woman named Sita. After some years of marriage, Sita runs off with (or is kidnapped by) Rama's enemy, Ravanna. Ravanna takes Sita by vimana to his capital city on an island called Lanka. Rama uses his own vimana, and a small army of friends, to fly to Lanka and get his troublesome wife back. He brings her back to his home city, Ayodhya, where she banishes herself in the forest for being unfaithful. Rama, after years of anguish, finally gets back together with her, and they live happily ever after.

The city of Ayodhya as mentioned in the *Ramayana* is thought to be the small town of Ayodhya in northern India. Every year a Hindu festival occurs in the town and a mock-vimana is paraded through the town. Recently, it was reported in the *Motilal Banarsidass Newsletter* on archeology (February 1998) on the Indian subcontinent that a retired geography professor named S.N. Pande has proposed that Rama's Ayodhya was actually in Afghanistan. Dr. Pande said that the current Ayodhya dated from only about 800 BC and the events of the Ramayana were much older.

Dr. Pande believes that the ancient city of Ayodhya was rebuilt as the city of Kushak, which was known after the tribe of Kashi and later by the name of Kusha, the son of Rama.

Thus Ayodhya and Kashi became synonymous in those days, says Pande. It is curious to think that many of the events of the *Ramayana* and *Mahabharata* occurred in Persia and Afghanistan, as well as in the Indian subcontinent. Considering the traditional connections between the eastern Mediterranean, Persia and India, this is not really so surprising. What is surprising is the tales of ancient flight and aerial warfare.

Rama ruled the earth for 11,000 years.
He gave a year-long festival
In this very Naimisha Forest.
All of this land was his kingdom then;
One age of the world ago;
Long before now, and far in the past.
Rama was king from the center of the world,
To the Four Oceans' shores.

—the beginning chapter of the
Ramayana by Valmiki

Fly the Friendly Skies in an Air India Vimana

Nearly every Hindu and Buddhist in the world—hundreds of millions of people—has heard of the ancient flying machines referred to in the *Ramayana* and other texts as vimanas. Vimanas are mentioned even today in standard Indian literature and media reports. An article called "Flight Path" by the Indian journalist Mukul Sharma appeared in the major newspaper *The Times of India* on April 8, 1999 which talked about vimanas and ancient warfare:

According to some interpretations of surviving texts, India's future it seems happened way back in its past. Take the case of the *Yantra Sarvasva,* said to have been written by the sage Maharshi Bhardwaj.

This consists of as many as 40 sections of which one, the Vaimanika Prakarana dealing with aeronautics, has eight chapters, a hundred topics and 500 sutras.

In it Bhardwaj describes vimana, or aerial craft, as being of three classes: (1) those that travel from place to place; (2) those that travel from one country to another; and (3) those that travel between planets. Of special concern among these were the military planes whose functions were delineated in some very considerable detail and which read today like something clean out of science fiction. For instance they had to be: impregnable, unbreakable, non-combustible and indestructible capable of coming to a dead stop in the twinkling of an eye; invisible to enemies; capable of listening to the conversations and sounds in hostile planes; technically proficient to see and record things, persons, incidents and situations going on inside enemy planes; know at every stage the direction of movement of other aircraft in the vicinity; capable of rendering the enemy crew into a state of suspended animation, intellectual torpor or complete loss of consciousness; capable of destruction; manned by pilots and co-travellers who could adapt in accordance with the climate in which they moved; temperature regulated inside; constructed of very light and heat absorbing metals; provided with mechanisms that could enlarge or reduce images and enhance or diminish sounds.

Notwithstanding the fact that such a contraption would resemble a cross between an American state-of-the-art Stealth Fighter and a flying saucer, does it mean that air and space travel was well known to ancient Indians and aeroplanes flourished in India when the rest of the world was just about learning the rudiments of agriculture? Not really [the perception of the absence of proof is no proof of the proof's absence. —Jai Maharaj], for the manufacturing processes described alongside are delightfully diffuse and deliberately vague.

But it does display a breathtaking expanse of imagination which, had it ever been implemented, would have propelled us even further than *Star Trek.*

It would seem from the above article that modern Indians now view their own past as something out of science fiction. Aerial battles and chases are common in ancient Hindu literature. Buck Rogers, Flash Gordon, and *Star Trek* all come mind when reading the ancient Indian epics.

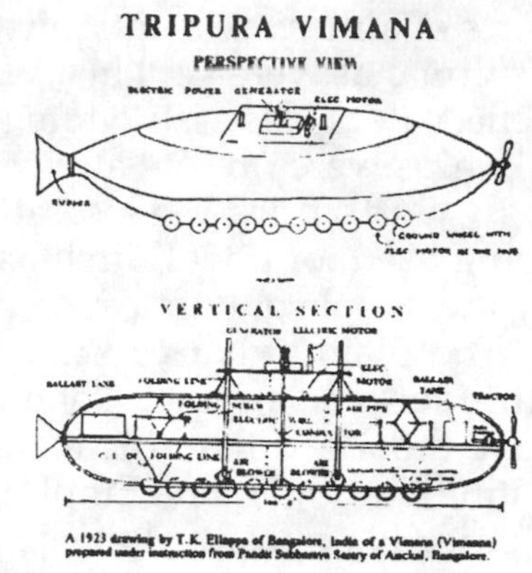

A 1923 drawing by T.K. Ellappa of Bangalore, India of a Vimana (Vimaana) prepared under instruction from Pandit Subbaraya Sastry of Anekal, Bangalore.

What did these airships look like? The ancient *Mahabharata* speaks of a vimana as "an aerial chariot with the sides of iron and clad with wings." The *Ramayana* describes a vimana as a double-deck, circular (cylindrical) aircraft with portholes and a dome. It flew with the "speed of the wind" and gave forth a "melodious sound" (a humming noise?). Ancient Indian texts on vimanas are so numerous it would take at least one entire book to relate what they have to say. See, among others, *Vimana Aircraft of Ancient India & Atlantis*[10] by this author. The ancient Indians themselves wrote entire flight manuals on the care and control of various types of vimanas. The *Samara Sutradhara* is a scientific treatise dealing with every possible facet of air travel in a vimana. There are 230 stanzas dealing with construction, take-off, cruising for thousands of miles, normal and forced landings, and even possible collisions with birds.[31, 15, 39] Would these texts exist (they do) without there being something to actually write about? Traditional historians and archaeologists simply ignore such writings as the imaginative ramblings of a bunch of stoned, ancient writers. After all, where are these vimanas that they write about? Perhaps they are seen every day around the world and are called UFOs!

Says Andrew Tomas, "There are two categories of ancient Sanskrit texts—the factual records known as the *Manusa,* and mythical and religious literature known as the *Daiva*. The *Samara Sutradhara,* which belongs to the factual type of records, treats of air travel from every angle. ...If this is the science fiction of antiquity, then it is the best that has ever been written."[24]

In 1875, the *Vaimanika Sastra,* a fourth-century BC text written by Maharshi Bhardwaj, was rediscovered in a temple in India. The book (taken from older texts, says the author) dealt with the operation of ancient vimanas and included information on steering, precautions for long flights, protection of the airships from storms and lightning, and how to switch the drive to solar energy, or some other "free energy" source, possibly some sort of "gravity drive." Vimanas were said to take off vertically, and were capable of hovering in the sky, like a modern helicopter or dirigible. Bhardwaj the Wise refers to no less than seventy authorities and ten experts of air travel in antiquity. These sources are now lost.[10, 33]

Vimanas were kept in a Vimana Griha, or hangar, were said to be propelled by a yellowish-white liquid, and were used for various purposes. Airships were present all over the world, if we are to believe these seemingly wild stories and look at archeological evidence accordingly. Besides being used for travel, airships unfortunately came to be used as warships by the people of Rama and Atlantis.

The plain of Nazca in Peru is very famous for appearing from high altitude to be a rather elaborate, if confusing, airfield. Some researchers have theorized that this was some sort of Atlantean outpost. It is worth noting that the Rama Empire had its outposts: Easter Island, almost diametrically opposite Mohenjo-Daro on the globe, astonishingly developed its own written language, an obscure script lost to the the present inhabitants, but found on tablets and other carvings. This odd script is found in only one other place in the world: Mohenjo-Daro and Harappa! Could it be that a trade network, operating even across the Pacific Ocean, was used by the Rama Empire and the Atlanteans?[52, 31]

Aerial Warfare in Ancient India

The ancient Indian epics go into considerable detail about aerial warfare over 10,000 years ago. So much detail that a famous Oxford professor included a chapter on the subject in a book on ancient warfare!

According to the Sanskrit scholar Ramachandra Dikshitar, the Oxford Professor who wrote *War in Ancient India* in 1944, "No question can be more interesting in the present circumstances of the world than India's contribution to the science of aeronautics. There are numerous illustrations in our vast Puranic and epic literature to show how well and wonderfully the ancient Indians conquered the air. To glibly characterize everything found in this literature as imaginary and summarily dismiss it as unreal has been the practice of both Western and Eastern scholars until very recently. The very idea indeed was ridiculed and people went so far as to assert that it was physically impossible for man to use flying machines. But today what with balloons, aeroplanes and other flying machines, a great change has come over our ideas on the subject."[100]

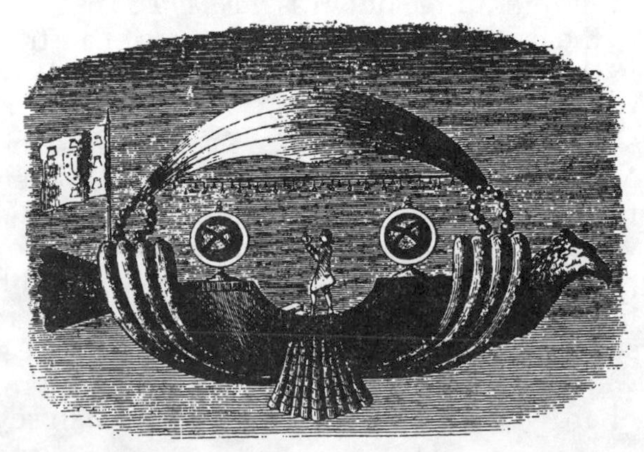

Says Dr. Dikshitar, "...the flying vimana of Rama or Ravana was set down as but a dream of the mythographer till aeroplanes and zeppelins of the present century saw the light of day. The mohanastra or the "arrow of unconsciousness" of old was until very recently a creature of legend till we heard the other day of bombs discharging Poisonous gases. We owe much to the energetic scientists and researchers who plod persistently and carry their torches deep down into the caves and excavations of old and dig out valid testimonials pointing to the misty antiquity of the wonderful creations of humanity."[100]

Dr. R. Dikshitar's 1944 book.

Dikshitar mentions that in Vedic literature, in one of the *Brahmanas,* occurs the concept of a ship that sails heavenwards. "The ship is the Agniliotra of which the Ahavaniya and Garhapatya fires represent the two sides bound heavenward, and the steersman is the *Agnihotrin* who offers milk to the three Agnis. Again in the still earlier *Rg Veda Samhita* we read that the Asvins conveyed the rescued Bhujya safely by means of winged ships. The latter may refer to the aerial navigation in the earliest times."

Commenting on the famous vimana text the *Vimanika Shastra,* he says:

> In the recently published *Samarangana Sutradhara* of Bhoja, a whole chapter of about 230 stanzas is devoted to the principles of construction underlying the various flying machines and other engines used for military and other purposes. The various advantages of using machines, especially flying ones, are given elaborately. Special mention is made of their attacking visible as well as invisible objects, of their use at one's will and pleasure, of their uninterrupted movements, of their strength and durability, in short of their capability to do in the air all that is done on earth. After enumerating and explaining a number of other advantages, the author concludes that even impossible things could be effected through them. Three movements are usually ascribed to these machines, ascending, cruising thousands of miles in the atmosphere and lastly descending. It is said that in an aerial car one can mount up to the Surya-mandala, 'solar region' and the Naksatra-mandala (stellar region) and also travel throughout the regions of air above the sea and the earth. These cars are said to move so fast as to make a noise that could be heard faintly from the ground. Still some writers have expressed a doubt and asked 'Was that true?' But the evidence in its favour is overwhelming.

> The making of machines for offense and defense to be used on the ground and in the air is described. Considering briefly some of the flying machines alone that find distinct mention in this work, we find that they were of different shapes like those of elephants, horses, monkeys, different kinds of birds, and chariots. Such vehicles were made usually of wood. We quote in this connection the following stanzas so as to give an idea of the materials and size, especially as we are in the days of rigid airships navigating the air for a very long time and at a long distance as well.
>
> An aerial car is made of light wood looking like a great bird with a durable and well-formed body having mercury inside and fire at the bottom. It has two resplendent wings, and is propelled by air. It flies in the atmospheric regions for a great distance and carries several persons along with it. The inside construction resembles heaven created by Brahma himself. Iron, copper, lead and other metals are also used for these machines. All these show how far art was developed in ancient India in this direction. Such elaborate descriptions ought to meet the criticism that the vimanas and similar aerial vehicles mentioned in ancient Indian literature should be relegated to the region of myth.[100]

The ancient texts also made the important distinction that vimanas were real machines, while contact with the spirit world, angels or fairies was a different matter. Says Dikshitar:

> The ancient writers could certainly make a distinction between the mythical which they designated *daiva* and the actual aerial wars designated *manusa*. Some wars mentioned in ancient literature belong to the *daiva* form, as distinguished from the *manusa*. An example of the daiva form is the encounter between Sumbha and the goddess Durga. Sumbha was worsted and he fell headlong to the ground. Soon he recovered and flew up again and fought desperately until at last he fell dead on the ground. Again, in the famous battle between "the celestials" and the Asuras elaborately described in the *Harivamsa,* Maya flung stones, rocks and trees from above, though the main fight took place in the field below. The adoption of such tactics is also mentioned in the war between Arjuna and the Asura Nivatakavaca, and in that between Karna and the Raksasa in both of which, arrows, javelins, stones and other missiles were freely showered down from the aerial regions.
>
> King Satrujit was presented by a Brahman Galava with a horse named Kuvalaya which had the power of conveying him to any place on the earth. If this had any basis in fact it must have been a

flying horse. There are numerous references both in the *Visnupurana* and the *Mahabharata* where Krishna is said to have navigated the air on the Garuda. Thither the accounts are imaginary or they are a reference to an eagle-shaped machine flying in the air. Subrahmanya used a peacock as his vehicle and Brahma a swan. Further, the Asura, Maya by name, is said to have owned an animated golden car with four strong wheels and having a circumference of 12,000 cubits, which possessed the wonderful power of flying at will to any place. It was equipped with various weapons and bore huge standards. ...After the great victory of Rama over Lanka, Vibhisana presented him with the Puspaka vimana which was furnished with windows, apartments, and excellent seats. It was capable of accommodating all the Vanaras besides Rama, Sita and Laksmana. Rama flew to his capital Ayodhya pointing to Sita from above the places of encampment, the town of Kiskindha and others on the way. Again Valmiki beautifully compares the city of Ayodhya to an aerial car.

"This is an allusion to the use of flying machines as transport apart from their use in actual warfare. Again in the *Vikramaurvasiya*, we are told that king Puraravas rode in an aerial car to rescue Urvasi in pursuit of the Danava who was carrying her away. Similarly in the *Uttararamacarita* in the fight between Lava and Candraketu (Act VI) a number of aerial cars are mentioned as bearing celestial spectators. There is a statement in the *Harsacarita* of Yavanas being acquainted with aerial machines. The Tamil work *Jivakacintamani* refers to Jivaka flying through the air."[100]

Mercury Engines and Vimana Texts

Perhaps the most valuable information that has been gotten from the *Vimaanika Shastra* of Bhardwaj is the description of what are known today as mercury vortex engines.

In chapter five of the *Vimaanika Shastra*, Bhardwaj describes from the ancient texts which are his reference, how to create a mercury vortex engine:

> Prepare a square or circular base of 9 inches width with wood and glass, mark its centre, and from about an inch and half thereof draw lines to edge in the 8 directions, fix 2 hinges in each of the lines in order to open shut. In the centre erect a 6 inch pivot and four tubes, made of *vishvodara* metal,

equipped with hinges and bands of iron, copper, brass or lead, and attach to the pegs in the lines in the several directions. The whole is to be covered.

Prepare a mirror of perfect finish and fix it to the *danda* or pivot. At the base of the pivot an electric *yantra* should be fixed. Crystal and glass beads should be fixed at the base, middle, and end of the pivot or by its side. The circular or goblet shaped mirror for attracting solar rays should be fixed at the foot of the pivot. To the west of it the image-reflector should be placed. Its operation is as follows:

First the pivot or pole should be stretched by moving the *keelee* or switch. The observation mirror should be fixed at its base. A vessel with mercury should be fixed at its bottom. In it a crystal bead with hole should be placed. Through the hole in the chemically purified bead, sensitive wires should be passed and attached to the end beads in various directions. At the middle of the pole, a mustard cleaned solar mirror should be fixed. At the foot of the pole a vessel should be placed with liquid *ruchaka* salt. A crystal should be fixed in it with hinge and wiring. In the bottom centre should be placed a goblet-like circular mirror for attracting solar rays. To the west of it a reflecting mechanism should be placed. To the east of the liquid salt vessel, the electric generator should be placed and the wiring of the crystal attached to it. The current from both the *yantras* should be passed to the crystal in the liquid *ruchaka* salt vessel. Eight parts of sun-power in the solar reflector and 12 parts of electric power should be passed through the crystal into the mercury and on to the universal reflecting mirror. And the that mirror should be focused in the direction of the region which has to be photographed. The image which appears in the facing lens will then be reflected through the crystal in the liquid salt solution. The picture which will appear in the mirror will be true to life, and enable the pilot to realize the conditions of the concerned region, and he can take appropriate action to ward off danger and inflict damage on the enemy.[33]

Two paragraphs later Bhardwaj says:

Two circular rods made of magnetic metal and copper should be fixed on the glass ball so as to cause friction when they revolve. To the west of it a globular ball made of *vaatapaa* glass with a wide open mouth should be fixed. Then a vessel made of *shaktipaa* glass, narrow at bottom, round in the middle, with narrow neck, and open mouth with 5 beaks should be fixed. Then a vessel made of *shaktipaa* glass, narrow at bottom, round in the middle, with narrow neck, and open mouth with 5 beaks should be fixed on the middle bolt. Similarly on the the end bolt should be paced a vessel sulfuric acid (*bhraajaswad-draavada*). On the pegs on southern side 3 interlocked wheels should be fixed. On the north side a liquefied mixture of load-stone, mercury, mica, and serpent-slough should be placed. And crystals should be placed at the requisite centres.

'Maniratnaakara' [here Bhardwaj is referring to an ancient authority, now lost—ed.] says that the *shaktyaakarshana* yantra should be equipped with 6 crystals known as *Bhaaradwaaja, Sanjanika, Sourrya, Pingalaka, Shaktipanjaraka,* and *Pancha-jyotirgarbha*.

The same work mentions where the crystals are to be located. The *sourrya mani* is to be placed in the vessel at the foot of the central pole. *Bhaaradwaaja mani* should be fixed at the foot of the central pole. *Sanjanika mani* should be fixed at the middle of the triangular wall. *Pingalaka mani* is to be fixed in the opening in the *naala-danda*. *Pancha-jyotirgarbha* mani should be fixed in the sulfuric acid vessel, and *Shakti-panjaraka mani* should be placed in the mixture of magnet, mercury, mica, and serpent-slough. All the five crystals should be equipped with wires passing through glass tubes.

Wires should be passed from the centre in all directions. Then the triple wheels should be set in revolving motion, which will cause the two glass balls inside the glass case, to turn with increasing speed rubbing each other the resulting friction generating a 100 degree power...[33]

From the text of the *Vimaanika Shastra* it is apparent that mercury, copper, magnets, electricity, crystals, gyros(?) and other pivots, plus antennas, are all part of at least one kind of vimana. The recent resurgence in the esoteric and scientific use of crystals is interesting in the context of the *Vimaanika Shastra*. Crystals (*mani* in Sanskrit), are apparently as integral a part of vimanas as they are today of a digital watch. It is interesting to note here that the familiar Tibetan prayer Om Mani Padme Hum, is an invocation to the "Crystal (or jewel) inside the Lotus (of the mind)."

While crystals are no doubt wondrous and important technological tools, it is mercury that concerns us here.

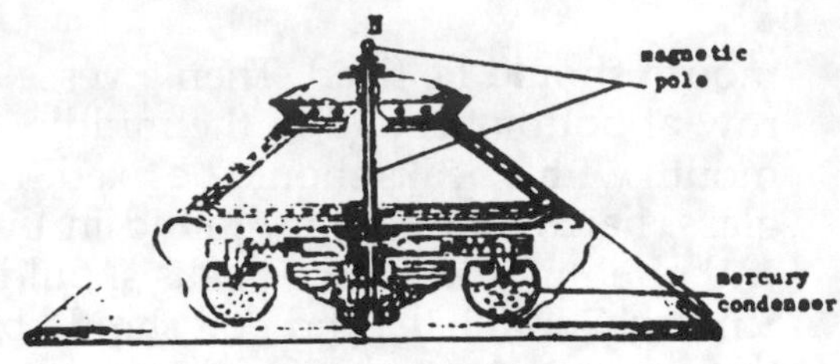

Mercury is an element and a metal. According to the *Concise Columbia Encyclopedia,* Mercury is a "metallic element, known to the ancient Chinese, Hindus, and Egyptians." The chief source of Mercury is cinnabar HgS, a mineral. According to *Van Nostrand's Scientific Encyclopedia,* Mercury was mined as early a 500 BC out of cinnabar crystals which are usually "small and often highly modified hexagonal crystals, usually of rhombohedral or tabular habit. Its name is supposed to be of Hindu origin."

Mercury was most certainly mined and used earlier than 500 BC; scientific encyclopedias and such are usually overly conservative. The metal was named after the messenger of the gods in Roman mythology. It is a heavy, silver-white liquid with the symbol *Hg*. The symbol for Mercury is derived from the Greek word *Hydrargos* meaning water, silver or liquid gyro. It is a liquid at ordinary temperatures and expands and contracts evenly when heated or cooled.

The liquid metal mercury, when heated by any means, gives forth a hot vapor that is deadly. Mercury is generally confined to glass tubes or containers that are sealed, and therefore harmless to the user. Present day mercury vapor turbine engines use large quantities of mercury, but little is required for renewal because of its closed circuit systems. Mercury and its vapor conduct electricity; its vapor is also a source of heat for power usage. Mercury amplifies sound waves and doesn't lose timbre in quality. Ultrasonics can be used for dispersing a metallic catalyst such as mercury in a reaction vessel or a boiler. High-frequency sound waves produce bubbles in liquid mercury. When the frequency of the bubbles grow to match that of the sound waves the bubbles implode, releasing a sudden burst of heat.

According to William Clendenon, well-known UFO investigator and the author of the book *Mercury: UFO Messenger of the Gods,*[105] a mercury-filled flywheel can be used for stabilization and propulsion in discoid aircraft/spacecraft. Liquid mercury proton gyroscopes, according to Clendenon, can be used as direction-sensing gyros if placed 120 degrees apart on the rotating stabilizer flywheel of a discoid craft.

Liquid mercury proton gyroscopes have several advantages, says Clendenon. Firstly, the heavy protons found in mercury atoms are very stable. Secondly, such gyros do not require a warm-up period as mechanical gyros do. Thirdly, the gyro using stable mercury protons is not affected by vibrations and shock. Fourthly, the liquid mercury proton gyroscope has no moving parts and can run forever. And lastly, the mercury atom offers the most stable gyro device in nature and has the additional advantages of saving space and weight. This is particularly valuable on long distance flights where all space and weight must be very carefully calculated and conserved.[105]

Ivan T. Sanderson mentions mercury engines and quotes from Bhardwaj's text: "Strong and durable must the body be made, like a great flying bird, of light material. Inside it one must place the *mercury-engine* with its iron heating apparatus beneath. By means of the *power latent in the mercury* which sets the driving *whirlwind* in motion, a man sitting inside may travel a great distance in the sky in a most marvelous manner.

"Similarly by using the prescribed processes one can build a vimana as large as the temple of the God-in-motion. Four strong *mercury* containers must be built into the interior structure. When these have been heated by controlled fire from iron containers, the vimana develops thunder-power through the mercury. And at once it becomes a pearl in the sky. Moreover, if this iron engine with properly welded joints be filled with mercury, and the fire be conducted to the upper part it develops power with the roar of a lion."[10]

Sanderson then goes on to make the basic observation that a circular dish of mercury revolves in a contrary manner to a naked flame circulated below it, and that it gathers speed until it exceeds the speed of revolution of said flame. Sanderson's observation of revolving mercury is one of the first references to what we now call mercury vortex engines.

The Caduceus

The mythical god Mercury (Hermes to the Greeks) was a messenger of the Gods; he flew through the air rapidly bearing important tidings and official news from kings, gods, or sovereign powers. It was said that if the gods wanted to communicate, carry on commerce, to move things swiftly from one place to another over a long distance safely, they made use of Mercury to accomplish their goals.

Mercury wore winged sandals and a winged hat which bore him over land and sea with great speed. He carried with him his magic wand or "caduceus"—the winged staff with which he could perform many wondrous feats. In one form or another, the ancient symbol has appeared throughout the world, though its actual origin remains a mystery. The caduceus staff was a rod entwined by two serpents and topped with a winged sphere. Today the caduceus is used by the medical profession as its symbol, a practice that apparently stems from the Middle Ages. Probably, the use of the caduceus as a medical symbol stems from the symbolism of the wings for speedy medical attention and the entwined snakes as chemical or medical symbols.

Clendenon, in *Mercury: Ufo Messenger of the Gods,*[105] states that he believes that the caduceus is an ancient symbol of "electromagnetic flight and cosmic energy." The entwined snakes are the vortex coils of the propellant, the rod the mercury boiler/starter/antenna, and the wings symbolic of flight.

Clendenon has done a great deal of experimentation with mercury vortex technology in the context of the ancient texts. His vimana, modeled after Adamski's "scout ship," consists of a circular air frame that is partly a powerful electromagnet though which is passed a rapidly pulsating direct current. It works basically like this:

• The electromagnetic field coil which consists of the closed circuit heat exchanger/condenser coil circuit containing the liquid metal mercury and/or its hot vapor, is placed with its core axis vertical to the craft.

• A ring conductor (directional gyro-armature) is placed around the field coil (heat exchanger) windings so that the core of the vertical heat exchanger coils protrudes through the center of the ring conductor.

• When the electromagnet (heat exchanger coils) is energized, the ring conductor is instantly shot into the air, taking the craft as a complete unit along with it.

• If the current is controlled by a computerized resistance (rheostat), the ring conductor armature and craft can be made to hover or float in the Earth's atmosphere.

• The electromagnet hums and the armature ring (or torus) becomes quite hot. In fact, if the electrical current is high enough, the ring will glow dull red or rust orange with heat.

• The phenomenon (outward sign of a working law of nature) is brought about by an induced current effect identical with an ordinary transformer.

• As the repulsion between the electromagnet and the ring conductor is mutual, one can imagine the craft being affected and responding to the repulsion phenomenon as a complete unit.

• Lift or repulsion is generated because of close proximity of the field magnet to the ring conductor. Clendenon says that lift would always be vertically opposed to the gravitational pull of the planet Earth, but repulsion can be employed to cause fore and aft propulsion.

Clendenon thus interprets the *Samaran Sutradhara* quite differently than most scholars, and voila—"By means of the power latent in the mercury which sets

the driving whirlwind in motion a man sitting inside may travel a great distance in the sky in a most marvelous manner."[105]

Clendenon's view of a great deal of discoid craft seen since 1947 is that many are vimanas, either of ancient manufacture, or modern manufacture. He believes that the famous scout ship observed by George Adamski (and later by other witnesses) is neither a hoax nor an interplanetary space craft. His mercury vortex engines are not capable of interplanetary flight, he says, but, like this version of a vimana, are for terrestrial flight only. He believes that a great number of UFO phenomena could be explained as effects of mercury vortex technology, and craft using it. He thought that some of these craft were ancient—flown by mysterious humans who lived for hundred of years—and that some of these craft were modern constructions, made by theAmericans, British, and Germans.

As to unusual UFO effects, he says that the ball of light that often surrounds the UFO-ship is the magneto-hydrodynamic plasma, a hot, continuously recirculating air flow through the the ship's gas turbine which is ionized (electrically conducting). According to Clendenon, at times a shimmering mirage-like effect caused by heat, accompanied by pulsations of the ball of light makes the craft appear to be alive and breathing. This has, at times, suggests Clendenon, made witnesses to certain UFOs think that they were seeing a living thing. For some of the above reasons, the ship may seem to suddenly disappear from view, though it is actually still there and not de-materialized. The ionized bubble of air surrounding the UFO may be controlled by a computerized rheostat so the ionization of the air may shift through every color of the spectrum, obscuring the aircraft from view.

Curiously, the following item appeared on the internet in 1998 concerning the U.S. government's secret aircraft called the TR-3B which was claimed to be powered by a mercury vortex drive as described in the *Vimanika Shastra*:

"The TR-3B Triangular Anti-Gravity Craft, by Ed Fouche: A very important speech was given by Ed Fouche to the *1998 Summer Sessions at the International UFO Congress,* describing the 200 foot across triangular UFO "anti-gravity" craft being built and tested in area S-4 inside Area 51 in Nevada. Supposedly uses a heated mercury vortex to offset gravity "mass."

Is mercury the element of the gods? Is the caduceus is a virtual diagram for a mercury vortex propulsion device? The ancient Indian civilization may truly have had the "technology of the gods."

The end of all learning is the recovery of the lost mind.
—Mencius, c. 282-301 BC

To Grand Teton in an Atlantean Airship

In 1899 an unusual book was published entitled *A Dweller on Two Planets.*[20] It was first dictated in 1884 by "Phylos the Thibetan" to a young Californian named Frederick Spencer Oliver who wrote the dictations down in manuscript form in 1886.

The book is a long and complicated history of a number of persons and the karma created by each of them during their many lives. It dwells especially on how the karmic relationships and events of the "amanuensis" (Frederick Spencer Oliver and his different lives as Rexdahl, Aisa and Mainin) intertwined with the many lives of "Phylos" as (Ouardl, Zo Lahm, Zailm and Walter Pierson).

A Dweller on Two Planets has remained a popular occult book for over a century, largely because it contains detailed descriptions of life in Atlantis plus devices and technology which were unquestionably well in advance of the time in which it was written. As the cover of one of the book's editions states, "One of the greatest wonders of our times is the uncanny way in which *A Dweller on Two Planets* predicted inventions which modern technology fulfilled after the writing of the book."[20]

Among the inventions and devices mentioned in the book are air conditioners (to overcome deadly and noxious vapors); airless cylinder lamps (tubes of crystal illuminated by the "night side forces"); electric rifles (guns employing electricity as a propulsive force—rail-guns are a similar, and very new invention); monorail transportation; water generators (an instrument for condensing water from the atmosphere); and the *vailx* (an aerial ship governed by forces of levitation and repulsion).

In *A Dweller on Two Planets,* the hero, Zailm (an earlier incarnation of Phylos and Walter Pierson), visits Caiphul, the capital of Atlantis, and views many wonderful electronic devices and the monorail system.

Later, the electromagnetic airships of Atlantis are introduced, along with radio and television (don't forget, this book was written in 1886). It is explained that the airships, similar to zeppelins, but more like a cigar-shaped craft, are electro-magnetic-gravitational in nature. They move through the air using a form of anti-gravity and are also capable of entering the water as submarines.

The book also contains a fascinating trip by one of these airships to a building on the summit of the Tetons. The main protagonist of the book, a young man named Zailm, visits "Umaur," a colony of Poseid. The description may be a rare psychic look at the ancient North American continent of 11,000 years ago.

"...From the city of Tolta, on the shore of Miti, our vailx arose and sped away north, across the lake Ui (Great Salt Lake) to its northwestern shore, hundreds of miles distant. On this far shore arose three lofty peaks, covered with snow, the Pitachi Ui, from which the lake at their feet took its name. On the tallest of these had stood, perhaps for five centuries, a building made of heavy slabs of granite. It had originally been erected for the double purpose of worship of Incal (the Sun, or God), and astronomical calculations, but was used in my day as a monastery. There was no path up the peak, and the sole means of access was by vailx."[10, 20]

An illustration from the 1884 book, *A Dweller on Two Planets.*

Then, in a break in the story, Frederick Spencer Oliver alleges that such massive, granite-slab walls were discovered in 1866 by a Professor Hayden, allegedly the first person to climb Grand Teton. Says the text, "In the neighborhood of twenty years ago, more or less, counting from 1886, an intrepid American explorer... went as far west as the Three Tetons. These mountain triplets were the Pitachi Ui, of Atl. Professor Hayden, having arrived at the base of these lofty peaks, succeeded, after indefatigable toil, in reaching the top of the greater peak, and made the first ascent known to modern times. On its top he found a roofless structure of granite slabs, within which, he said, that 'the granite detritus was of a depth indicating that for eleven thousand years it had been undisturbed.' His inference was that this period had elapsed since the construction of the granite walls. Well, the professor was right, as I happen to know. He was examining a structure made by Poseid hands one hundred and twenty-seven and a half thousand years ago, and it was because Professor Hayden was once a Poseida and held a position under the Atlan Government, as an attache of the government body of scientists stationed at Pitachi Ui, that he was karmically attracted to return to the scene of his labors long ago. Perhaps knowledge of this fact would have increased the interest he felt in the Three Tetons."

The narrative then resumes the journey: "Our vailx alighted upon the ledge without the temple Ui just as nightfall came on. It was very cold there, so far north, and at such an altitude...The primary cause of our visit was our desire to pay devotion to Incal as He arose next morning... Next morning after sunrise our vessel lifted and departed for the east, that we might visit our copper mines in the present Lake Superior region. We were conducted in electric trams through the labyrinths of galleries and tunnels. When we were about to leave, the government overseer of the mines presented each of our company with various articles of tempered copper."[20]

The group then returns to Poseid, making part of the journey underwater.

The book is curious and the statements quite interesting, to say the least. Do massive granite slabs in the form of walls exist on the top of Grand Teton? If they did they would certainly be in poor condition and if they exist, they might be thought to be natural. It would be nice to have the idea, begun by the book, that Atlantean ruins exist somewhere in the Grand Tetons, either proven or disproven.

The ancient copper mines of the Lake Superior region do indeed exist and are a mysterious archaeological fact. They were known in the mid-1800s, and are a source of pure copper. It is estimated that hundreds of thousands of tons of pure copper was taken out of the open pit mines of Lake Superior starting over 5,000 years ago. The civilization that mined this copper—and where it went—is still a mystery.

A Dweller on Two Planets is an odd book, one that seems to go beyond mere fiction. If the vimanas of the ancient Rama Empire flew around the world, surely they would transport cargo and passengers just like today's airships. One might enter a vimana port in Ayodhya, India circa 12000 BC and fly over the Pacific Ocean and then on to South America. The next stop of your round-the-world trip might be the Atlantean fortress on top of Grand Teton, before returning home to Poseid. As we walk down the long corridors of the airline terminal on the way to our next flight, should we wonder—is there nothing new under the sun?

Top: A gunner using a hot iron to discharge his cannon, ca. 1400 AD. Bottom: The Chinese inventor Wan Hoo and his rocket-powered car.

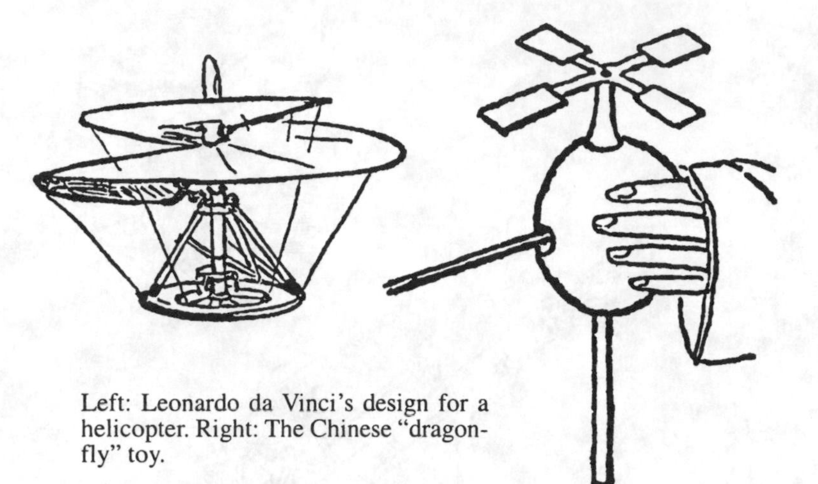

Left: Leonardo da Vinci's design for a helicopter. Right: The Chinese "dragon-fly" toy.

Robert Goddard and his rocket, 1925.

The airplane-glider model from Egypt.

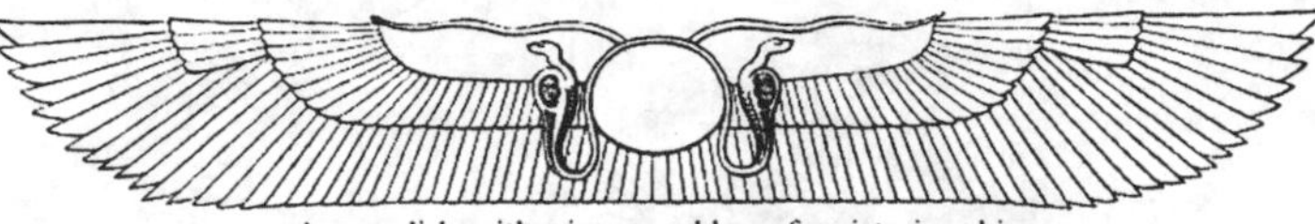

the sun disk, with wings—emblem of a victorious king

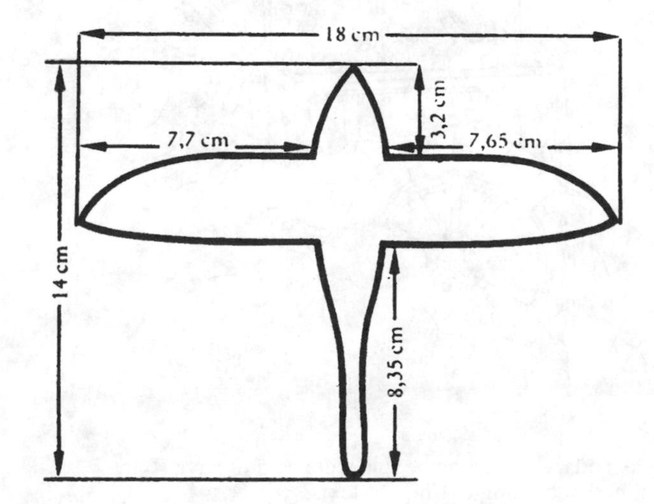

The legendary Chi-Kung people were said to have flying machines, according to Chinese texts.

Top: Woodcut of the legendary Chi-Kung people in a flying machine. Left: Fanciful diagram for Besnier's 1678 "flying engine."

Left, Above, Below: As we entered into our own era of flight, we first developed lighter-than-air balloons with propeller blades, and later, wings.

Jim Woodman's hot-air balloon flies over the Nazca Plain in Peru.

An Assyrian cylinder seal depiction of a winged disk.

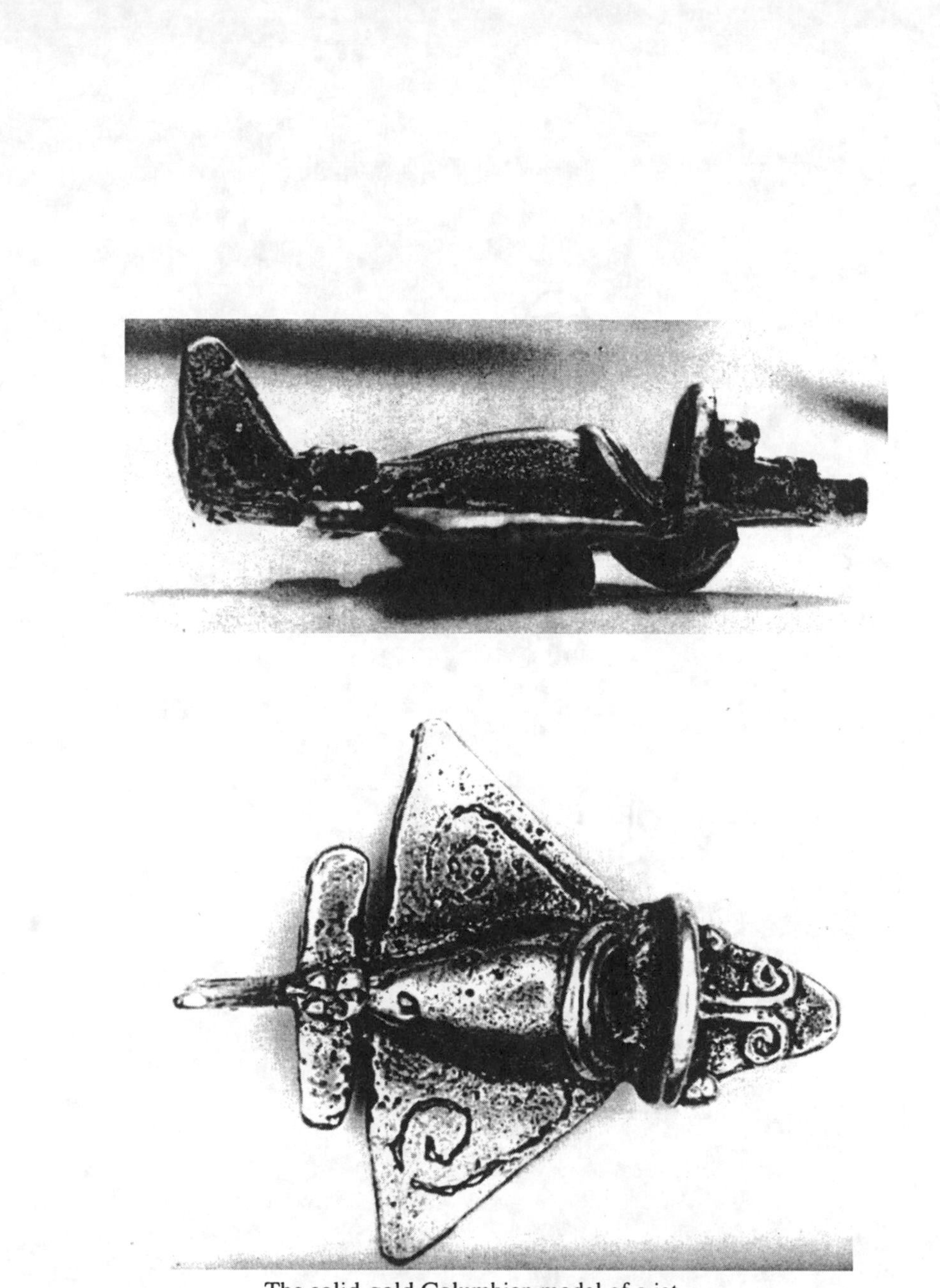

The solid-gold Columbian model of a jet.

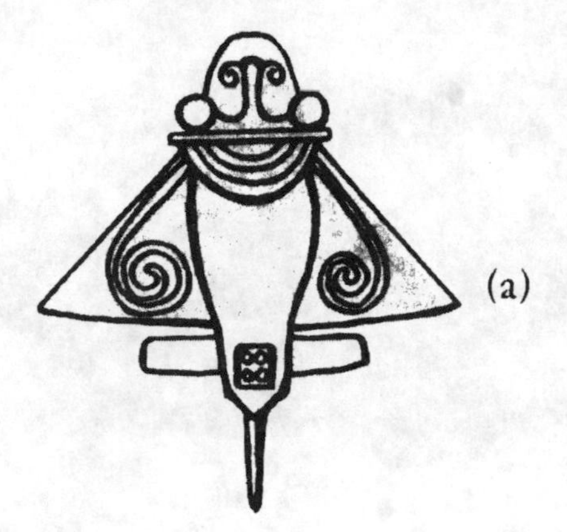

(b)

(c)

The little gold trinket from the Columbian National Collection: (a) as seen from above, (b) the side, (c) in front, (d) from behind.

Left: The gold "flying fish" in Chicago. Right: Top view of one of the jets.

The solid-gold Columbian models in a jet formation.

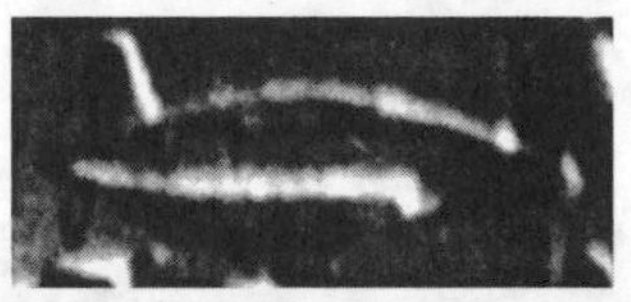

A close-up of the Abydos jet.

The lintel at the Abydos Temple in Egypt.

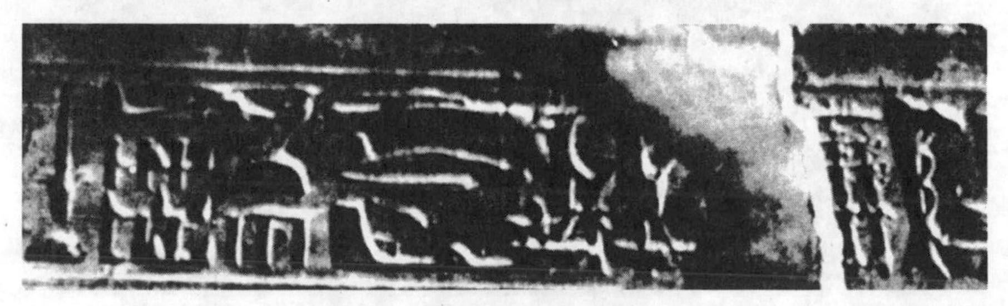

The symbols at Abydos look identical to a modern helicopter, a rocket, a flying saucer-type craft and a jet.

An Assyrian cylinder seal depiction of three men in a winged disk.

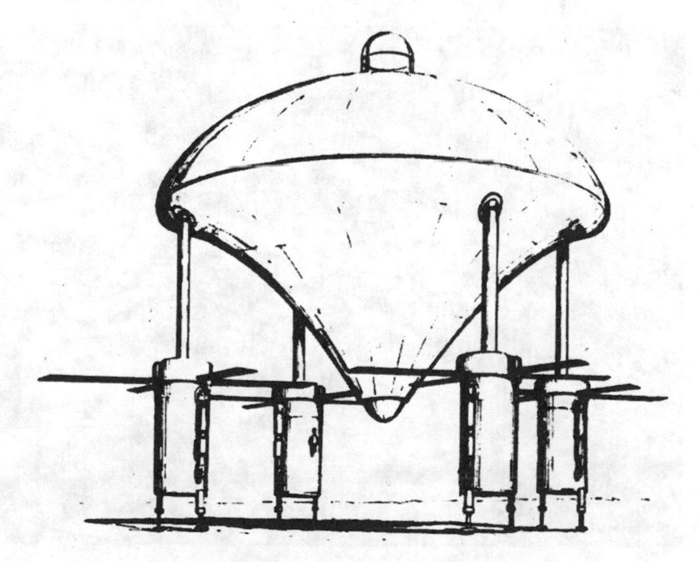

Ezekiel's Biblical vision as interpreted by NASA engineer Joseph F. Blumrich. Was it a helicopter-like vimana?

Solomon visits Sheba in this cartoon.

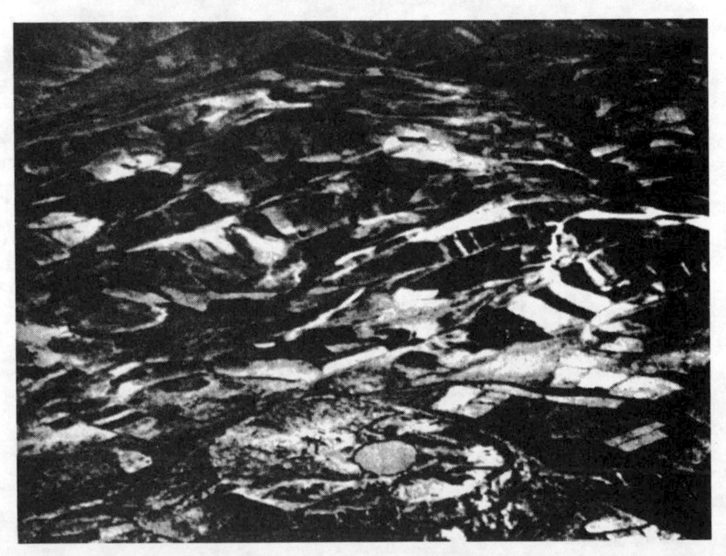

The strange hill-top fortress in northern Iran known as Tacht-i-Suleiman. It is thought to have been a landing place for King Solomon's famous airship.

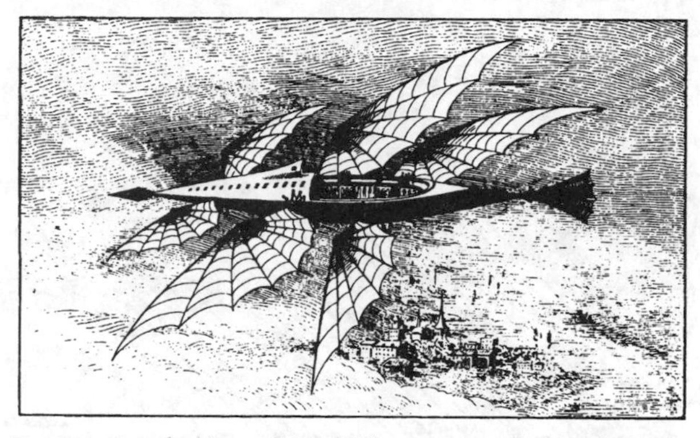

Designs for airships of the 1800s were similar to vimanas.

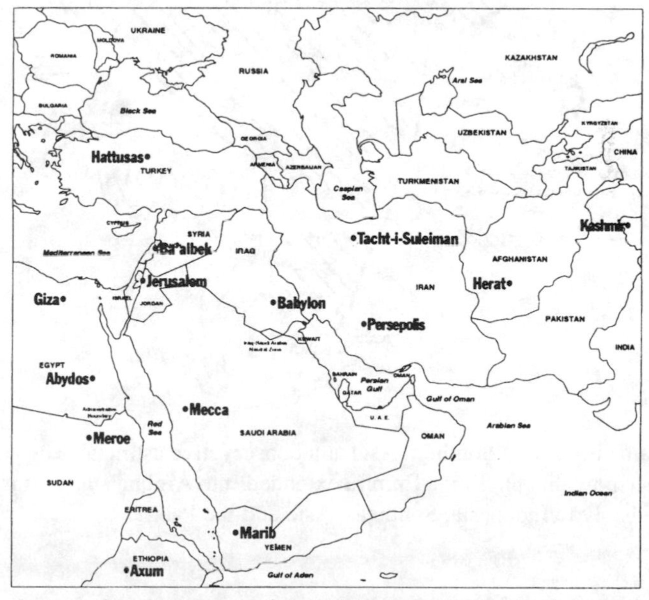

A map showing some of the places mentioned in the text, including the places visited by Solomon in his airship.

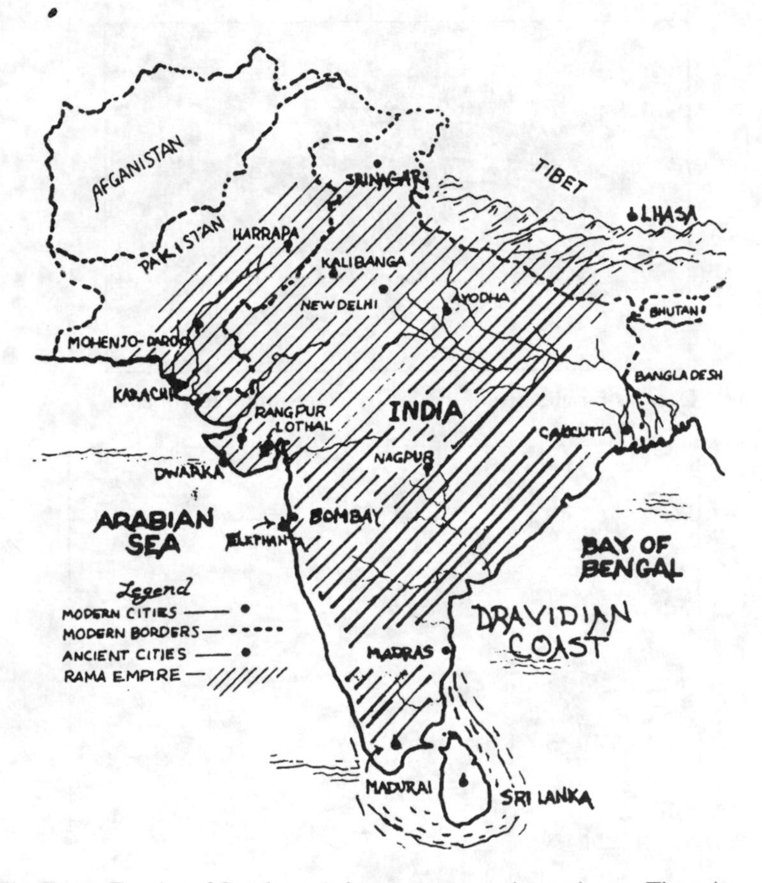

The Rama Empire of Southern Asia at a conservative estimate. There is evidence now that the Rama Empire extended into Afghanistan and Iran, and probably to Indonesia, Southeast Asia and the Pacific.

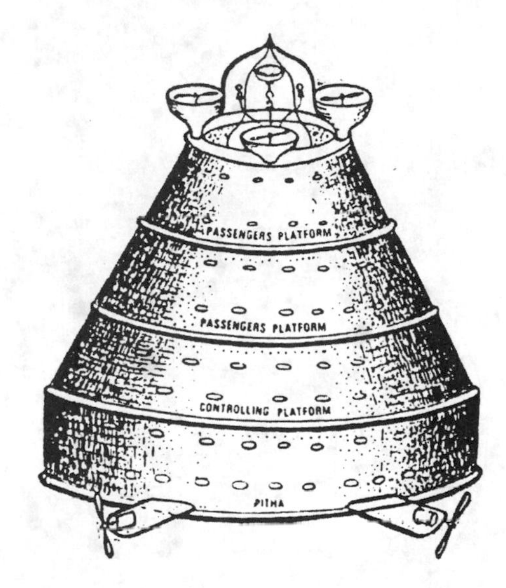

According to the Vimana texts, there were 4 kinds of vimanas.

1. The Rukma Vimana, a disc or circular craft.
2. The Sundara Vimana, also circular and pointed, like a rocket.
3. The Shakuna Vimana, a winged craft with a central tower.
4. The Tripura Vimana, a tubular or cigar-shaped craft.

RUKMA VIMANA

Plan of Top Floor

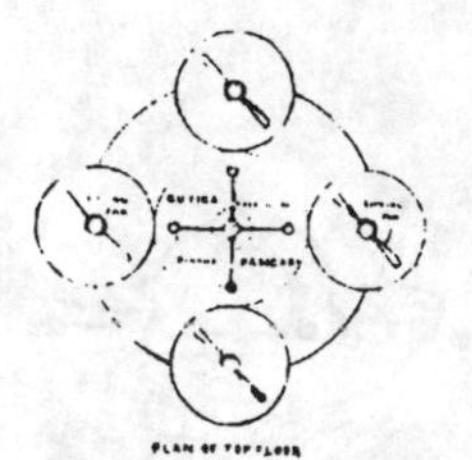

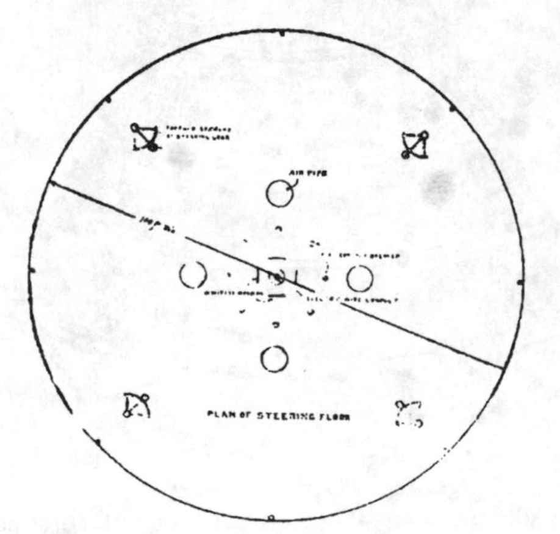

PLAN OF STEERING FLOOR

Drawn by
T. K. ELLAPPA.
Bangalore.
2-12-1923.

Prepared under instruction of
Pandit SUBBARAYA SASTRY,
of Anekal, Bangalore.

Drawings done in 1923 from the vimana texts.

RUKMA VIMANA

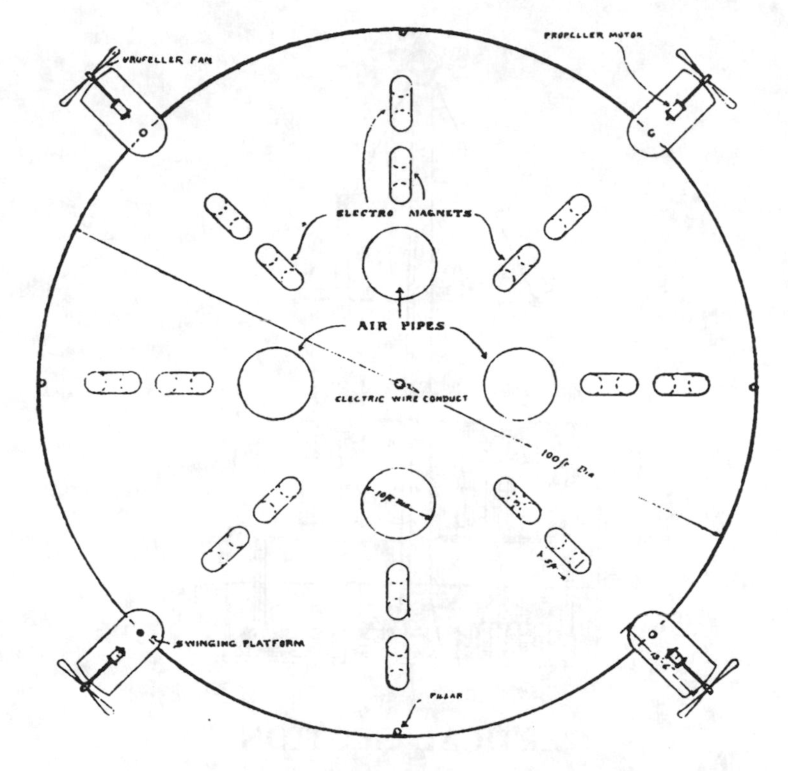

PLAN OF BASE OR PITHA

Drawn by
T. K. ELLAPPA,
Bangalore.
2-12-1923.

Prepared under instruction of
Pundit SUBBARAYA SASTRY,
of Anekal, Bangalore.

Drawings done in 1923 from the vimana texts.

SUNDARA VIMANA

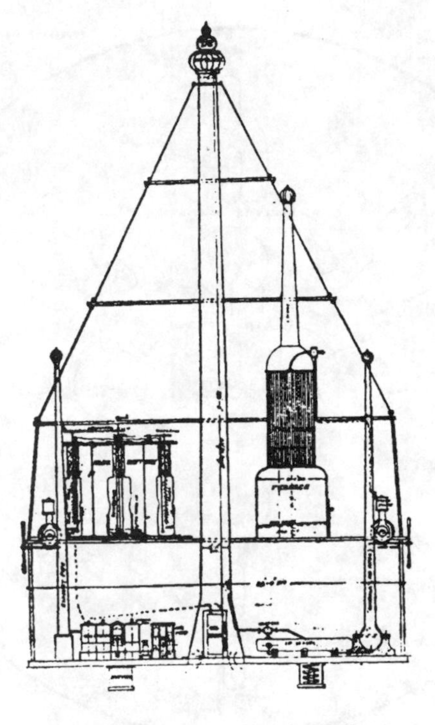

Drawn by
T. K. ELLAPPA,
Bangalore.
2-12-1923.

Prepared under instruction of
Pandit SUBBARAYA SASTRY,
of Anekal, Bangalore

Drawings done in 1923 from the vimana texts.

A vimana depicted in a temple relief at Ellora Caves, India.

At Borobodur in Indonesia, Buddha figures are seen riding in "flying saucers."

SHAKUNA VIMANA

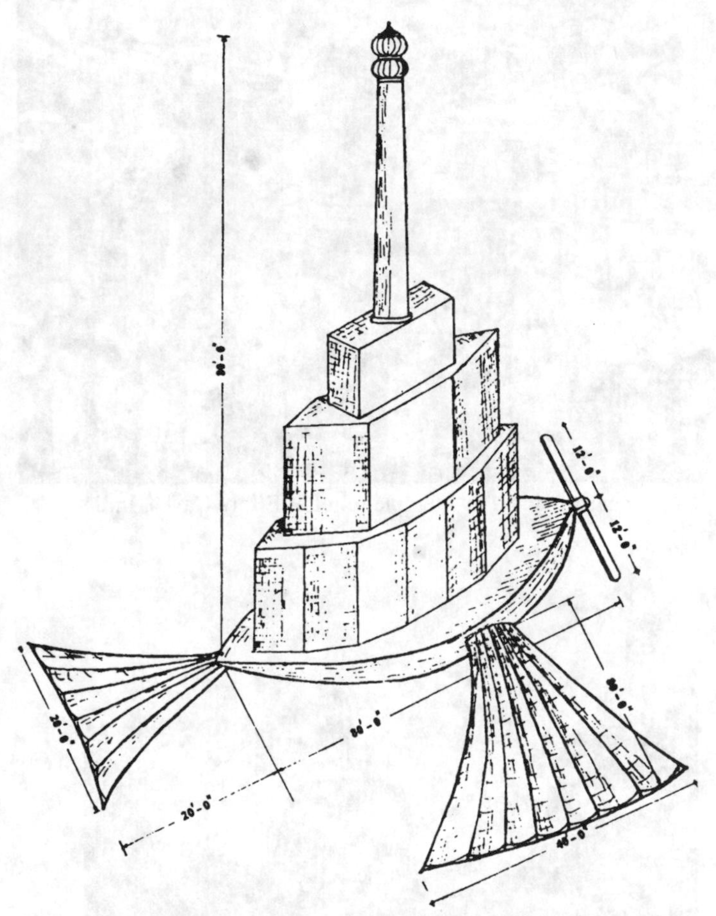

PERSPECTIVE VIEW

Drawn by
T. K. ELLAPPA,
Bangalore.
2-12-1923.

Prepared under instruction of
Pandit SUBBARAYA SASTRY,
of Anekal, Bangalore.

Drawings done in 1923 from the vimana texts.

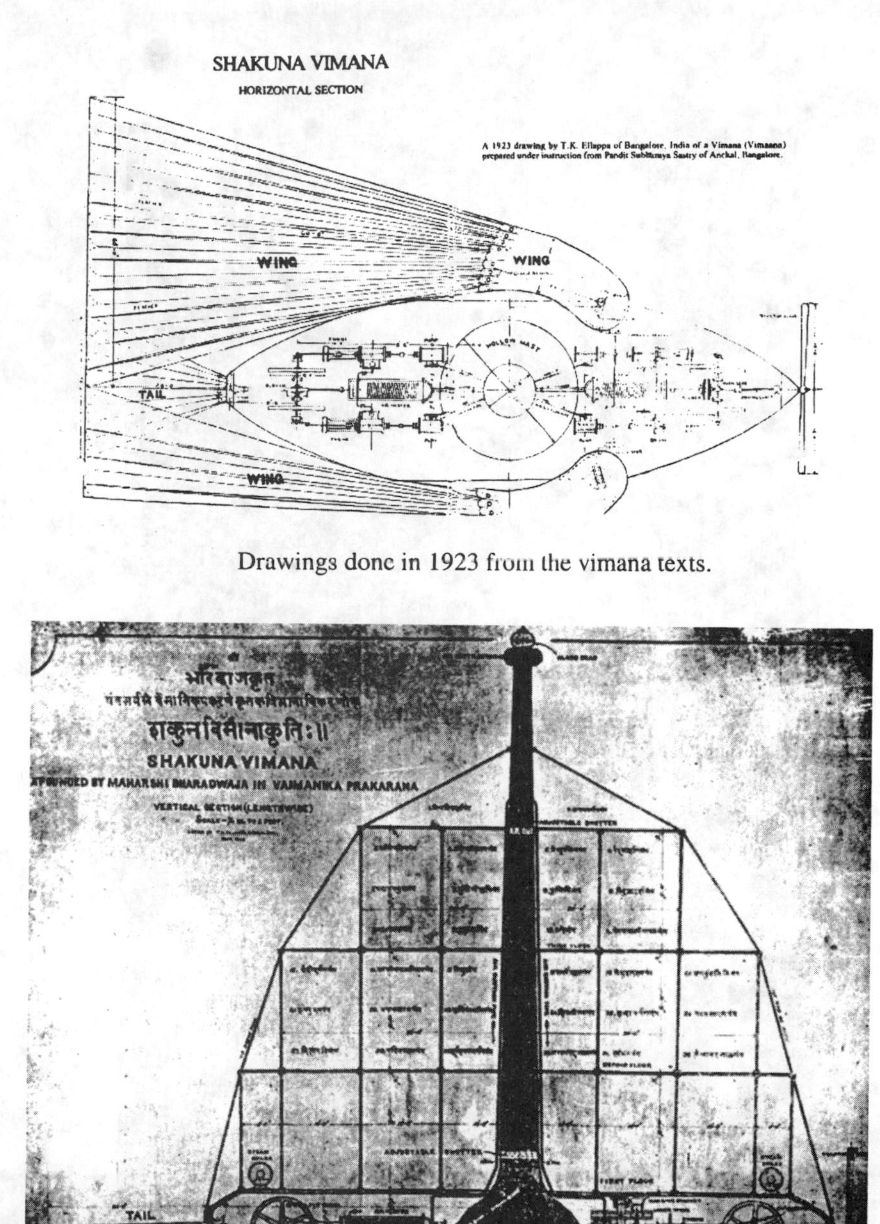

Drawings done in 1923 from the vimana texts.

Drawings done in 1923 from the vimana texts.

Mercury, the messenger god who can fly.

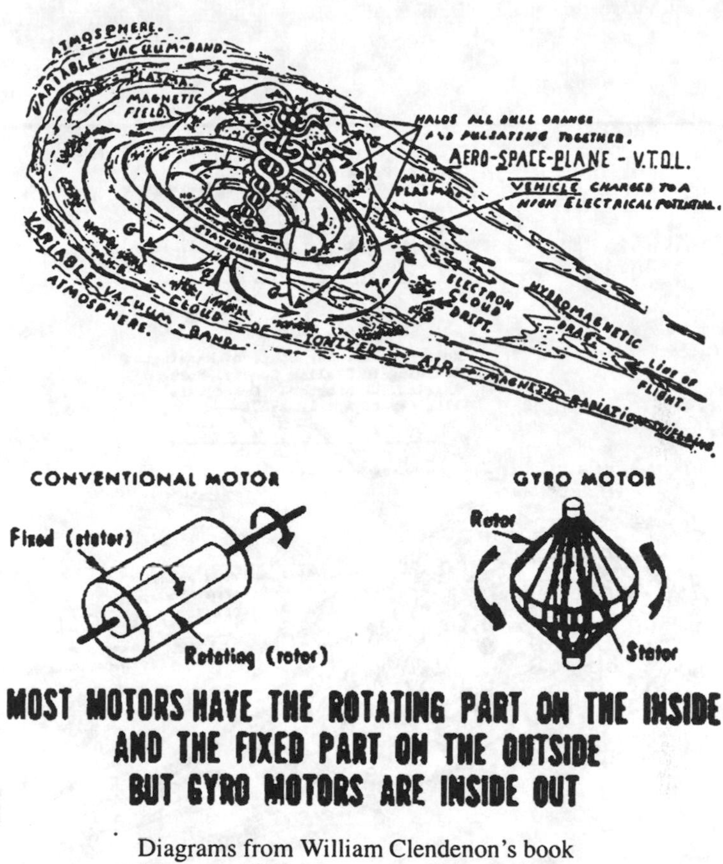

Diagrams from William Clendenon's book *Mercury: UFO Messenger of the Gods.*

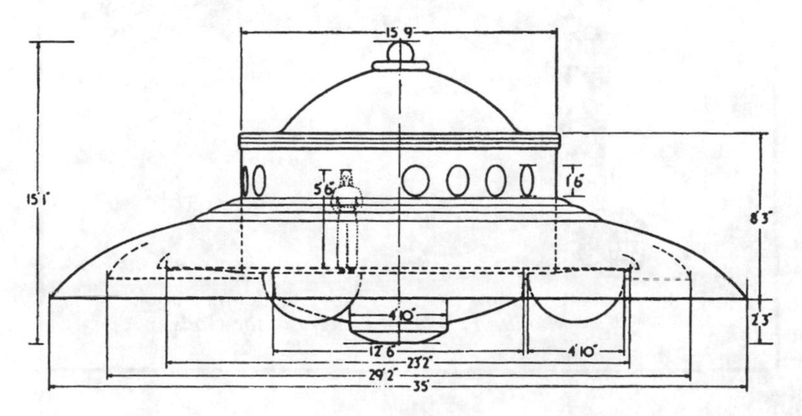

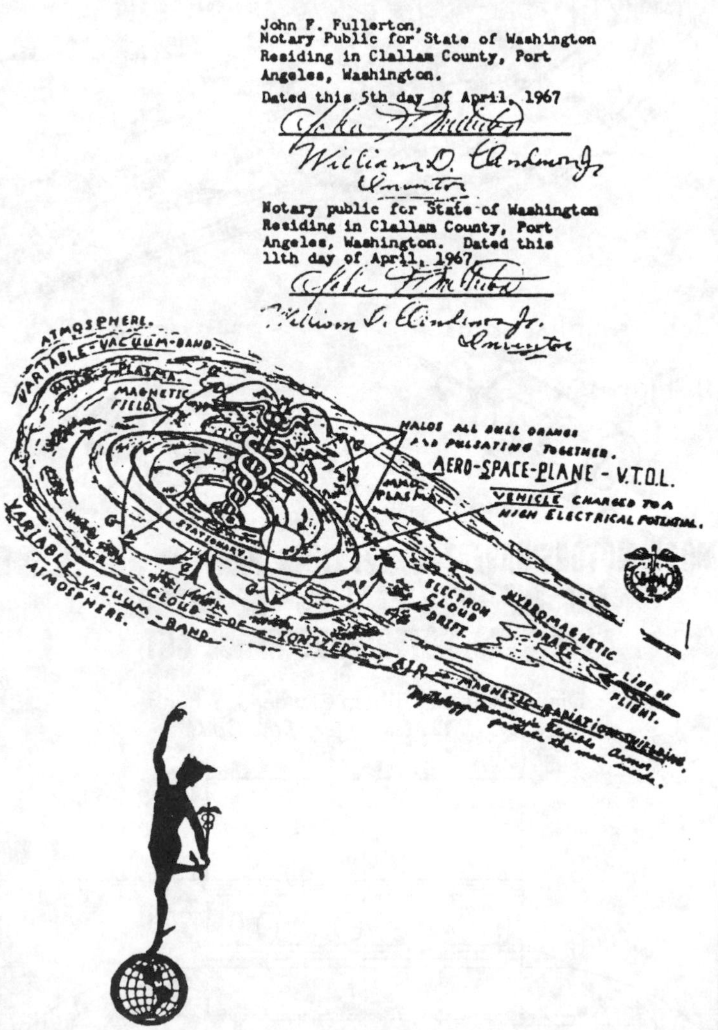
John F. Fullerton,
Notary Public for State of Washington
Residing in Clallam County, Port
Angeles, Washington.
Dated this 5th day of April, 1967
William D. Clendenon Jr.
Inventor
Notary public for State of Washington
Residing in Clallam County, Port
Angeles, Washington. Dated this
11th day of April, 1967
William D. Clendenon Jr.
Inventor
ATMOSPHERE.
VARIABLE-VACUUM-BAND.
PLASMA.
MAGNETIC FIELD
HALOS ALL DULL ORANGE AND PULSATING TOGETHER.
AERO-SPACE-PLANE - V.T.O.L.
VEHICLE CHARGED TO A HIGH ELECTRICAL POTENTIAL.
STATIONARY.
ELECTRON CLOUD DRIFT.
HYDROMAGNETIC DRAG.
LINE OF FLIGHT.
VARIABLE-VACUUM-BAND.
ATMOSPHERE.
CLOUD OF IONIZED AIR
MAGNETIC-RADIATION-SHIELDING.

Diagrams from William Clendenon's book
Mercury: UFO Messenger of the Gods.

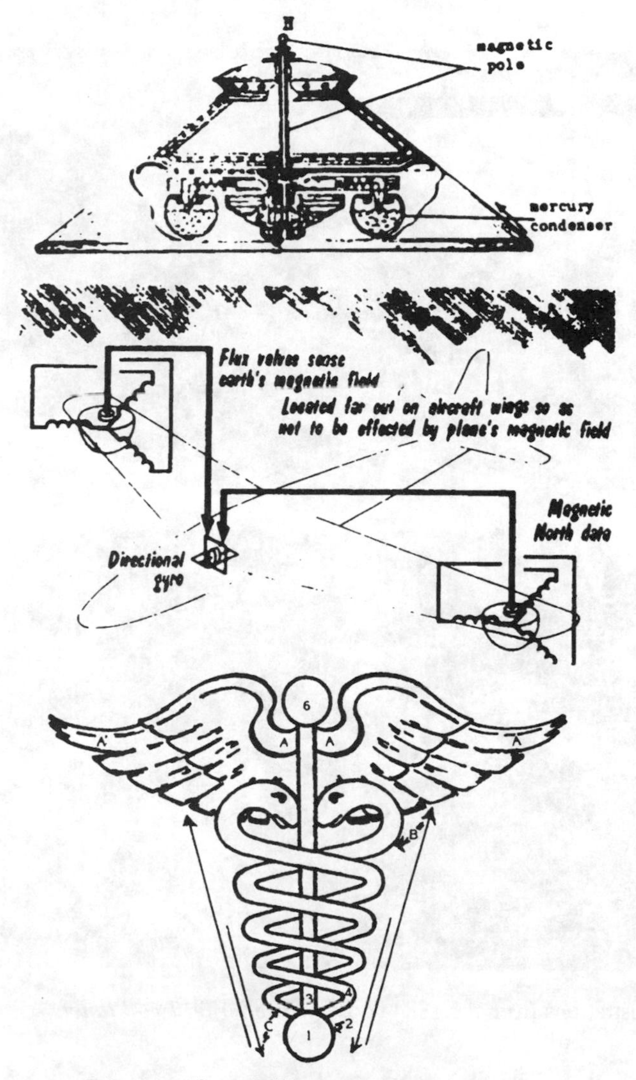

Diagrams from William Clendenon's book
Mercury: UFO Messenger of the Gods.

Two illustrations from the 1884 book, *A Dweller on Two Planets.*

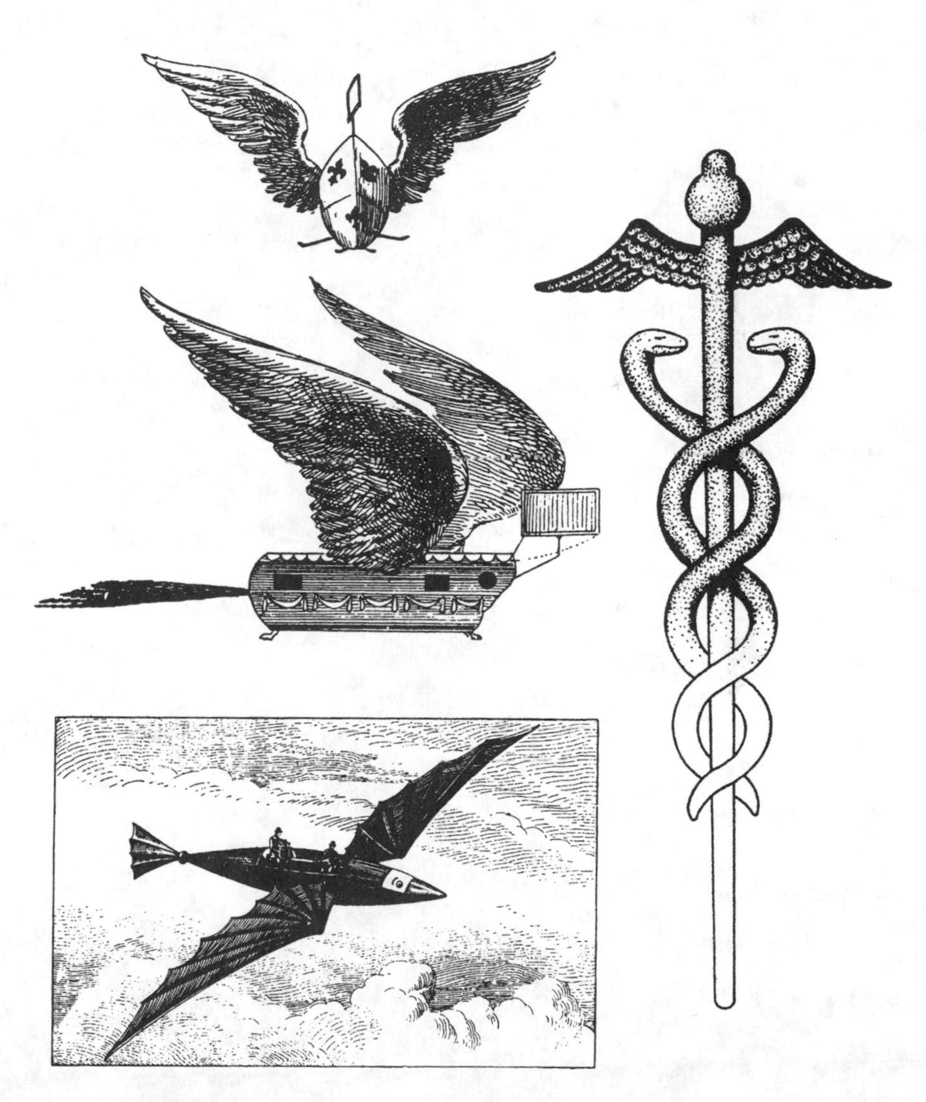

Various different versions of ancient flight and vimanas.

6.
Ancient Atomic Warfare

We learn from history
that we do not learn anything from history.
—Mark Twain

War is an instrument entirely inefficient toward redressing wrong;
and multiplies, instead of indemnifying losses.
—Thomas Jefferson

Incredible Evidence for an Ancient Atomic War

The following item appeared in the New York *Herald Tribune* on February 16, 1947 (and was repeated by Ivan T. Sanderson in the January, 1970 issue of his magazine *Pursuit*):

> When the first atomic bomb exploded in New Mexico, the desert sand turned to fused green glass. This fact, according to the magazine *Free World,* has given certain archaeologists a turn. They have been digging in the ancient Euphrates Valley and have uncovered a layer of agrarian culture 8000 years old, and a layer of herdsman culture much older, and a still older caveman culture. Recently, they reached another layer... of fused green glass. Think it over, brother.

It is well known that atomic detonations on or above a sandy desert will melt the silicon in the sand and turn the surface of the earth into a sheet of glass. But if sheets of ancient desert glass can be found in various parts of the world, does it mean that ancient atomic wars were fought in the past, or at the very least, that atomic testing occurred in the dim ages of history?

This is a startling theory, but one that is not lacking in evidence, as such ancient sheets of desert glass are a geological fact. Lightning strikes can sometimes fuse sand, meteorologists contend, but this is always in a distinctive, root-like pattern. These strange geological oddities are called fulgurites and manifest as branched, tubular forms, rather than as flat sheets of fused sand. Therefore, lightning is largely ruled out as the cause of such finds by geologists, who prefer to hold onto the theory of a meteor or comet strike as the cause. The problem with this theory is that there is usually no crater associated with these anomalous sheets of glass.

Brad Steiger and Ron Calais report in their book *Mysteries of Time and Space*[16] that Albion W. Hart, one of the first engineers to graduate from Massachusetts Institute of Technology, was assigned an engineering project in the interior of Africa. While he and his men were traveling to an almost inaccessible region, they had first to cross a great expanse of desert.

"At the time he was puzzled and quite unable to explain a large expanse of greenish glass which covered the sands as far as he could see," writes Margarethe Casson in an article on Hart's life in the magazine *Rocks and Minerals* (No. 396, 1972).

She then goes on to mention that, "Later on, during his life... he passed by the White Sands area after the first atomic explosion there, and he recognized the same type of silica fusion which he had seen fifty years earlier in the African desert."[16]

The Mystery of Tektites

Large desert areas strewn with mysterious globules of "glass"—known as tektites—are occasionally discussed in geological literature. These blobs of "hardened glass" (glass is a liquid, in fact) are thought to come from meteorite impacts in most cases, but the evidence shows that in many cases there is no impact crater.

Another explanation is that tektites have a terrestrial explanation—one that includes atomic war or high-tech weapons capable of melting sand. The tektite debate was summed up in an article in *Scientific American* (August 1978) by John O'Keefe entitled *The Tektite Problem.* Said O'Keefe:

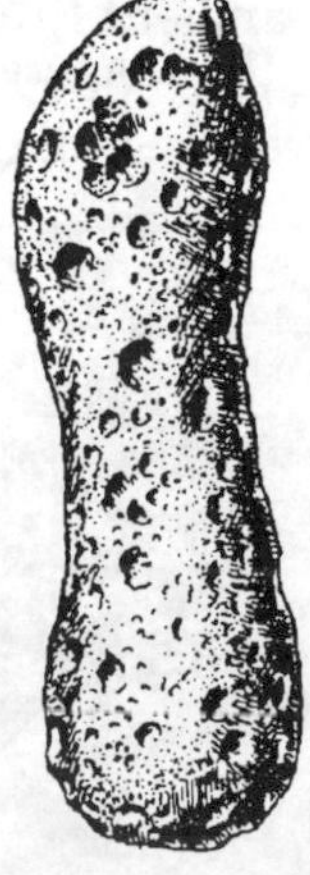

Indonesian Tektites

If tektites are terrestrial, it means that some process exists by which soil or common rocks can be converted in an instant into homogeneous, water-free, bubble-free glass and be propelled thousands of miles above the atmosphere. If tektites come from the moon, it seems to follow that there is at least one powerful volcano somewhere on the moon that has erupted at least as recently as 750,000 years ago. Neither possibility is easy to accept. Yet one of them must be accepted, and I believe it is feasible to pick the more reasonable one by rejecting the more unlikely.

The key to solving the tektite problem is an insistence on a physically reasonable hypothesis and a resolute refusal to be impressed by mere numerical coincidences such as the similarity of terrestrial sediments to tektite material. I believe that the lunar-volcanism hypothesis is the only one physically possible, and that we have to accept it. If it leads to unexpected but not impossible conclusions, that is precisely its utility.

To cite just one example of the utility, the lunar origin of tektites strongly supports the idea that the moon was formed by fission of the earth. Tektites are indeed much more like terrestrial rocks than one would expect of a chance assemblage. If tektites come from a lunar magma, then deep inside the moon there must be material that is very much like the mantle of the earth—more like the mantle than it is like the shallower parts of the moon from which the lunar surface basalts have originated. If the moon was formed by fission of the earth, the object that became the moon would have been heated intensely, and from the outside, and would have lost most of its original mass, and in particular the more volatile elements. The lavas constituting most of the moon's present surface were erupted early in the moon's history, when its heat was concentrated in the shallow depleted zone quite near the surface. During the recent periods represented by tektite falls the sources of lunar volcanism have necessarily been much deeper, so that any volcanoes responsible for tektites have drawn on the lunar material that suffered least during the period of ablation and is therefore most like unaltered terrestrial mantle material. Ironically, that would explain why tektites are in some ways more like terrestrial rocks than they are like the rocks of the lunar surface.[37]

Mysterious Glass in the Egyptian Desert

One of the strangest mysteries of ancient Egypt is that of the great glass sheets that were only discovered in 1932. In December of that year, P. Clayton, a surveyor for the Egyptian Geological Survey, was driving among the dunes of the Great Sand Sea near the Saad Plateau in the virtually uninhabited area just north of the southwestern corner of Egypt, when he heard his tires crunch on something that wasn't sand. It turned out to be large pieces of marvelously clear yellow-green glass.

In fact, this wasn't just any ordinary glass, but ultra-pure glass that was an astonishing 98% silica. Clayton wasn't the first person to come across this field of glass, as various "prehistoric" hunters and nomads had obviously also found the now-famous Libyan Desert Glass or LDG. The glass had been used in the past to make knives and other sharp-edged tools as well as other objects. A carved scarab of LDG was even found in Tut-Ankh-Amen's tomb, indicating that the glass was sometimes used for jewelry.

An article in the British science magazine *New Scientist* (July 10, 1999) by Giles Wright entitled "The Riddle of the Sands," says that LDG is the purest natural silica glass ever found. Over a thousand tons of it are strewn across hundreds of kilometers of bleak desert. Some of the chunks weigh 26 kilograms, but most LDG exists in smaller, angular pieces looking like shards left when a giant green bottle was smashed by colossal forces.

According to the article, pure as it is, LDG does contain tiny bubbles, white wisps and inky black swirls. The whitish inclusions consist of refractory minerals, such as cristobalite. The ink-like swirls, though, are rich in iridium, which is diagnostic of an extraterrestrial impact such as a meteorite or comet, according to conventional wisdom. The general theory is that the glass was created by the searing, sand-melting impact of a cosmic projectile.

However, there are serious problems with this theory, says Wright, and many mysteries concerning this stretch of desert containing the pure glass. The main problem: Where did this immense amount of widely dispersed glass shards come from? There is no evidence of an impact crater of any kind; the surface of the Great Sand Sea shows no sign of a giant crater and neither do microwave probes made deep into the sand by satellite radar.

Furthermore, LDG seems to be too pure to be derived from a messy cosmic collision. Wright mentions that known impact craters, such as the one at Wabar in Saudi Arabia, are littered with bits of iron and other meteorite debris. This is not the case with the Libyan Desert Glass sites. What is more, LDG is concentrated in two areas, rather than one. One area is oval-shaped; the other is a circular ring 6 kilometers wide and 21 kilometers in diameter. The ring's wide center is devoid of LDG.

One theory is that there was a "soft" projectile impact, which would mean that a meteorite, perhaps 30 meters in diameter, detonated about 10 kilometers or so above the Great Sand Sea. The searing blast of hot air might have melted the sand beneath. Such a craterless impact is thought to have occurred in the 1908 Tunguska Event in Siberia, at least as far as mainstream science is concerned. That event, like the pure desert glass, remains a mystery.

Another theory has a meteorite glancing off the desert surface leaving a glassy crust and a shallow crater that was soon filled in. But there are two known areas of LDG. Were there two cosmic projectiles in tandem?

It is possible that the vitrified desert is the result of the atomic wars of the ancient past. It is also possible that it was affected by a Tesla-type beam weapon which melted the desert, perhaps in a test.

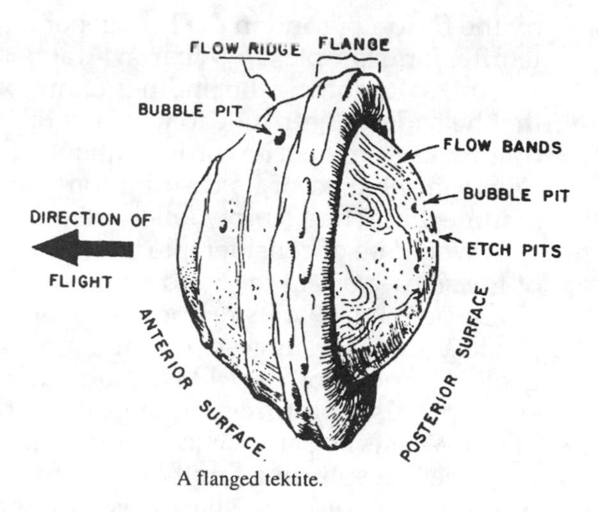

A flanged tektite.

An article by Kenneth Oakley appeared on the subject of the mysterious desert glass in the British journal *Nature*, (No. 170, 1952) entitled "Dating the Libyan Desert Silica-Glass." Said Oakley:

> Pieces of natural silica-glass up to 16 lb. in weight occur scattered sparsely in an oval area measuring 130 km. north to south and 53 km. from east to west, in the Sand Sea of the Libyan Desert. This remarkable material, which is almost pure (97 per cent silica), relatively light (sp. gin. 2. 21), clear and yellowish-green in colour, has the qualities of a gemstone. It was discovered by the Egyptian Survey Expedition under Mr. P. A. Clayton in 1932, and was thoroughly investigated by Dr. L. J. Spencer, who joined a special expedition of the Survey for this purpose in 1934.
>
> The pieces are found in sand-free corridors between north-south dune ridges, about 100 m. high and 2—5 km. apart. These corridors or 'streets' have a rubbly surface, rather like that of a 'speedway' track, formed by angular gravel and red loamy weathering debris overlying Nubian Sandstone. The pieces of glass lie on this surface or partly embedded in it. Only a few small fragments were found below the surface, and none deeper than about one metre. All the pieces on the surface have been pitted or smoothed by sand-blast. The distribution of the glass is patchy… While undoubtedly natural, the origin of the Libyan silica-glass is uncertain. In its constitution it resembles the tektites of supposed cosmic origin, but these are much smaller. Tektites are usually black, although one variety found in Bohemia and Moravia and known as moldavite is clear deep green. The Libyan silica-glass has also been compared with the glass formed

by the fusion of sand in the heat generated by the fall of a great meteorite, for example, at Wabar in Arabia and at Henbury in Central Australia. Reporting the findings of his expedition, Dr. Spencer said that he had not been able to trace the Libyan glass to any source; no fragments of meteorites or indications of meteorite craters could be found in the area of its distribution. He said: "It seemed easier to assume that it had simply fallen from the sky."

It would be of considerable interest if the time of origin or arrival of the silica-glass in the Sand Sea could be determined geologically or archaeologically. Its restriction to the surface or top layer of a superficial deposit suggests that it is not of great antiquity from the geological point of view. On the other hand, it has clearly been there since prehistoric times. Some of the flakes were submitted to Egyptologists in Cairo, who regarded them as "late Neolithic or predynastic." In spite of a careful search by Dr. Spencer and the late Mr. A. Lucas, no objects of silica-glass could be found in the collections from Tut-Ankh-Amen's tomb or from any of the other dynastic tombs. No potsherds were encountered in the silica-glass area, but in the neighbourhood of the flakings some "crude spear-points of glass" were found, also some quartzite implements, "quernstones" and ostrich-shell fragments.[37]

Oakley is apparently incorrect when he says that LDG was not found in Tutankhamen's tomb, as according to O'Keefe, a piece was. At any rate, it appears that areas of the Libyan Desert are yet to be explained. Are the vitrified areas of the North Africa desert evidence of an ancient war, a war that may have turned North Africa and Arabia into the desert that it is today?

The Vitrified Forts of Scotland

One of the great mysteries of classical archaeology is the existence of many vitrified forts in Scotland. Are they also evidence of some ancient atomic war? Maybe, but maybe not.

There are said to be at least sixty such forts throughout Scotland. Among the most well-known are Tap O'Noth, Dunnideer, Craig Phadrig (near Inverness), Abernathy (near Perth), Dun Lagaidh (in Ross), Cromarty, Arka-Unskel, Eilean na Goar, and Bute-Dunagoil on the sound of Bute on Arran Island. Another well-known vitrified fort is the Cauadale hill fort in Argyll, West Scotland.

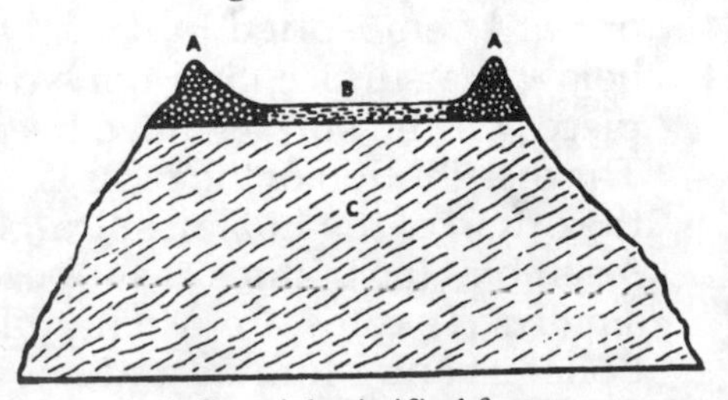

A Scottish vitrified fort.

One of the best examples of a vitrified fort is Tap O'Noth, which is near the village of Rhynie in northeastern Scotland. This massive fort from prehistory is on the summit of a mountain of the same name (1,859 ft.; 560 m.), which commands an impressive view of the Aberdeenshire countryside.

At first glance it seems that the walls are made of a rubble of stones, but on closer look, it is apparent that they are made not of dry stones, but of melted rocks! What were once individual stones are now black and cindery masses, fused together by heat that must have been so intense that molten rivers of rock once ran down the walls.

Reports on vitrified forts were made as far back as 1880 when Edward Hamilton wrote an article entitled "Vitrified Forts on the West Coast of Scotland" in the *Archaeological Journal* (Number 37, pages 227-243, 1880).

In his article, Hamilton describes several sties in detail, including Arka-Unskel: "At the point where Loch na Nuagh begins to narrow, where the opposite shore is about one-and-a-half to two miles distant, is a small promontory connected with the mainland by a narrow strip of sand and grass, which evidently at one time was submerged by the rising tide. On the flat summit of this promontory are the ruins of a vitrified fort, the proper name for which is Arka-Unskel.

"The rocks on which this fort are placed are metamorphic gneiss, covered with grass and ferns, and rise on three sides almost perpendicular for about 110 feet from the sea level. The smooth surface on the top is divided by a slight depression into two portions. On the largest, with precipitous sides to the sea, the chief portion of the fort is situated, and occupies the whole of the flat surface. It is of somewhat oval form. The circumference is about 200 feet, and the vitrified walls can be traced in its entire length... We dug under the vitrified mass, and there found what was extremely interesting, as throwing some light on the manner in which the fire was applied for the purpose of vitrification. The internal part of the upper or vitrified wall for about a foot or a foot-and-a-half was untouched by the fire, except that some of the flat stones were slightly agglutinated together, and that the stones, all feldspatic, were placed in layers one upon another.

"It was evident, therefore, that a rude foundation of boulder stones was first formed upon the original rock, and then a thick layer of loose, mostly flat stones of feldspatic sand, and of a different kind from those found in the immediate neighborhood, were placed on this foundation, and then vitrified by heat applied externally. This foundation of loose stones is found also in the vitrified fort of Dun Mac Snuichan, on Loch Etive."[5]

Hamilton describes another vitrified fort that is much larger, situated on the island at the entrance of Loch Ailort. "This island, locally termed Eilean na Goar, is the most eastern and is bounded on all sides by precipitous gneiss rocks; it is the abode and nesting place of numerous sea birds. The flat surface on the top is 120 feet from the sea level, and the remains of the vitrified forts are situated on this, oblong in form, with a continuous rampart of vitrified wall five feet thick; attached at the S.W. end to a large upright rock of gneiss. The space enclosed by this wall is 420 feet in circumference and 70 feet in width. The rampart is continuous and about five feet in thickness. At the eastern end, is a great mass of wall in situ, vitrified on both sides. In the center of the enclosed space is a deep depression in which are masses of the vitrified wall strewed about, evidently detached from their original site."[5]

Hamilton naturally asks a few obvious questions about the forts: Were these structures built as a means of defense? Was the vitrification the result of design or accident? How was the vitrification produced?

In the vitrification process, huge blocks of stones are fused with smaller rubble and form a hard and glassy mass. Explanations for vitrification are few and far between, and none of them are universally accepted. One early theory was that these forts are located on ancient volcanoes (or the remains of them) and that the people used molten stone ejected from eruptions to build their settlements.

This idea was replaced with the theory that the builders of the walls had designed the forts in such a way that the vitrification was purposeful, in order to strengthen the walls. This theory postulated that fires had been lit, and flammable material added, to produce walls strong enough to resist the dampness of the local climate or the invading armies of the enemy.

It is an interesting theory, but one that presents several problems. For starters, there is really no indication that such vitrification actually strengthens the walls of the fortress; rather, it seems to weaken them. In many cases, the walls of the forts seem to have collapsed because of the fires. Also, since the walls of many Scottish forts are only partially vitrified, this would hardly have proved an effective building method.

Julius Caesar described a type of wood and stone fortress known as a *murus Gallicus* in his account of the Gallic Wars. This was interesting to those seeking solutions to the vitrified fort mystery because these forts were made of a stone wall filled with rubble, with wooden logs inside for stability. It seemed logical to suggest that perhaps the burning of such a wood-filled wall might create the phenomenon of vitrification.

Some researchers are sure that the builders of the forts caused the vitrification. Arthur

A vitrified tower in Iraq.

C. Clarke quotes one team of chemists from the Natural History Museum in London who were studying the many forts: "Considering the high temperatures which have to be produced, and the fact that possibly sixty or so vitrified forts are to be seen in a limited geographical area of Scotland, we do not believe that this type of structure is the result of accidental fires. Careful planning and construction were needed."[41]

However, one Scottish archaeologist, Helen Nisbet, believes that the vitrification was not done on purpose by the builders of the forts. In a thorough analysis of rock types used, she reveals that most of the forts were built of stone easily available at the chosen site and not chosen for their property of vitrification.[64]

The vitrification process itself, even if purposely set, is quite a mystery. A team of chemists on Arthur C. Clarke's *Mysterious World* subjected rock samples from eleven forts to rigorous chemical analysis, and stated that the temperatures needed to produce the vitrification were so intense—up to 1,100°C—that a simple burning of walls with wood interlaced with stone could not have achieved such temperatures.[41]

Nevertheless, experiments carried out in the 1930s by the famous archaeologist V. Gordon Childe and his colleague Wallace Thorneycroft showed that forts could be set on fire and generate enough heat to vitrify the stone. In 1934, these two created a test wall that was 12 feet long, six feet wide and six feet high which was built for them at Plean Colliery in Stirlingshire. They used old fireclay bricks for the faces, pit props as timber, and filled the cavity between the walls with small cubes of basalt rubble. Finally, they covered the top with turf. Then, they piled about four tons of scrap timber and brushwood against the walls and set fire to them. Because of a snowstorm in progress, a strong wind fanned the blazing mixture of wood and stone so that the inner core did attain some vitrification of the rock.

In June of 1937, Childe and Thorneycroft duplicated their test vitrification at the ancient fort of Rahoy in Argyllshire, using rocks found at the site. Their experiments did not resolve any of the questions surrounding vitrified forts, however, because they had only proven that it was theoretically possible to pile enough wood and brush on top of a mixture of wood and stone to vitrify the mass of stone. One criticism of Childe is that he seems to have used a larger proportion of wood to stone than many historians believe made up the ancient wood and stone fortresses.

An important part of Childe's theory was that it was invaders, not the builders, who were assaulting the forts and then setting fire to the walls with piles of brush and wood; however, it is hard to understand why people would have repeatedly built defenses that invaders could destroy with fire, when great ramparts of solid stone would have survived unscathed.[41]

Critics of the assault theory point out that in order to generate enough heat by a natural fire, the walls would have to have been specially constructed to create the heat necessary. It seems unreasonable to suggest the builders would specifically create forts to be burned or that such a great effort would be made by invaders to create the kind of fire it would take to vitrify the walls—at least with traditional techniques.

One problem with all the many theories is their assumption of a primitive state of culture associated with ancient Scotland.

It is astonishing to think of how large and well coordinated the population or army must have been that built and inhabited these ancient structures. Janet and Colin Bord in their book *Mysterious Britain*[65] speak of Maiden Castle to give an idea of the vast extent of this marvel of prehistoric engineering. "It covers an area of 120 acres, with an average width of 1,500 feet and length of 3,000 feet. The inner circumference is about 1-1/2 miles round, and it has been estimated, as mentioned earlier, that it would require 250,000 men to defend it! It is hard, therefore, to believe that this construction was intended to be a defensive position.

"A great puzzle to archaeologists has always been the multiple and labyrinthine east and west entrances at each end of the enclosure. Originally they may have been built as a way for processional entry by people of the Neolithic era. Later, when warriors of the Iron Age were using the site as a fortress they probably found them useful as a means of confusing the attacking force trying to gain entry. The fact that so many of these 'hillforts' have two entrances—one north of east and the other south of west—also suggests some form of Sun ceremonial."

With 250,000 men defending a fort, we are talking about a huge army in a very organized society! This is not a bunch of fur-wearing Picts with spears defending a fort from marauding bands of hunter-gatherers.

The questions remain though as to what huge army might have occupied these cliffside forts by the sea or lake entrances? And what massive maritime power were these people unsuccessfully defending themselves against?

The forts on the western coast of Scotland are reminiscent of the mysterious clifftop forts in the Aran Islands on the west coast of Ireland. Here we truly have shades of the Atlantis story with a powerful naval fleet attacking and conquering its neighbors in a terrible war. It has been theorized that the terrible battles of the Atlantis story took place in Wales, Scotland, Ireland and England—however, in the case of the Scottish vitrified forts it looks as if these were the losers of a war, not the victors. And defeat can be seen across the land: the war dikes in Sussex, the vitrified forts of Scotland, the utter collapse and disappearance of the civilization that built these things. What long-ago Armageddon destroyed ancient Scotland?

In ancient times there was a substance known through writings as "Greek fire." This was some sort of ancient napalm bomb that was hurled by catapult and could not be put out. Some forms of Greek fire were even said to burn under water and were therefore used in naval battles. (The actual composition of Greek fire is unknown, but it must have contained chemicals such as phosphorus, pitch, sulfur or other flammable chemicals.)

Could a form of Greek fire have been responsible for the vitrification? While ancient astronaut theorists may believe that extraterrestrials with their atomic weapons vitrified these walls, it seems more likely that they are the result of a manmade apocalypse of a chemical nature. With siege machines, battleships and Greek fire, did a vast flotilla storm the huge forts and eventually burn them down in a hellish blaze?

The evidence of the vitrified forts is clear: some hugely successful and organized civilization was living in Scotland, England and Wales in prehistoric times, circa 1000 BC or more, that was building gigantic structures, including forts. This apparently was a maritime civilization that prepared itself for naval warfare as well as other forms of attack.

More Vitrified Ruins

Other vitrified ruins can be found in France, Turkey and other areas of the Middle East. Vitrified forts in France are discussed in the *American Journal of Science* (Volume 3, Number 22, pages 150-151, 1881), which carried an article entitled "On the Substances Obtained from Some 'Forts Vitrifies' in France" by M. Daubree.

The author mentions several forts in Brittany and northern France whose granite blocks have been vitrified. He cites the "partially fused granitic rocks from the forts of Chateau-vieux and of Puy de Gaudy (Creuse), also from the neighborhood of Saint Brieuc (Cotes-du-Nord)."[17] Daubree, understandably, could not readily find an explanation for the vitrification.

Similarly, the ruins of Hattusas in central Turkey, an ancient Hittite city, are partially vitrified. The Hittites are said to be the inventors of the chariot, and horses were of great importance to them. It is on the ancient Hittite stelae that we first see the chariot being used. However, it seems unlikely that horsemanship and wheeled chariots were first invented by the Hittites. It is highly likely, for instance, that chariots were in use in ancient China at the same time.

The Hittites used the unusual double-headed eagle motif, a symbol Germany still uses today. The Hittites are also related to the amazing world of ancient India. Proto-Indic writing is found at Hattusas and scholars now admit that the civilization of India, as the ancient Indian texts like the

Ramayana have said, goes back many millennia.

In his 1965 book, *The Bible as History*,[29] the German historian Werner Keller cites some of the mysteries concerning the Hittites. According to Keller, the Hittites are first mentioned in the *Bible* in connection with the Biblical patriarch Abraham who (in *Genesis* 23) acquires from the Hittites a burial place in Hebron for his wife Sarah. Conservative classical scholar Keller is confused by this, because the time period of Abraham was circa 2000-1800 BC, while the Hittites are traditionally said to have appeared in the 16th century BC.

Even more confusing to Keller is the Biblical statement that the Hittites are the founders of Jerusalem (*Numbers* 13:29-30).[29] This is a fascinating statement, as it would mean that the Hittites also occupied Ba'albek, which lies between their realm and Jerusalem. As we have seen, the Temple Mount at Jerusalem is built on a foundation of huge ashlars like Ba'albek. The Hittites definitely used the gigantic megalithic construction known as cyclopean—huge odd-shaped polygonal blocks perfectly fitted together. The massive walls and gates of Hattusas are eerily similar in construction to those in the high Andes and other megalithic sites around the world. The difference at Hattusas is that parts of the city are vitrified, and the walls of rock are partly melted.

If the Hittites were the builders of Jerusalem, it would mean that the ancient Hittite empire existed for several thousand years with frontiers with Egypt. Indeed, the Hittite hieroglyphic script is undeniably similar to Egyptian hieroglyphs, probably more so than any other language.

Just as Egypt goes back many thousands of years BC and is ultimately connected to Atlantis, so does the ancient Hittite empire. Like the Egyptians, the Hittites carved massive granite sphinxes, built on a cyclopean scale, and worshiped the Sun. The Hittites also used the common motif of a winged disc for their Sun God, just as the Egyptians did. The Hittites were well known in the ancient world, because they were the main manufacturer of iron and bronze goods. The Hittites were metallurgists and seafarers. Their winged disks may, in fact, have been representations of the flying machines called vimanas.

Some of the ancient ziggurats of Iran and Iraq also contain vitrified material, sometimes thought by archaeologists to be caused by the "Greek Fire." For instance, the vitrified remains of the ziggurat at Birs Nimrod (Borsippa), south of Hillah, were once confused with the 'Tower of Babel.' The ruins are crowned by a mass of vitrified brickwork, actual clay bricks fused together by intense heat. This may be due to the horrific ancient wars described in the *Ramayana* and *Mahabharata*, although early archaeologists attributed the effect to lightning.

Destruction cometh; and they shall seek peace,
and there shall be none.
—Ezekiel(7: 25)

Greek Fire, Plasma Guns, and Atomic Warfare

If one were to believe the great Indian epic of the *Mahabharata,* fantastic battles were fought in the past with airships, particle beams, chemical warfare and, presumably, atomic weapons. Just as battles in this century have been fought with incredibly devastating weapons, it may well be that battles in the latter days of Atlantis were fought with highly sophisticated and high-tech weapons.

The mysterious Greek fire was a "chemical fireball." Incendiary mixtures go back at least to the 5th century BC when Aineias the Tactician wrote a book called *On the Defense of Fortified Positions.* Said he: "And fire itself which is to be powerful and quite inextinguishable is to be prepared as follows. Pitch, sulfur, tow, granulated frankincense, and pine sawdust in sacks you should ignite if you wish to set any of the enemy's works on fire."[27]

L. Sprague de Camp mentions in his book *The Ancient Engineers,*[27] that at some point it was found that petroleum, which seeps out of the ground in Iraq and elsewhere, made an ideal base for incendiary mixtures because it could be squirted from syringes of the sort then used in fighting fires. Other substances were added to it, such as sulfur, olive oil, rosin, bitumen, salt, and quicklime.

Some of these additives may have helped—sulfur at least made a fine stench—but others did not, although it was thought that they did. Salt, for instance, may have been added because the sodium in it gave the flame a bright orange color. The ancients, supposing that a brighter flame was necessarily a hotter flame, mistakenly believed that salt made the fire burn more fiercely. Such mixtures were put in thin wooden casks and thrown from catapults at hostile ships and at wooden siege engines and defense works.

According to de Camp, in 673 AD, the architect Kallinikos fled ahead of Arab invaders from Helipolis-Ba'albek to Constantinople. There he revealed to the emperor Constantine IV an improved formula for a liquid incendiary. This could not only be squirted at the foe but could also be used with great effect at sea, because it caught fire when it touched the water and floated flaming on the waves.

De Camp says that Byzantine galleys were armed with a flame-throwing apparatus in the bow, consisting of a tank of this mixture, a pump, and a nozzle. With the help of this compound, the Byzantines broke the Arab sieges of AD 674-76 and AD 715- 18, and also beat off the Russian attacks of AD 941 and 1043. The incendiary liquid wrought immense havoc; of 800 Arab ships that attacked Constantinople in 716 AD, only a handful returned home.

The formula for the wet version of Greek fire has never been discovered. Says de Camp, "By careful security precautions, the Byzantine Emperors succeeded in keeping the secret of this substance, called "wet fire" or "wild fire," so dark that it never did become generally known. When asked about it, they blandly replied that an angel had revealed the formula to the first Constantine.

"We can, therefore, only guess the nature of the mixture. According to one disputed theory, wet fire was petroleum with an admixture of calcium phosphide, which can be made from lime, bones, and urine. Perhaps Kallinikos stumbled across this substance in the course of alchemical experiments."[27]

Vitrification of brick, rock and sand may have been caused by any number of high-tech means. New Zealand author Robin Collyns suggests in his book *Ancient Astronauts: A Time Reversal?*[26] that there are five methods by which the ancients or "ancient astronauts" might have waged war on various societies on planet Earth. He outlines how these methods are again on the rise in modern society.

The five methods are: plasma guns, fusion torches, holes punched in the ozone layer, manipulation of weather processes, and the release of immense energy such as an atomic blast. As Collyns' book was published in Britain in 1976, the mention of holes in the ozone and weather warfare seem strangely prophetic.

Explaining the plasma gun, Collyns says, "The plasma gun has already been developed experimentally for peaceful purposes: Ukrainian scientists from the Geotechnical Mechanics Institute have experimentally drilled tunnels in iron ore mines by using a plasmatron, i.e., a plasma gas jet which delivers a temperature of 6,000°C."[26]

A plasma, in this case, is an electrified gas. Electrified gases are also featured in the ancient book from India on vimanas called the *Vymanika Shastra*,[33] which cryptically talks of using for fuel the liquid metal mercury, which could be a plasma if it were electrified.

Collyns goes on to describe a fusion torch: "This is still another possible method of warfare used by spacemen, or ancient advanced civilizations on Earth. Perhaps the solar mirrors of antiquity really were fusion torches? The fusion torch is basically a further development of the plasma jet. In 1970 a theory to develop a fusion torch was presented at the New York aerospace science meeting by Drs. Bernard J. Eastlund and William C. Cough. The basic idea

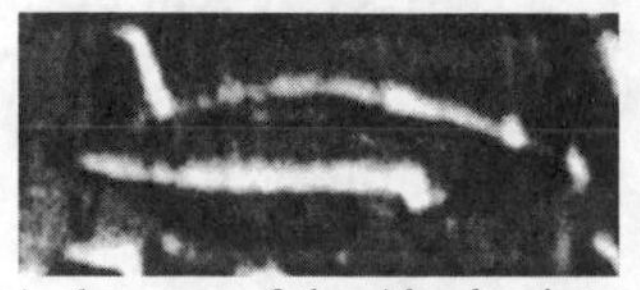

A close-up of the Abydos jet.

is to generate a fantastic heat of at least fifty million degrees Celsius which could be contained and controlled. That is, the energy released could be used for many peaceful applications with zero radioactive waste products to avoid contaminating the environment, or zero production of radioactive elements which would be highly dangerous, such as Plutonium which is the most deadly substance known to Man. Thermonuclear fusion occurs naturally in stellar processes, and unnaturally in man-made H-bomb explosions.

"The fusion of a deuterium nuclei (a heavy hydrogen isotope which can be easily extracted from sea water) with another deuterium nuclei, or with tritium (another isotope of hydrogen) or with helium, could be used. The actual fusion torch would be an ionized plasma jet which would vaporize anything and everything that the jet was directed at—if it was to be used for harmful purposes—while for peaceful applications, one use of the torch could be to reclaim basic elements from junk metals.

"University of Texas scientists announced in 1974 that they had actually developed the first experimental fusion torch which gave an incredible heat output of ninety-three degrees Celsius. This is five times the previous hottest temperature for a contained gas and is twice the minimum heat needed for fusion, but it was held only for one fifty-millionth of a second instead of the one full second which would be required."[26]

It is curious to note here that Dr. Bernard Eastlund is the patent holder of another unusual device, one that is associated with the HAARP project at Gakona, Alaska. HAARP (High Altitude Aerial Research Project) is allegedly linked to weather manipulation, one of the ways that Collyns thinks that the ancients waged warfare.

As far as holes in the ozone and weather manipulation go, Collyns says, "Soviet scientists have discussed and proposed at the United Nations a ban on developing new warfare ideas such as creating holes or 'windows' in the ozone layer to bombard specific areas of the Earth with increased natural ultra-violet radiation which would kill all life-forms, and turn the land into barren desert.

"Other ideas discussed at the meeting were the use of 'infrasound' to demolish ships by creating acoustic fields on the sea, and hurling a huge chunk of rock into the sea with a cheap atomic device. The resultant tidal wave could demolish the coastal fringe of a country. Other tidal waves could be created by detonating nuclear devices at the frozen poles. Controlled floods, hurricanes, earthquakes and droughts directed towards specific targets and cities are other possibilities."

"Finally," says Collyns, "although not a new method of warfare, incendiary weapons are now being developed to the point where 'chemical fireballs' will be produced which radiated thermal energy similar to that of an atomic bomb."[26]

Does California's Death Valley Show Evidence of an Atomic War?

In *Secrets of the Lost Races,*[32] Rene Noorbergen discusses the evidence for a cataclysmic war in the remote past that included the use of airships and weapons that vitrified stone cities. "The most numerous vitrified remains in the New World are located in the Western United States. In 1850 the American explorer Captain Ives William Walker was the first to view some of these ruins, situated in Death Valley. He discovered a city about a mile long, with the lines of the streets and the positions of the buildings still visible. At the center he found a huge rock, between 20 to 30 feet high, with the remains of an enormous structure atop it. The southern side of both the rock and the building was melted and vitrified. Walker assumed that a volcano had been responsible for this phenomenon, but there is no volcano in the area. In addition, tectonic heat could not have caused such a liquefication of the rock surface.

"An associate of Captain Walker who followed up his initial exploration commented, 'The whole region between the rivers Gila and San Juan is covered with remains. The ruins of cities are to be found there which must be most extensive, and they are burnt out and vitrified in part, full of fused stones and craters caused by fires which were hot enough to liquefy rock or metal. There are paving stones and houses torn with monstrous cracks... [as though they had] been attacked by a giant's fire-plough.'"[32]

These vitrified ruins in Death Valley sound fascinating—but do they really exist?

There certainly is evidence of ancient civilizations in the area. In Titus Canyon, petroglyphs and inscriptions have been scratched into the walls by unknown prehistoric hands. Some experts think the graffiti might have been made by people who lived here long before the Indians we know of, because extant Indians know nothing of the glyphs and, indeed, regard them with superstitious awe.

Says Jim Brandon in *Weird America,*[34] "Piute legends tell of a city beneath Death Valley that they call *Shin-au-av.* Tom Wilson, an Indian guide in the 1920s, claimed that his grandfather had rediscovered the place by wandering into a miles-long labyrinth of caves beneath the valley floor.

"Eventually the Indian came to an underworld city where the people spoke an incomprehensible language and wore clothing made of leather. Wilson told this story after a prospector named White claimed he had fallen through the floor of an abandoned mine at Wingate Pass and into an unknown tunnel. White followed this into a series of rooms, where he found hundreds of leather-clad humanoid mummies. Gold bars were stacked like bricks and piled in bins.

"White claimed he had explored the caverns on three occasions. On one his wife accompanied him and on another his partner, Fred Thomason.

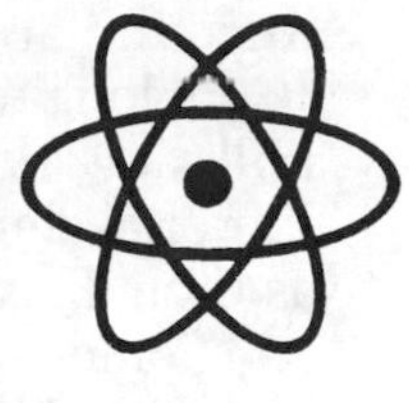

However, none of them were able to relocate the opening to the cavern when they tried to take a group of archaeologists on a tour of the place."

It seems one local character knew how to find the place. Brandon relates that "Death Valley Scotty," an eccentric who spent millions building a castle-estate in the area, was known to go "prospecting" when funds ran low. Death Valley Scotty would check out for a few days of wandering in the nearby Grapevine Mountains, bringing back suspiciously refined-looking gold that he claimed he had prospected. Many believe that he got his gold from the stacked gold bars in the tunnel system beneath Death Valley.[34]

Evidence of a lost civilization in Death Valley came in a bizarre report of caves and mummies in the *Hot Citizen,* a Nevada paper, on August 5, 1947. The story ran as follows:

EXPEDITION REPORTS NINE-FOOT SKELETONS

A band of amateur archaeologists announced today they have discovered a lost civilization of men nine feet tall in Californian caverns. Howard E. Hill, spokesman for the expedition said the civilization may be "the fabled lost continent of Atlantis."

The caves contain mummies of men and animals and implements of a culture 80,000 years old but "in some respects more advanced than ours," Hill said. He said the 32 caves covered a 180-square-mile area in California's Death Valley and Southern Nevada.

ARCHAEOLOGISTS SKEPTICAL

"This discovery may be more important than the unveiling of King Tut's tomb," he said.

Professional archaeologists were skeptical of Hill's story. Los Angeles County Museum scientists pointed out that dinosaurs and tigers which Hill said lay side by side in the caves appeared on earth 10,000,000 to 13,000,000 years apart.

Hill said the caves were discovered in 1931 by Dr. F. Bruce Russell, Beverly Hills physician, who literally fell in while sinking a shaft for a mining claim.

"He tried for years to interest people in them," Hill said, "but nobody believed him."

Russell and several hobbyists incorporated after the war as Amazing Explorations, Inc. and started digging. Several caverns contained mummified remains of "a race of men eight to nine feet tall," Hill said, "they apparently wore a prehistoric zoot suit—a hair garment of medium length, jacket and knee length trousers."

CAVERN TEMPLE FOUND

Another cavern contained their ritual hall with devices and markings similar to the Masonic order, he said.

"A long tunnel from this temple took the party into a room where," Hill said, "well-preserved remains of dinosaurs, saber-toothed tigers, imperial elephants and other extinct beasts were paired off in niches as if on display.

"Some catastrophe apparently drove the people into the caves," he said. "All of the implements of their civilization were found," he said, "including household utensils and stoves which apparently cooked by radio waves."

"I know," he said, "that you won't believe that."

While of doubtful authenticity, this is an interesting story, to say the least. The last comment about cooking food with radio waves being unbelievable is ironic. That is the one thing that modern readers of the story could certainly believe was true, considering the widespread use of microwave ovens today—who had heard of them in 1947?

Sodom and Gomorrah meet Hiroshima and Nagasaki

Probably the most famous of all ancient "nukem" stories is the well-known Biblical tale of Sodom and Gomorrah.

> "And the Lord said, Because the cry of Sodom and Gomorrah is great, and because their sin is very grievous... Then the Lord rained upon Sodom and upon Gomorrah brimstone and fire from the Lord out of heaven; And he overthrew those cities, and all the plain, and all the inhabitants of the cities, and that which grew upon the ground. But his [Lot's] wife looked back from behind him, and she became a pillar of salt... and, lo, the smoke of the country went up as the smoke of a furnace." (*Genesis* 18:20; 19:24-26, 28)

This Biblical passage has come to epitomize the destructive power of God's wrath, visited on those places which sin. The *Bible* is very specific about the site of Sodom and Gomorrah, plus several other towns; they were in the Vale of Siddim which was located at the southern end of the salt sea (now called the Dead Sea). Other towns in the area, according to the *Bible*, were Zoar, Admah and Zeboiim (*Gen.* 14:2). As late as the Middle Ages, a town called Zoar existed in the area.

The Dead Sea is 1,280 feet below sea level and an incredible 1,200 feet deep. The bottom of the sea is therefore about 2,500 feet below the level of the Mediterranean. Approximately 30% of the water of the Dead Sea consists of solid ingredients, mostly sodium chloride, i.e., cooking salt. Normal ocean water is only 3.3% to 4% salt. The Jordan and many smaller rivers empty themselves into this basin, which has not a solitary outlet.

What its tributaries bring to it in the way of chemical substances remains deposited in the Dead Sea's 500 square miles. Evaporation under the broiling sun takes place on the surface of the sea at a rate of over 230 million cubic feet per day. Arab tradition has it that so many poisonous gases come out of the lake that birds could not fly across it, as they would die before reaching the other side.

The Dead Sea was first explored in modern times in 1848 when W.F. Lynch, an American geologist, led an expedition. He brought ashore from his government research ship two metal boats which he fastened onto large-wheeled carts. Pulled by a long team of horses, Lynch's expedition reached the Dead Sea some months later. Lynch's team discovered that the traditions were correct in that a man could not sink in the sea. They also surveyed the lake, noting its unusual depth, and the shallow area, or "tongue" at the southern end of the lake. This area is thought to be where the Vale of Siddim was located and the five cities existed. It is possible to see entire forests of trees encrusted with salt beneath the water in this southern part of the lake.

Standard historical theory on the destruction of Sodom and Gomorrah, such as in *The Bible as History* by Werner Keller,[29] holds that the cities of the Vale of Siddim were destroyed when a plate movement caused the Great Rift Valley, of which the Dead Sea is a part, to shift, and the area at the southern end of the Dead Sea subsided. In the great earthquake, there were probably explosions, natural gases issuing forth, and brimstone falling like rain. This is likely to have happened about 2000 BC, the time of Abraham and Lot, thinks Keller, though geologists place the event many thousands of years before this.[29]

Says Keller, "The Jordan Valley is only part of a huge fracture in the earth's crust. The path of this crack has meantime been accurately traced. It begins far north, several hundred miles beyond the borders of Palestine, at the foot of the Taurus mountains in Asia Minor. In the south it runs from the south shore of the Dead Sea through the Wadi el-Arabah to the Gulf of Aqabah and only comes to an end beyond the Red Sea in Africa. At many points in this vast depression signs of earlier volcanic activity are obvious. In the Galilean mountains, in the highlands of Transjordan, on the banks of the Jabbok, a tributary of the Jordan, and on the Gulf of Aqabah are black basalt and lava.

"The subsidence released volcanic forces that had been lying dormant deep down along the whole length of the fracture. In the upper valleys of the Jordan near Bashan there are still the towering craters of extinct volcanoes; great stretches of lava and deep layers of basalt have been deposited on the limestone surface. From time immemorial the area around this depression has been subject to earthquakes. There is repeated evidence of them and the *Bible* itself records them. Did Sodom and Gomorrah sink when perhaps a part of the base of this huge fissure collapsed still further to the accompaniment of earthquakes and volcanic eruptions?"[29]

As for the pillars of salt, Keller says, "To the west of the southern shore and in the direction of the Biblical "Land of the South," the Negev, stretches a ridge of hills about 150 feet high and 10 miles from north to south. Their slopes sparkle and glitter in the sunshine like diamonds. It is an odd phenomenon of nature. For the most part this little range of hills consists of pure rock salt. The Arabs call it Jebel Usdum, an ancient name, which preserves in it the word 'Sodom.' Many blocks of salt have been worn away by the rain and have crashed downhill. They have odd shapes and some of them stand on end, looking like statues. It is easy to imagine them suddenly seeming to come to life.

"These strange statues in salt remind us vividly of the Biblical description of Lot's wife who was turned into a pillar of salt. And everything in the neighborhood of the Salt Sea is even to this day quickly covered with a crust of salt."[29]

However, Keller himself admits that there is a very serious problem with this theory of a cataclysm sending the Vale of Siddom to the bottom of the Dead Sea—it must have happened many hundreds of thousands, even millions, of years ago, at least according to most geologists. Says Keller, "In particular, we must remember there can be no question that the Jordan fissure was formed before about 4000 BC. Indeed, according to the most recent presentation of the facts, the origin of the fissure dates back to the Oligocene, the third oldest stage of the Tertiary Period. We thus have to think in terms not of thousands, but of millions of years. Violent volcanic activity connected with the Jordan fissure has been shown to have occurred since then, but even so we do not get any further than the Pleistocene which came to an end approximately ten thousand years ago. Certainly we do not come anywhere near to the third, still less the second millennium before Christ, the period that is to say, in which the patriarchs are traditionally placed."[29]

In short, Keller is saying that any geological catastrophe that would have destroyed Sodom and Gomorrah would have had to have happened a million years ago, or so geologists have told him. Keller says that geologists have not found any evidence of a recent catastrophe at the southern end of the Dead Sea, at least not for about 10,000 years.

Says Keller, "In addition, it is precisely to the south of the Lisan peninsula, where Sodom and Gomorrah are reported to have been annihilated, that the traces of former volcanic activity cease. In short, the proof in this

area of a quite recent catastrophe which wiped out towns and was accompanied by violent volcanic activity is not provided by the findings of the geologists."[29]

So here is the problem: the Dead Sea area may have had a cataclysm that could be the origin of the Old Testament story, however, conservative uniformitarian geologists have said that any such earth changes must have occurred long before any sort of collective memory of the event could have occurred.

In late 1999 a new theory was proposed by British *Bible* scholar Michael Sanders and an international team of researchers, who discovered what appear to be the salt-encrusted remains of ancient settlements on the seabed after several fraught weeks diving in a mini-submarine.

Sanders told a television crew from BBC Channel 4, who made a documentary about the expedition, "There is a good chance that these mounds are covering up brick structures and are one of the lost cities of the plains, possibly even Sodom or Gomorrah, though I would have to examine the evidence. These *Bible* stories were handed down by word of mouth from generation to generation before they were written down, and there seems to be a great deal in this one."

Mr. Sanders unearthed a map dating from 1650 which reinforced his belief that the sites of the two cities could be under the north basin, rather than on the southern edge of the Dead Sea. He recruited Richard Slater, an American geologist and expert in deep sea diving, to take him to the depths of the Dead Sea in the two-man Delta mini-submarine that was involved in the discovery of the sunken ocean liner, the *Lusitania*.

Sanders' location for Sodom and Gomorrah, in the deep northern part of the Dead Sea, is even more at odds with history and geology than Keller's theories of the cities being at the shallow southern end. Therefore, we come back to the popular theory that these cities were were not destroyed in a geological cataclysm, but in a man-made (or extraterrestrial-made) apocalypse that was technological in nature. Were Sodom and Gomorrah attacked with atomic weapons, as were Hiroshima and Nagasaki?

Researcher L.M. Lewis in his book *Footprints on the Sands of Time*[30] maintains that Sodom and Gomorrah were both destroyed by atomic weapons and says that the salt pillars and high salt content around the Dead Sea are evidence of a nuclear blast.

Says Lewis, "When Hiroshima was being rebuilt, stretches of sandy soil were found to have been atomically changed into a substance resembling a glazed silicon permeated by a saline crystalloid. Little blocks of this were cut from the mass and sold to tourists as souvenirs of the town—and of atomic action.

"Had an even larger explosion pulverized every stone of every building—and had the complete city disappeared into thin air—there would still have been tell-tale indications of what had occurred on the outskirts of the area of devastation. At some points there would surely be a marked difference in the soil or an atomic change in some object of note."[30]

Lewis maintains that if the pillars of salt at the end of the Dead Sea were ordinary salt, they would have disappeared with the periodic rains. Instead, these pillars are of a special, harder salt, only created with a nuclear reaction such as an atomic explosion.

These pillars of salt have indeed lasted a long time. Not only were they present in ancient times, but are still standing today. Lewis quotes from the well-known Roman historian Josephus who says in his *History of the Jews*, "... but Lot's wife continually turning back to view the city as she went from it, although God had forbidden her so to do, was changed into a pillar of salt: for I have seen it, and it remains to this day."

Comments Lewis, "It should be emphasized that Flavius Josephus lived from 37 to approximately 100 AD. As previously stated, Sodom was disintegrated in 1898 BC. How amazing, then that Josephus should actually have seen the human 'pillar of salt' after it had stood for almost 2,000 years! If it had been ordinary salt, it would have disappeared with the first rains."[29]

Though there may have been many pillars of salt throughout history, Lewis thinks that the evidence supports an atomic blast. "The atomic change of the soil upon which Lot's wife stood and that of the shore of Hiroshima have a similarity that cannot be denied! Both had undergone a sudden atomic conversion which could only have been caused by the instant action of nuclear fission. As those things which equal the same thing must be equal to one another, it is difficult to escape the conviction that as Hiroshima was destroyed, so, by similar means, Sodom was disintegrated and Lot's wife at the same moment atomically changed.

"Relying on the veracity of Josephus, the only conclusion that can be reached is that Sodom was destroyed by nuclear fission."[29]

The story of Sodom and Gomorrah is puzzling not just because of the destruction, but also because of the personalities involved, such as the angel warning Lot to leave the doomed cities.

Was Lot warned before the cities were going to be "nuked" by extraterrestrials or humans with high-tech weapons? They warned Lot to get his family out, but his wife looked back, and was blinded by the atomic flash. Perhaps her body was even atomically changed.

At the southern end of the Dead Sea today is a modern chemical plant that looks like an alien base. Strange towers shoot up out of the desert. Bizarre buildings with domes and spires are covered with multi-colored lights. One expects to see a flying saucer land at any moment. It is the Dead Sea Chemical Works. During the day, it looks more normal, like an oil refinery or something, but at night, the lights that are strung about the facility make it seem otherworldly.

This huge chemical plant is said to have an endless supply of valuable minerals with which to work, including radioactive salts. Are some of these chemicals the result of an ancient atomic blast?

Atomic Devastation, Indian Style

(Quotes from the *Mahabharata*)

"Various omens appeared among the gods—winds blew, meteors fell in thousands, thunder rolled through a cloudless sky."

"There he saw a wheel with a rim as sharp as a razor whirling around the soma... Then taking the soma, he broke the whirling machine..."

"Drona called Arjuna and said: ... 'Accept from me this irresistible weapon called Brahmasira. But you must promise never to use it against a human foe, for if you did it might destroy the world. If any foe who is not a human attacks you, you may use it against him in battle... None but you deserves the celestial weapon that I gave you." (This is a curious statement, as what other kind of foe, different from a human might there have been? Are we talking about an interplanetary war?)

"I shall fight you with a celestial weapon given to me by Drona. He then hurled the blazing weapon..."

"At last they came to blows, and seizing their maces struck each other... they fell like falling suns."

"These huge animals [elephants] like mountains, struck by Bhima's mace fell with their heads broken, fell upon the ground like cliffs loosened by thunder."

"Bhima took him by the arm and dragged him away to an open place where they began to fight like two elephants mad with rage. The dust they raised resembled the smoke of a forest fire; it covered their bodies so that they looked like swaying cliffs wreathed in mist."

"Arjuna and Krishna rode to and fro in their chariots on either side of the forest and drove back the creatures which tried to escape. Thousands of animals were burnt, pools and lakes began to boil... The flames even reached Heaven... Indra without loss of time set out for Khandava and covered the sky with masses of clouds; the rain poured down but it was dried in mid-air by the heat."

These verses are from the *Mahabharata* (written in ancient Dravidian, then later in Sanskrit) describing horrific wars fought long before the recorder's lifetime. Several historical records claim that Indian culture has been around for literally tens of thousands of years. Yet, until 1920, all the "experts" agreed that the origins of the Indian civilization should be placed within a few hundred years of Alexander the Great's expedition to the subcontinent in 327 BC. However that was before several great cities like Harappa and Mohenjo-Daro (Mound of the Dead), were discovered and excavated, including Kot Diji, Kalibanga, and Lothal. Lothal, a former port city now miles from the ocean, was discovered in Gujerat, western India, just in the late 20th century.[31]

The discoveries of these cities forced archeologists to push the dates for the origin of Indian civilization back thousands of years, just as Indians themselves insisted. A wonder to modern-day researchers, the cities are highly developed and advanced. The way that each city is laid out in regular blocks, with streets crossing each other at right angles, and the entire city laid out in sections, causes archaeologists to believe that the cities were conceived as a whole before they were built: a remarkable early example of city planning. Even more remarkable is that the plumbing-sewage systems throughout the large cities are so sophisticated, they are superior to those found in Pakistan, India, and many Asian countries today. Sewers were covered, and most homes had private toilets and running water. Furthermore, the water and sewage systems were kept well separated.[31, 15, 39]

This advanced culture had its own writing, which has never been deciphered. The people used personalized clay seals, much as the Chinese still do today, to officialize documents and letters. Some of the seals found contain figures of animals that are unknown to us today, including an extinct form of the brahma bull.

Archaeologists really have no idea who the builders were, but attempts to date the ruins (which they ascribe to the "Indus Valley Civilization," also called "Harappan") have come up with something like 2500 BC, and older. Radiation from the wars apparently fought in the area may have thrown off any dating techniques.

The Rama Empire, described in the *Mahabharata* and *Ramayana*, was supposedly contemporaneous with the great cultures of Atlantis and Osiris in the west. Atlantis, well-known from Plato's writings and ancient Egyptian records, apparently existed in the mid-Atlantic and was a very highly technological and patriarchal civilization. As we have noted, the Osirian civilization existed in the Mediterranean basin and North Africa, according to esoteric doctrine and archeological evidence. The Osirian civilization is generally known as pre-dynastic Egypt, and was flooded when Atlantis sank and the Mediterranean began to fill up with water.

A clay seal of an extinct type of bull.

The Rama Empire flourished during the same period, according to esoteric tradition, fading out in the millennium after the destruction of the Atlantean continent. As noted above, the ancient Indian epics describe a series of horrific wars, wars which could have been between ancient India and Atlantis, or perhaps a third party in the Gobi region of western China. The *Mahabharata* and the *Drona Parva* (another ancient Indian epic) speak of the war and of the weapons used: great fireballs that could destroy a whole city, "Kapilla's Glance" which could burn fifty thousand men to ashes in seconds, and flying spears that could ruin whole "cities full of forts."[31, 15, 39]

The Rama Empire was started by the Nagas (Naacals) who had come into India from Burma and ultimately from "the Motherland to the east," or so Col. James Churchward was told. After settling in the Deccan Plateau in northern India, they made their capital in the ancient city of Deccan, where today the modern city of Nagpur stands.

The empire of the Nagas apparently began to extend all over northern India to include the cities of Harappa, Mohenjo-Daro, and Kot Diji (now in Pakistan), and Lothal, Kalibanga, Mathura, and possibly other cities such as Benares, Ayodha, and Pataliputra.

These cities were led by "Great Teachers" or "Masters" who were the benevolent aristocracy of the Rama civilization. Today, they are generally called "Priest-Kings" of the Indus Valley Civilization, and a number of statues of these so-called gods have been discovered. In reality, these were apparently men whose mental-psychic powers were of a degree that seems incredible to most people of today. It was at the height of power for both the Rama Empire and Atlantis that the war allegedly broke out, seemingly because of Atlantis' attempt to subjugate Rama.

According to the Lemurian Fellowship lesson materials, the populace surrounding Mu (Lemuria, which predated the other civilizations) had eventually split into two opposing factions: those who prized practicality and those who prized spirituality. The citizenry, or educated elite, of Mu itself, were balanced equally in these two qualities. The citizenry encouraged the other groups to emigrate to uninhabited lands. Those that prized practicality emigrated to the Poseid Island Group (Atlantis) and those that prized spirituality eventually ended up in India. The Atlanteans, a patriarchal civilization with an extremely materialistic, technology-oriented culture deemed themselves the "Masters of the World" and eventually sent a well-equipped army to India in order to subjugate the empire and bring it under the suzerainty of Atlantis. One account of the battle related by the Lemurian Fellowship in its lessons tells how the Rama Empire Priest-Kings defeated the Atlanteans.

Equipped with a formidable force and a "fantastic array of weapons," the Atlanteans landed in their vailixi outside of one of the Rama cities, got their troops in order and sent a message to the ruling Priest-King of the city that he should surrender. The Priest-King sent word back to the Atlantean general: "We of India have no quarrel with you of Atlantis. We ask only that we be permitted to follow our own way of life."

Regarding the ruler's mild request as a confession of weakness, and expecting an easy victory, as the Rama Empire did not possess the technology of war nor the aggressiveness of the Atlanteans, the Atlantean general sent another message: "We shall not destroy your land with the mighty weapons at our command provided you pay sufficient tribute and accept the rulership of Atlantis."

The Priest-King of the city responded humbly again, seeking to avert war: "We of India do not believe in war and strife, peace being our ideal. Neither would we destroy you or your soldiers who but follow orders. However, if you persist in your determination to attack us without cause and merely for the purpose of conquest, you will leave us no recourse but to destroy you and all of your leaders. Depart, and leave us in peace."

A Rama Empire "Priest King"?

Arrogantly, the Atlanteans did not believe that the Indians had the power to stop them, certainly not by technical means. At dawn, the Atlantean army began to march on the city. From a high viewpoint,the Priest-King sadly watched the army advance. Then he raised his arms heavenward and (using a mental technique perhaps known today by certain very knowledgeable persons) he caused the General, and then each officer in order of rank, to drop dead in his tracks, probably of some sort of heart failure. In a panic, and without leaders, the remaining Atlantean force fled to the waiting vailixi and retreated in terror to Atlantis! Of the sieged Rama city, not one man was lost.

While this may be nothing but fanciful conjecture, the Indian epics go on to tell the rest of the horrible story, and things do not turn out well for Rama. Atlantis, assuming the above story is true, was not pleased at the humiliating defeat, and therefore used its most powerful and destructive weapon, probably an atomic weapon! These are verses from the ancient *Mahabharata:*

...(it was) a single projectile
Charged with all the power of the Universe.
An incandescent column of smoke and flame
As bright as the thousand suns
Rose in all its splendor...
...it was an unknown weapon,
An iron thunderbolt,
A gigantic messenger of death,
Which reduced to ashes
The entire race of the
Vrishnis and the Andhakas.
...The corpses were so burned
As to be unrecognizable.
The hair and nails fell out;
Pottery broke without apparent cause,
And the birds turned white...
...After a few hours
All foodstuffs were infected...
...to escape from this fire
The soldiers threw themselves in streams
To wash themselves and their equipment."[44]

In the way we traditionally view ancient history, it seems absolutely incredible that there was an atomic war approximately ten thousand years ago. And yet, of what else could the *Mahabharata* be speaking? Perhaps this is just a poetic way to describe cavemen clubbing each other to death; after all, that is what we are told the ancient past was like. Until the bombing of Hiroshima and Nagasaki, modern mankind could not imagine any weapon as horrible and devastating as those described in the ancient Indian texts. Yet they very accurately described the effects of an atomic explosion. Radioactive poisoning will make hair and nails fall out. Immersing one's self in water is the only respite, though not a cure.

Interestingly, Dr. J. Robert Oppenheimer, the "Father of the H-Bomb," is known to be familiar with ancient Sanskrit literature. In an interview conducted after he watched the first atomic test, he quotes from the *Bhagavad Gita;* "'Now I've become death—the Destroyer of Worlds.' I suppose we all felt that way." When asked in an interview at Rochester University seven years after the Alamogordo nuclear test whether that was the first atomic bomb ever to be detonated, his reply was, "Well, yes, in modern history." 44

The Doom of Mohenjo-Daro

Incredible as it may seem, archaeologists have found evidence in India indicating that some cities were destroyed in atomic explosions. When excavations of Mohenjo-Daro and Harappa reached the street level, they discovered scattered skeletons about the cities, many holding hands and sprawling in the streets, as if some instant, horrible doom had taken place. I mean, people are just lying, unburied, in the streets of the city. And these skeletons are thousands of years old, even by traditional archaeological standards! What could cause such a thing? Why did the bodies not decay or get eaten by wild animals? Furthermore, there is no apparent cause of a violent death (heads hacked off, bashed in, etc.).

These skeletons are among the most radioactive ever found, on par with those at Nagasaki and Hiroshima. Soviet scholars have found a skeleton at one site that had a radioactive level fifty times greater than normal.[44] The Russian archaeologist A. Gorbovsky mentions the high incidence of radiation associated with the skeletons in his 1966 book, *Riddles of Ancient History*.[45] Furthermore, thousands of fused lumps, christened "black stones," have been found at Mohenjo-Daro. These appear to be fragments of clay vessels that melted together in extreme heat.

Other cities have been found in northern India that show indications of explosions of great magnitude: one such city, found between the Ganges and the mountains of Rajmahal, seems to have been subjected to intense heat. Huge masses of walls and foundations of the ancient city are fused together, literally vitrified! Since there is no indication of a volcanic eruption at Mohenjo-Daro, or at the other cities, the intense heat to melt clay vessels can only be explained by an atomic blast or some other unknown weapon.[15,24,45]

The cities were wiped out entirely. If we accept the Lemurian Fellowship stories as fact, then Atlantis wanted to waste no more time with the Priest-Kings of Rama and their mental tricks. In terrifying revenge, they utterly destroyed the Rama Empire, leaving no country to even pay tribute to them. The area around the cities of Harappa and Mohenjo-Daro are desolate deserts, though agriculture takes place to a limited extent in the vicinity today.

It is said in esoteric literature that Atlantis at the same time, or shortly afterwards, also attempted to subjugate a civilization extant in the area of the Gobi Desert, which was then a fertile plain. By using so-called "Scalar Wave Weaponry" and firing through the center of the earth, they wiped out their adversaries, and, possibly at the same time, did themselves in! Much speculation naturally exists in connection with remote history. We may never actually know the complete truth, though ancient texts are certainly a good start.

Atlantis met its own doom, according to Plato, by sinking into the ocean in a mighty cataclysm; not too long after the war with the Rama Empire, I imagine.

Kashmir is also connected with the fantastic war in ancient times that destroyed the Rama Empire. The massive ruins of a temple called Parshaspur can be found just outside Srinagar. It is a scene of total destruction; huge blocks of stone are scattered about a wide area giving the impression of explosive annihilation.[42] Was Parshaspur destroyed by some fantastic weapon during one of the horrendous battles detailed in the *Mahabharata*?

Another curious sign of an ancient nuclear war in India is a giant crater near Bombay. India's nearly circular 2,154-meter diameter Lonar crater, located 400 kilometers northeast of Bombay and aged at less than 50,000 years old, could be related to nuclear warfare of antiquity. No trace of any meteoric, etc., material has been found at the site or in the vicinity, while it is the world's only known 'impact' crater in basalt. Indications of great shock (from a pressure exceeding six hundred thousand atmospheres) and intense abrupt heat (indicated by basalt glass spherules) can be ascertained at the site.

Orthodoxy cannot, of course, concede nuclear possibilities for such craters even in the absence of any material meteorite, or related evidence. If such geologically recent craters as the Lonar were of meteoric origin, why then do such tremendous meteorites not fall today? The Earth's atmosphere fifty thousand years ago probably was no different from today's, so a lighter atmosphere cannot be advanced as a hypothesis to attempt to explain an immense meteorite size, which of course would be considerably reduced by heat oxidization within a gaseously heavier atmosphere. A theory was advanced by American space consultant Pat Frank to the effect that some of the huge craters on the Earth may in fact be scars from ancient nuclear explosions![40]

The echoes of ancient atomic warfare in south Asia continue to this day with India and Pakistan currently threatening each other. Modern India is proud of their nukes, likening them to "Rama's Arrow." Similarly, Pakistan would love to use its Islamic bomb on India. Ironically, Kashmir, possibly the site of an earlier atomic war, is the focus of this conflict. Will the past repeat itself in Pakistan and India?

In the crazy world of the new millennium and its underground tunnels, secret bases, UFOs, and nuclear weapons, there is always the possibility that this has all happened before. Maybe it has. Dejá Vu!

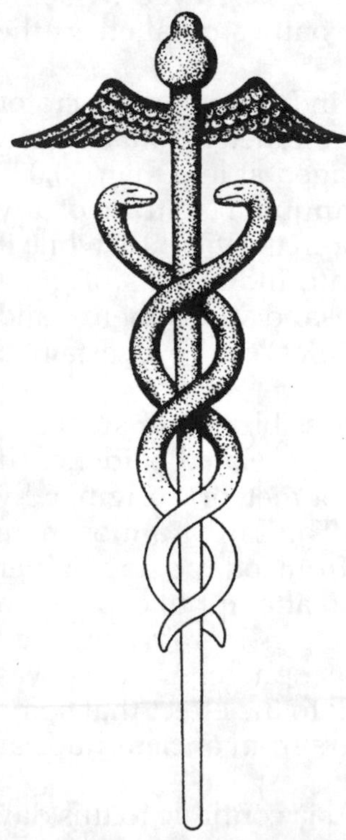

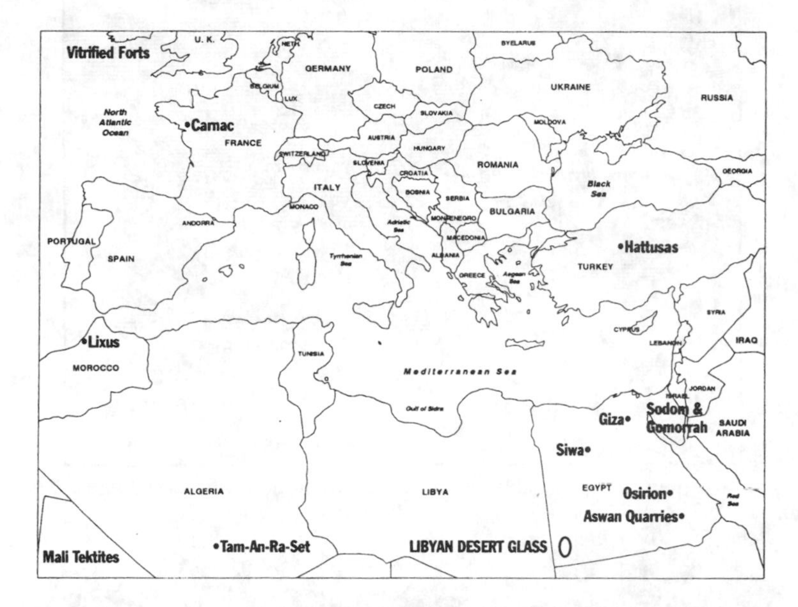

Top: A map of the Libyan Desert Glass and other sites. Bottom: Does history repeat itself?

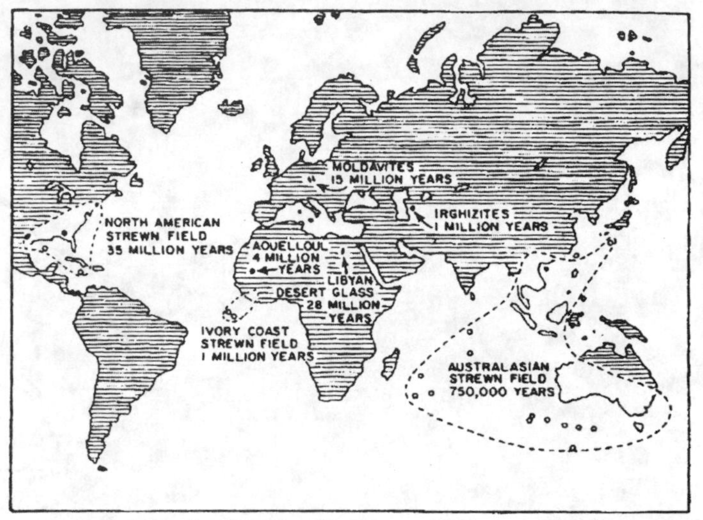

A map of the distribution of tektites with their theoretical ages.

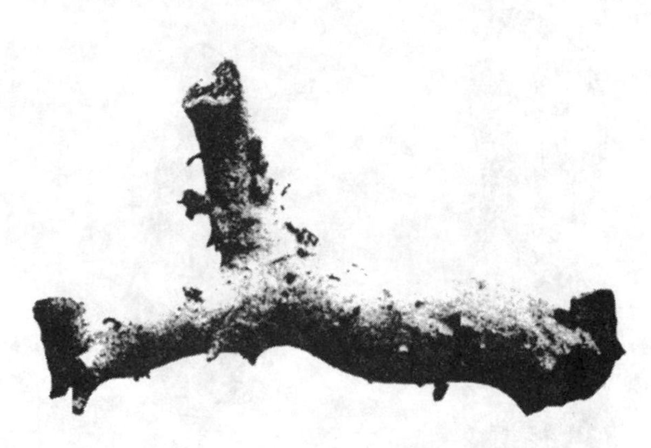

A fulgurite: fused sand created by lightning.

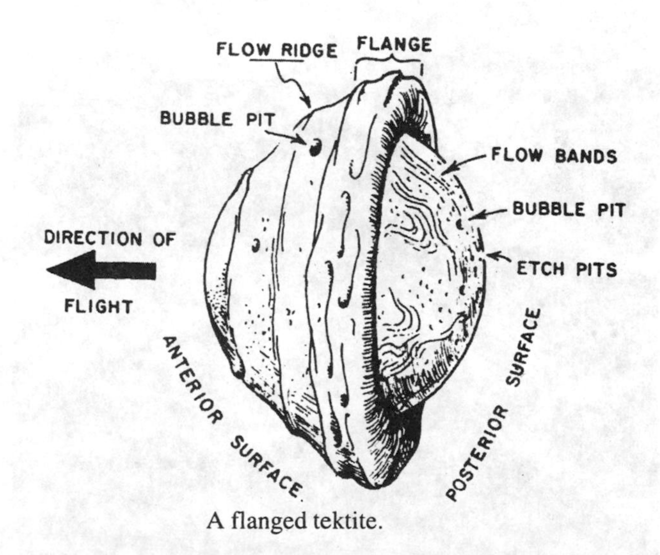

A flanged tektite.

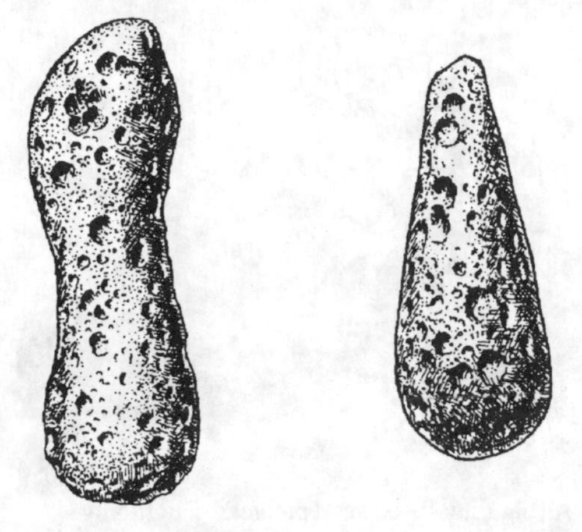

Indonesian tektites.

One of the vitrified forts of Scotland, Tap O'Noth.

Remains of the ziggurat at Birs Nimrod (Borsippa), south of Hillah, Iraq, once confused with the 'Tower of Babel.' The ruins are crowned by a mass of vitrified brickwork, fused together by intense heat. This may be due to an ancient atomic war, although early archaeologists attributed the effect to lightning.

Fire From Heaven: What was Greek Fire?

The Four Horsemen of the Apocalypse.

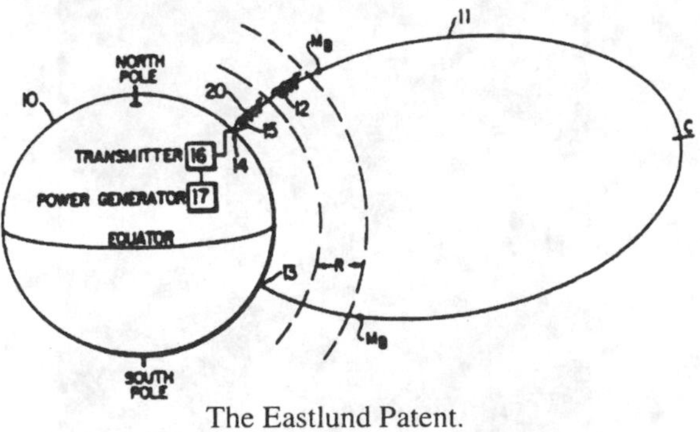

The Eastlund Patent.

The Alamogordo atomic test at .016 seconds.

The Alamogordo atomic test at 15 seconds.

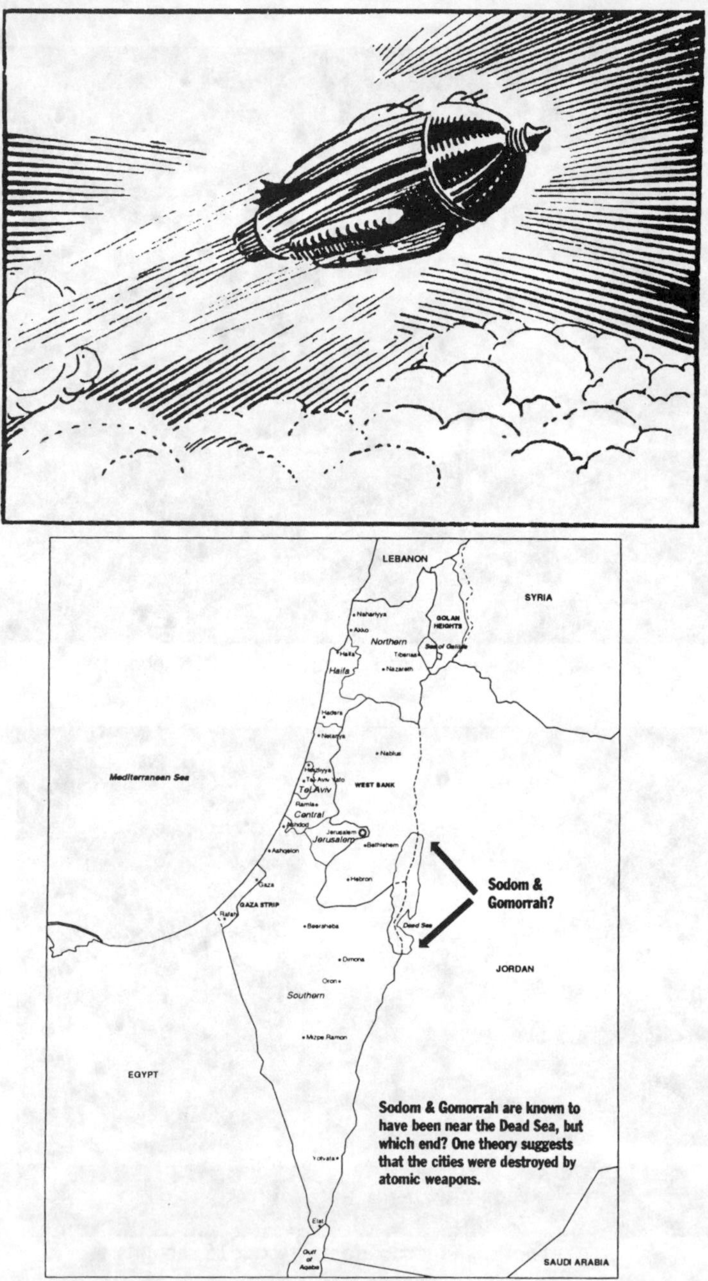

Top: The ancient epics speak of airships and devastating wars.
Bottom: A map showing the location of Sodom and Gommorah.

Lynch explores the Dead Sea in 1848.

Lot and his family leave Sodom.

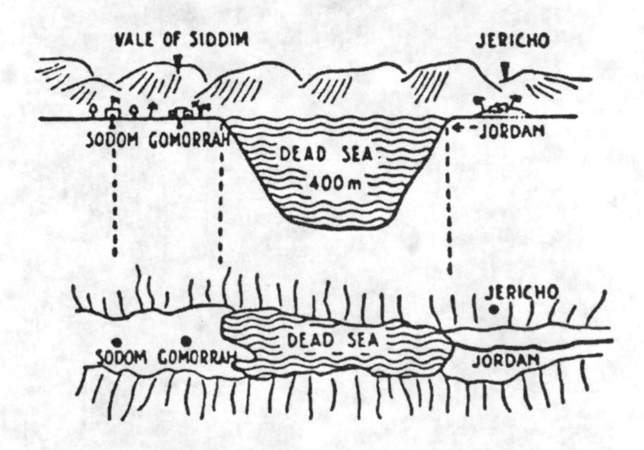

The Dead Sea area before and after subsidence.

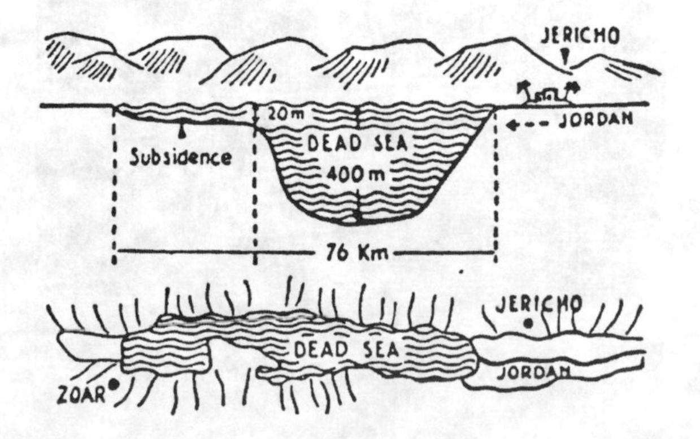

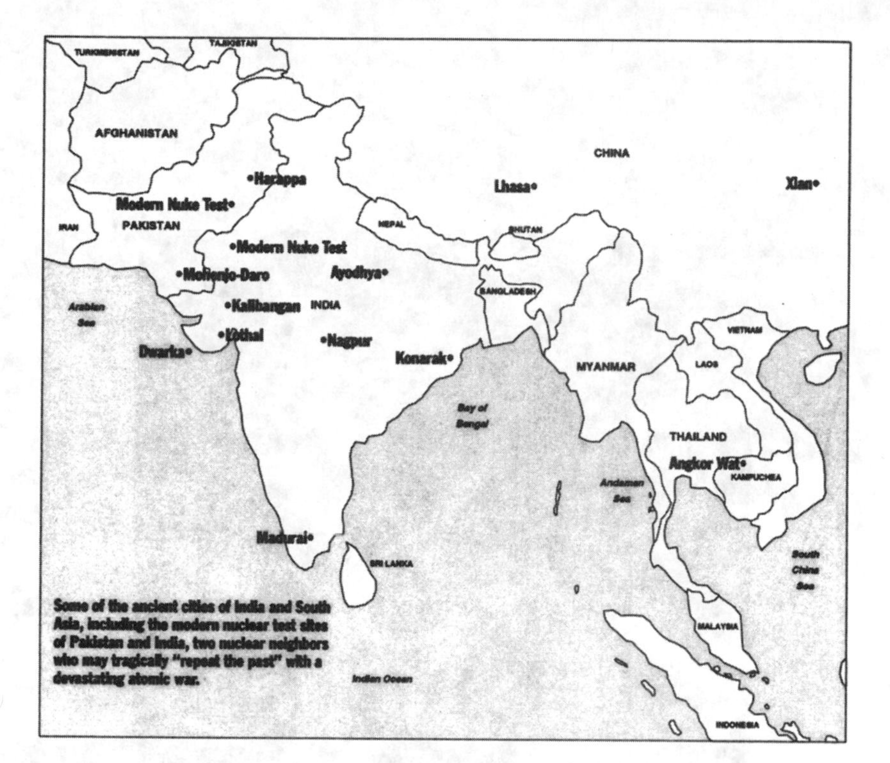

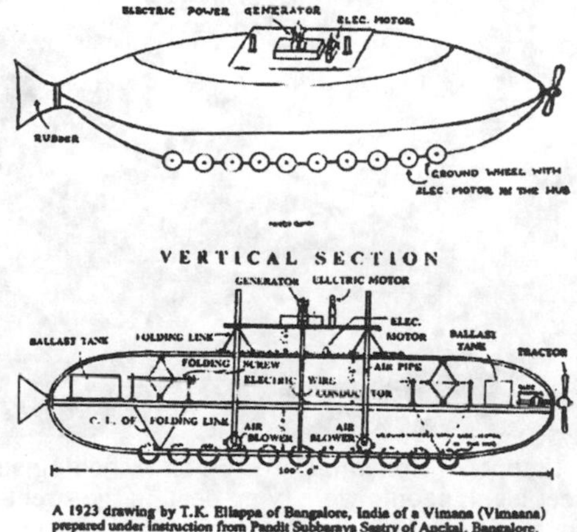

A diagram of one kind of vimana.

A street scene at Mohenjo Daro: When archeologists got to the street level, people were lying dead in the street—after thousands of years.

The ancient port city of Lothal, now miles from the ocean.

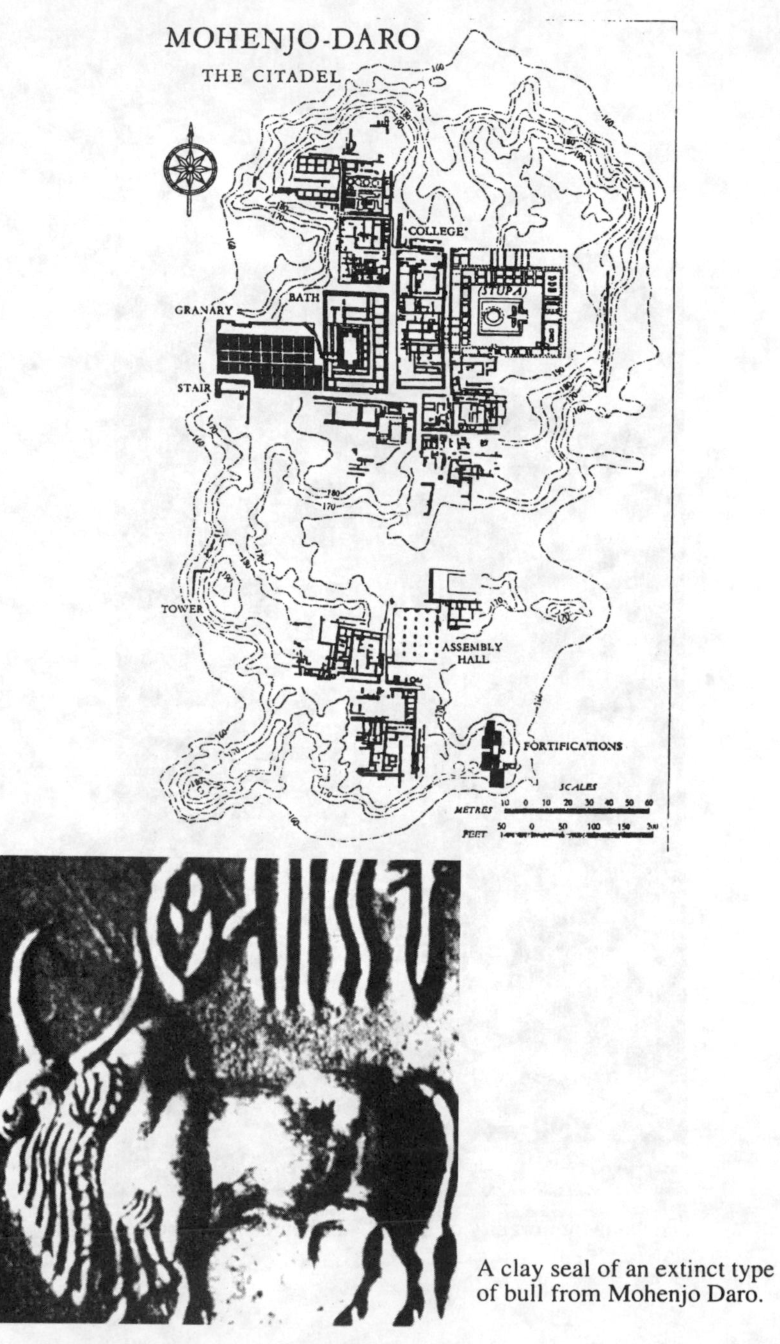

A clay seal of an extinct type of bull from Mohenjo Daro.

Top: The great bath at Mohenjo Daro. Bottom: One of the "Priest Kings" of the Rama Empire?

The first Duke-Nukem: the Hindu Sudershan Chakra's cosmic wheel.
Does it signify the vimanas and destructive ancient wars?

The ruins of Parshaspur in Kashmir.

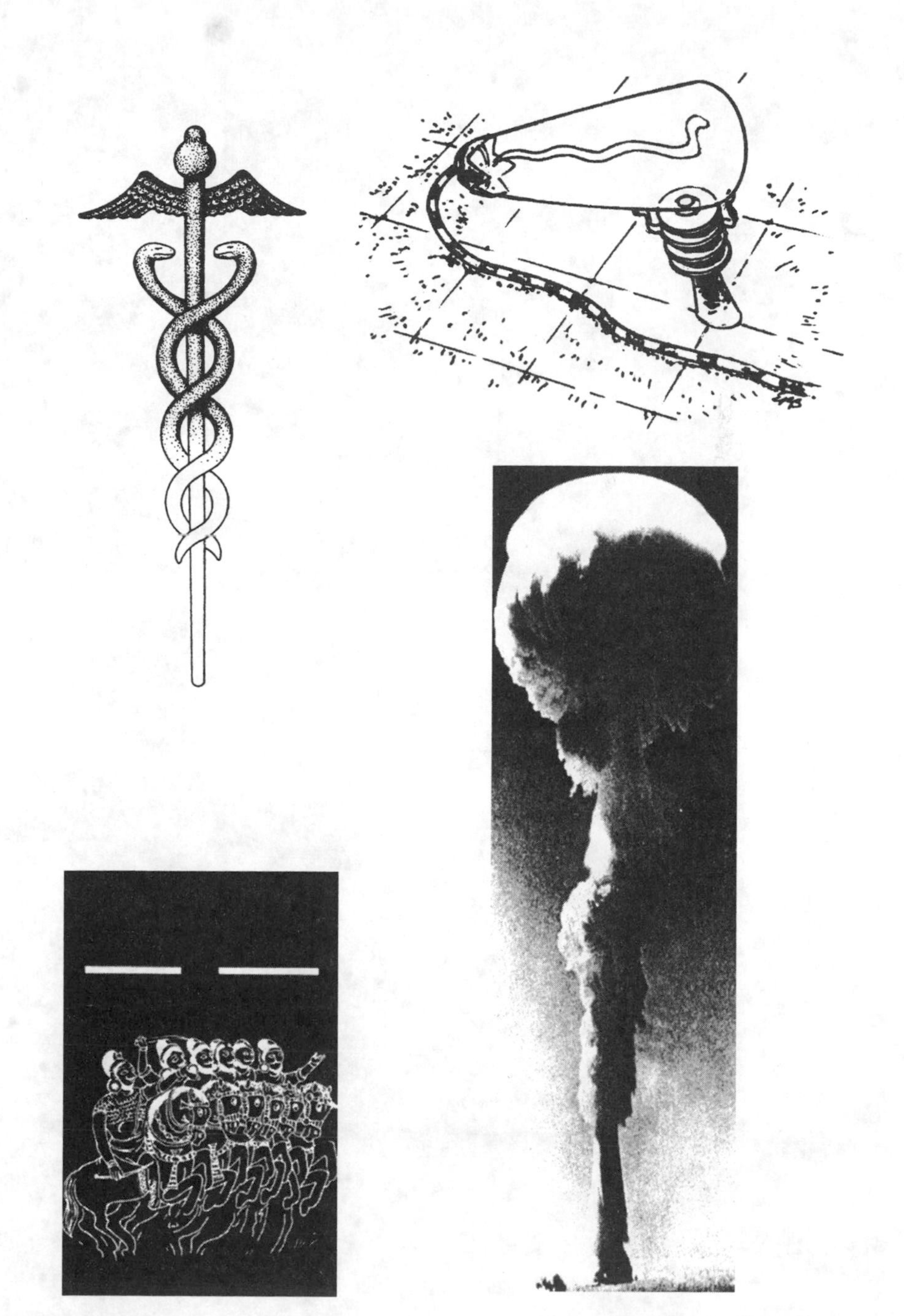

7.
The Earth As A Giant Power Plant

The priests told me that the Great Pyramid embodied all the wonders of Physics.
—Herodotus (350 BC)

You gotta know the rules before you can break em'. Otherwise, it's just no fun.
—Sonny Crockett, *Miami Vice*

The Giza Pyramid Complex

The structures of the Giza Plateau in Egypt are perhaps the most famous examples of technological wonders built by the ancients using technologies we can neither understand nor duplicate today. Who built these monuments, how and why, has been the focus of immense speculation over the years.

The Sphinx is one of the three most controversial structures in Egypt, along with the Great Pyramid and the Osirion at Abydos. Carved out of solid rock, the Sphinx seems to typify the mystery of Egypt as it gazes silently out away from the pyramids. The age of the Sphinx is a matter of great debate. The body is severely eroded, though the Egyptian government is now reconstructing it.

What could have caused this heavy decay? The controversial German Egyptologist Schwaller de Lubicz observed that the the dramatically severe erosion on the body could not be the result of wind and sand, as universally assumed, but was rather the result of water. Geologists agree that in the not so distant past Egypt was subjected to severe flooding. This period is usually held to coincide with the melting of the ice from the last Ice Age, circa 15000 to 10000 BC![112]

This would indicate that the Sphinx was already constructed by this time, and would make it easily the oldest structure in Egypt, built long before the recognized Egyptian civilization began. Suddenly, we are back to the tales of the ancient Osirian Empire, Atlantis, and the cataclysmic pole shifts that have rocked our planet every ten thousand years or so.

The Sphinx is often said to be the likeness of the Pharaoh Chephren, of whom several statues, one in the form of a sphinx, were found upside down in the temple next to the sphinx. The sphinx, in fact, is said to have been recarved at least once, and its head is unnaturally small for its body, indicating that it once had a much larger head. Perhaps the Pharaoh Chephren had the Sphinx recarved in his likeness during his lifetime.

The Valley Temple of Chephren next to the Sphinx is also an unusual structure. It is constructed of huge granite and limestone blocks weighing up to 100 tons apiece. No inscriptions of any kind are to be found in the temple, and the blocks are perfectly fitted together in a curious jigsaw pattern that interlocks the blocks. As has been noted previously, this is a trademark of "The Builders," a type of megalithic construction that is not only extremely difficult to make, but also difficult to tear down. Because the blocks are interlocked, they cannot shear like bricks or square blocks of masonry. It is especially interesting to compare the construction technique found at the Valley Temple of Chephren to that found at Cuzco, Sacsayhuaman, Ollantaytambo and even Machu Picchu.

There are also said to be secret passages beneath the Giza plateau. These passages go to the pyramids, allegedly starting from the Sphinx, and are part of the ancient Mystery Schools of Egypt. A strange shaft in the Giza plateau between the Pyramid of Chephren and the Sphinx is known as Campbell's Tomb or Campbell's Well. This shaft is now blocked by a grate, but one can still look down it. The shaft is about 15 feet square on each side and about a 100 feet deep. On each side of the walls, one can see numerous tunnels, passages and doors cut into the solid rock. These passageways are part of the tunnel system that goes beneath the Giza plateau. It is purposely dangerous to attempt to reach the pyramids or the secret underground rooms that can be found in the tunnels. Their existence, and what lies in them, is a matter of legend and prophecy.

It has been suggested that a secret library from Atlantis is hidden somewhere inside, underneath or nearby the pyramids of Giza; this library is generally called the Ancient Hall of Records. According to some, the Hall of Records preserved ancient knowledge in the form of encoded quartz crystals, much like a hologram might be encoded by a laser today. Also in these secret chambers, sealed from the rest of mankind during the dark age of Egyptian history when the evil priesthoods sought to control mankind, are supposedly machines and devices from that forgotten age. There is the belief by some that the Ark of the Covenant was contained inside the Great Pyramid for some time, and then taken by Moses when the Israelites split for the Promised Land.

The pyramids of the Giza Plateau have been recognized as feats of engineering wonder since time immemorial. Herodotus (the Greek historian of the 1st century BC) claims that he was told by priests two thousand years or more after the Great Pyramid's construction that teams of ten thousand men each labored for ten years to build a ramp for the blocks; then another twenty years to build the pyramid; and finally a further ten years to put the casing stones on the pyramid starting from the top down. Herodotus claimed that Cheops financed the construction by having his own daughter work as a prostitute. An inscription read to Herodotus at the base of the pyramid by priests told of the number of onions and radishes it took to feed the laborers.

Yet, it appears that Herodotus was being told a story. No trace of a ramp has actually been discovered. Most scholars believe that the ramp Herodotus refers to is the causeway leading up from the Nile, past the Sphinx. All the pyramids had this causeway leading up to them, but it apparently had nothing to do with their construction. There are no actual wall drawings of a pyramid being constructed, but there are drawings that portray the transport of gigantic obelisks and giant statues weighing more than a hundred tons by men pulling sleds.[112]

According to John Anthony West, even though it was possible to muster up enough labor to build the pyramids over time, some sort of lifting device would have to have been employed, which so far no one has come up with. Other engineers claim that no lifting device was needed and that the ramp merely had to reach the top of the pyramid. However, a Danish engineer named P. Garde-Hanson has calculated that such a ramp would require seventeen and a half million cubic yards of material, seven times more than that used in the building of the pyramid itself! Garde-Hanson believes that a ramp going halfway up the pyramid would be better, but it would still be necessary to use a lifting device, which brings us back to the old problem.[112]

The placing of the limestone casing stones, which weigh up to 10 tons or more, is an even bigger problem, as they are cut and fitted with such precision. Cheops did not even sign his own pyramid—the only marks in the pyramid are quarry marks on granite blocks on the inside of the pyramid. These were only discovered with the tearing apart of the pyramid, and were never meant to be seen.

One possibility is the ingenious theory that the pyramid itself was a hydraulic pump, and the blocks were floated into place on rafts from nearby

Lake Moeris. Another theory that has a certain attraction to mystics is that the blocks were levitated, using what the Egyptians called Ma-at, a force of mind power akin to the Sanskrit Mana power of the mind.

Was the Great Pyramid Poured into Place?

One curious theory of the pyramids is postulated by an authority on ancient construction techniques named Dr. Joseph Davidovits. Davidovits has been saying over the last several years that the Great Pyramid of Egypt, as well as other pyramids in Egypt, was not constructed out of cut stone as has always been assumed. Davidovits believes that the large blocks were actually poured into place, and that they are an advanced and ingenious form of synthetic stone that was cast on the spot like concrete.

Davidovits reported on his research at a meeting of the American Chemical Society in the mid-1980s. He is the founder and director of the Institute for Applied Archaeological Sciences located near Miami. He is also the author of the 1988 book *The Pyramids: An Enigma Solved.*[114] Davidovits claims that a new deciphering of an ancient hieroglyphic text has provided some direct information about pyramid construction and that it supports his theory that synthetic stone was the construction material.

The text, called the "Famine Stele," was discovered 100 years ago on an island near Elephantine, Egypt. It consists of 2,600 hieroglyphs, about 650 of which have been interpreted as dealing with stone-fabrication techniques. The text claims that an Egyptian god gave the instructions for making synthetic stone to Pharaoh Zoser, who is said to have built the first pyramid in 2750 BC.

Included were a list of 29 minerals that could be processed with crushed limestone and other natural aggregates into a synthetic stone for use in the building of temples and pyramids. Like the chemists of the 17th and 18th centuries, the Egyptians named these minerals according to their physical properties. The materials were called "onion ore," "garlic ore" and "horseradish ore" because of their distinctive smells.

Davidovits believes the minerals in the ores contained arsenic. Other ingredients for making synthetic stone—phosphates from bones or dung, Nile silt, limestone and quartz—were also readily available.

According to the theory, the ingredients were mixed with water and placed into wooden forms similar to those used for concrete. Davidovits said the cement used in the pyramid stone binds the aggregate and other ingredients together chemically in a process similar to that involved in the formation of natural stone. Thus, pyramid stone is extremely difficult to distinguish from natural stone. Portland cement, by contrast, involves mechanical rather than molecular bonding of its ingredients. Also, "Egyptian cement" would last for thousands of years, while ordinary cement has an average life span of only 150 years. Organic fibers, having accidentally fallen into the mixture, have been found in the stone blocks of the Great Pyramid, claims Davidovits.[114]

What Was the Function of the Pyramids?

Actually, though it is usually said that the pyramids were tombs for the pharaohs, the evidence is generally against this theory. Startling as it may seem, no Egyptian mummy was ever found in a pyramid. Many Egyptian mummies have been found, but not inside pyramids. Rather, they have been found in underground vaults and tunnels, such as those in the Valley of the Kings where Tutankhamen was found.

As the normally traditional archaeologist Kurt Mendelssohn says in his book *The Riddle of the Pyramids:*[111] "While the funerary function of the pyramids cannot be doubted, it is rather more difficult to prove that the pharaohs were ever buried inside them...

"Leaving out Zoser's Step Pyramid, with its unique burial chambers, the nine remaining pyramids contain no more than three authentic sarcophagi. These are distributed over no fewer than fourteen tomb chambers. Petrie has shown that the lidless sarcophagus in the Khufu [Cheops] Pyramid had been put into the King's Chamber before the latter was roofed over since it is too large to pass through the entrance passage... One would like to know what has happened to the missing sarcophagi. The robbers might have smashed their lids but they would hardly have taken the trouble to steal a smashed sarcophagus. In spite of careful search no chips of broken sarcophagi have been found in any of the pyramid passages or chambers. Moreover, it has to be remembered that from the Meidum pyramid onward the entrance was well above ground level. At the Bent Pyramid even the lower corridor is located 12 meters above the base and bringing a heavy sarcophagus in or out would have necessitated the use of a substantial ramp...

"The fact that sarcophagi in the Khufu and Khafre [Chephren] pyramids were found empty is easily explained as the work of intruders, but the empty sarcophagi of Sekhemket, Queen Hetepheres and a third one in a shaft under the Step Pyramid, pose a different problem. They were all left undisturbed since early antiquity. As these were burials without a corpse, we are almost driven to the conclusion that something other than a human body may have been ritually entombed.

"We have already referred to the fact that Snefru seems to have had two, or even three, large pyramids, and he can hardly have been buried in all of them...

"While very few people will dispute that the pyramids had some connection with the afterlife of the pharaoh, the general statement that the pharaohs were buried in them is by no means undisputable...Quite possibly each pyramid once housed the body of pharaoh, but there also exists

…an unpleasantly large number of factors that speak against it. It is on the basis of these complexities and contradictions that Egyptologists had to try and find a solution to the most difficult problem of all: why were these immense pyramids built in the first place?"[111]

If the pyramids weren't tombs, what were they?

There is the theory that they were astronomical observatories. Another idea is that the pyramids, especially the Great Pyramid, were geodetic markers and "time capsules" in the sense that higher knowledge, such as sophisticated geometry and mathematics were incorporated into the structures. Others have claimed that the pyramids were centers of initiation. Then, of course, there is always "pyramid energy." The word "pyramid" is in fact Greek for "fire in the center."

The Giza Power Plant

The belief that the pyramids were a device for tapping the energy of the Van Allen Belt, where the bulk of the pyramid served as a protective baffle, like insulation around an electric cable, is perhaps the most incredible suggestion of all. This theory is now championed by British engineer Christopher Dunn. Dunn is the author of the 1998 book *The Giza Power Plant: Technologies of Ancient Egypt*.[98] In this book, Dunn outlines his theories, and gives evidence for advanced machining and engineering knowledge in ancient Egypt.

Dunn claims that the Earth may be a giant power plant, and the many pyramids, obelisks, and megalithic standing stones may be part of this great "energy system." He says that the Great Pyramid was a giant power plant and that harmonic resonators were housed in slots above the King's Chamber. He also theorized that there was a hydrogen explosion inside the King's Chamber that shut down the power plant's operation.

In August 1984, *Analog* magazine published Dunn's article "Advanced Machining in Ancient Egypt?" It was a study of the book *Pyramids and Temples of Gizeh*, written by Sir William Flinders Petrie. Dunn is convinced that sophisticated machining was used in certain cases. Says Dunn, "Since the article's publication, I have visited Egypt twice, and with each visit I leave with more respect for the ancient pyramid builders. While in Egypt in 1986, I visited the Cairo museum and gave a copy of my article, along with a business card, to the director of the museum. He thanked me kindly, threw it in a drawer to join other sundry material, and turned away. Another Egyptologist led me to the "tool room" to educate me in the methods of the ancient masons by showing me a few cases that

housed primitive copper tools. I asked my host about the cutting of granite, for this was the focus of my article. He explained that the ancient Egyptians cut a slot in the granite, inserted wooden wedges, and then soaked them with water. The wood swelled creating pressure that split the rock. Splitting rock is vastly different than machining it and he did not explain how copper implements were able to cut granite, but he was so enthusiastic with his dissertation that I did not interrupt. To prove his argument, he walked me over to a nearby travel agent encouraging me to buy airplane tickets to Aswan, where, he said, the evidence is clear. I must, he insisted, see the quarry marks there, as well as the unfinished obelisk.

"Dutifully, I bought the tickets and arrived at Aswan the next day. After learning some of the Egyptian customs, I got the impression that this was not the first time that my Egyptologist friend had made that Quarry Marks at Aswan trip to the travel agent. The quarry marks I saw there did not satisfy me that the methods described were the only means by which the pyramid builders quarried their Drill Hole at Aswan rock. There is a large round hole drilled into the bedrock hillside, that measures approximately 12 inches in diameter and 3 feet deep that is located in the channel, which runs the length of the estimated 3,000 ton obelisk. The hole was drilled at an angle with the top intruding into the channel space. The ancients may have used drills to remove material from the perimeter of the obelisk, knocked out the webs between the holes, and then removed the cusps."[98]

Dunn says that archaeology is largely the study of history's toolmakers, and archaeologists understand a society's level of advancement with tools and artifacts. The hammer was probably the first tool ever invented, and hammers have forged some elegant and beautiful artifacts. Ever since man first learned that he could effect profound changes in his environment by applying force with a reasonable degree of accuracy, the development of tools has been a continuous and fascinating aspect of human endeavor. Dunn says that the Great Pyramid leads a long list of artifacts that have been misunderstood and misinterpreted by archaeologists, who have promoted theories and methods that are based on a collection of tools that they struggle with to replicate the most simple aspects of the work.

Says Dunn, "For the most part, primitive tools that are discovered are considered contemporaneous with the artifacts of the same period. Yet during this period in Egyptian history, artifacts were produced in prolific number with no tools surviving to explain their creation. The ancient Egyptians created artifacts that cannot be explained in simple terms. These tools do not fully represent the 'state of the art' that is evident in the artifacts. There are some intriguing objects that survived after this civilization, and in spite of its most visible and impressive monuments, we have only a sketchy understanding of the full scope of its technology. The tools displayed by Egyptologists as instruments for the creation of many of these incredible artifacts are physically incapable of reproducing them. After standing in awe before these engineering marvels, and then being shown a paltry collection of copper implements in the tool case at the Cairo Museum, one comes away bemused and frustrated."

Dunn claims that the British Egyptologist Sir William Flinders Petrie also recognized that these tools were insufficient. He explored this anomaly thoroughly in *Pyramids and Temples of Gizeh,* and expressed amazement about the methods the ancient Egyptians used to cut hard igneous rocks. He credited them with methods that "...we are only now coming to understand."[98]

Says Dunn, "I'm not an Egyptologist, I'm a technologist. I do not have much interest in who died when, whom they may have taken with them and where they went to. No lack of respect is intended for the mountain of work and the millions of hours of study conducted on this subject by intelligent scholars (professional and amateur), but my interest, thus my focus, is elsewhere. When I look at an artifact to investigate how it was manufactured, I am not concerned about its history or chronology. Having spent most of my career working with the machinery that actually creates modern artifacts, such as jet-engine components, I am able to analyze and determine how an artifact was created. I have also had training and experience in some non-conventional manufacturing methods, such as laser processing and electrical discharge machining. Having said that, I should state that contrary to some popular speculations, I have not seen evidence of laser for cutting on the Egyptian rocks. Still, there is evidence for other non-conventional machining methods, as well as more sophisticated, conventional type sawing, lathe and milling practices. Undoubtedly, some of the artifacts that Petrie was studying were produced using lathes. There is also evidence of clearly defined lathe tool marks on some "sarcophagi" lids. The Cairo Museum contains enough evidence that will prove that the ancient Egyptians used highly sophisticated manufacturing methods once it's properly analyzed.

"[T]here are several artifacts that almost undeniably indicate machinery power being used by the pyramid builders. These artifacts, scrutinized by William Flinders Petrie, are all fragments of extremely hard igneous rock. These pieces of granite and diorite exhibit marks that are the same as those resulting from cutting hard igneous rock with modern machinery. It is shocking that Petrie's studies of these fragments have not attracted greater attention, for there is unmistakable evidence of machine tooling methods. It will probably surprise many people to know that evidence proving that the ancient Egyptians used tools such as straight saws, circular saws, and even lathes has been recognized for over a century. The lathe is the father of all machine tools in existence, and Petrie submits evidence showing that not only were lathes used by the ancient Egyptians, but they performed tasks which would, by today's standards, be considered impossible without highly developed specialized techniques, such as cutting concave and convex spherical radii without splintering the material.

"While digging through the ruins of ancient civilizations, would archeologists instantly recognize the work of machine tools by the kind of marks made on the material or the configuration of the piece at which they were looking? Fortunately, one archeologist had the perception and knowledge to recognize such marks, and, although at the time Petrie's findings were published the machining industry was in its infancy, the growth in the industry since then warrants a new look at his findings."

Continues Dunn, "Having worked with copper on numerous occasions, and having hardened it in the manner suggested above, this statement struck me as being entirely ridiculous. You can certainly work-harden copper by striking it repeatedly or even by bending it. However, after a specific hardness has been reached, the copper will begin to split and break apart. This is why, when working with copper to any great extent, it has to be periodically annealed, or softened, if you want to keep it in one piece. Even after hardening in such a way, the copper will not be able to cut granite. The hardest copper alloy in existence today is beryllium copper. There is no evidence to suggest that the ancient Egyptians possessed this alloy, but even if they did, this alloy is not hard enough to cut granite. Copper has predominantly been described as the only metal available at the time the Great Pyramid was built. Consequently, it would follow that all work must have sprung from the able use of this basic metallic element. We may be entirely wrong, however, even in the basic assumption that copper was the only metal available to the ancient Egyptians. For another little known fact about the pyramid builders is that they were iron makers as well.

"Without going back in time and interviewing the craftsmen who worked on the pyramids, we will probably never know for sure what materials their tools were made of. Any debate of the subject would be futile, for until the proof is at hand, no satisfactory conclusion can be reached. However, the manner in which the masons used their tools can be discussed, and, perhaps if we compare current methods of cutting granite with the finished product (i.e. the granite coffers), there may be some basis on which to draw a parallel.

"Today's granite cutting methods include the use of wire-saws and an abrasive, usually silicon-carbide, which has a hardness comparable with diamond, and, therefore, is hard enough to cut through the quartz crystal in the granite. The wire is a continuous loop that is held by two wheels, one of the wheels being the driver. Between the wheels, which can vary in distance depending on the size of the machine, the granite is cut by being pushed against the wire or by being held firmly and allowing the wire to feed through it. The wire does not cut the granite, but is designed to effectively hold the silicon carbide grit that does the actual cutting.

"By looking at the shapes of the cuts that were made in the basalt items 3b, and 5b, one could certainly speculate that a wire saw had been used and left its imprint in the rock. The full radius at the bottom of the cut is exactly the shape that would be left by such a saw.

"Mr. John Barta, of the John Barta Company informed me, that the wire saws used in quarry mills today cut through granite with great rapidity. Mr. Barta told me that the wire saws with silicon-carbide cut through the granite like it is butter. Out of interest, I asked Mr. Barta what he thought of the copper chisel theory. Mr. Barta, possessing an excellent sense of humor, came forth with some jocular remarks regarding the practicality of such an idea.

"If the ancient Egyptians had indeed used wire saws for cutting hard rock, were these saws powered by hand or machine? With my experience in machine shops and the countless number of times I have had to use a saw (both handsaws and power saws), there appears to be strong evidence that, in at least some instances, the latter method was used."

Sir William Petrie's observations bear out what Dunn is saying. The following are his notes on the coffer inside the King's Chamber in the Great Pyramid:

"On the N. end (of the coffer) is a place, near the west side, where the saw was run too deep into the granite, and was backed out again by the masons; but this fresh start they made still too deep, and two inches lower they backed out a second time, having cut out more than .10 inch deeper than they had intended..."

The following concerns the coffer inside the Second Pyramid:

"The coffer is well polished, not only inside but all over the outside; even though it was nearly all bedded into the floor, with the blocks plastered against it. The bottom is left rough, and shows that it was sawn and afterwards dressed down to the intended height; but in sawing it the saw was run too deep and then backed out; it was, therefore, not dressed down all over the bottom, the worst part of the sawing being cut .20 inch deeper than the dressed part. This is the only error of workmanship in the whole of it; it is polished all over the sides in and out, and is not left with the saw lines visible on it like the Great Pyramid coffer."

Petrie estimated that a pressure of one to two tons on jeweled tipped bronze saws would have been necessary to cut through the extremely hard granite. If we agree with these estimates as well as with the methods proposed by Egyptologists regarding the construction of the pyramids, then a severe inequity can be discerned between the two.

Says Dunn, "So far, Egyptologists have not given credence to any speculation that suggests that the builders of the pyramid might have used machines instead of manpower in this massive construction project. In fact, they do not give the pyramid builders the intelligence to have developed and used the simple wheel. It is quite remarkable that a culture, which possessed sufficient technical ability to make a lathe and progressed from there to develop a technique that enabled them to machine radii in hard diorite, would not have thought of the wheel before this.

"Petrie logically assumes that the granite coffers found in the Giza Pyramids were marked prior to being cut. The workmen were given a guideline with which to work. The accuracy exhibited in the dimensions of the coffers confirms this, plus the fact that guidelines of some sort would have been necessary to alert the masons of their error.

"While no one can say with certainty how the granite coffers were cut, the saw marks in the granite have certain characteristics, which suggests that they were not the result of hand sawing. If there was not evidence to the contrary, I might agree that the manufacturing of the granite coffers in the Great Pyramid and the Second Pyramid could quite possibly have been achieved using pure manpower, and a tremendous amount of time.

"It is extremely unlikely that a team of masons operating a 9-foot handsaw would be cutting through hard granite fast enough that they would pass their guideline before noticing the error. To then back the saw out and repeat the same error, as they did on the coffer in the King's Chamber, does nothing to confirm the speculation that this object was the result of hand work.

"When I read Petrie's passage concerning these deviations, a flood of memories came to me of my own history with saws, both power and manual driven. With these experiences, plus those observed in others, it seems inconceivable to me that manpower was the motivating force behind the saw which cut the granite coffers. While cutting steel with handsaws, an object that has a long workface, and certainly one with such dimensions as the coffers, would not be cut with great rapidity, and the direction the saw may turn can be seen well in advance of a serious mistake being made; the smaller the workpiece, naturally, the faster the blade would cut through it.

"On the other hand, if the saw is mechanized and is cutting rapidly through the workpiece, the saw could 'wander' from its intended course and cut through the guideline at a certain point at such a speed that the error is made before the condition can be corrected. This is not uncommon.

"This does not mean that a manually operated saw cannot 'wander,' but that the speed of the operation would determine the efficiency in discovering any deviation that the saw may have from its intended course.

"...Along with the evidence on the outside further evidence of the use of high speed machine tools can be found on the inside of the granite coffer in the King's Chamber. The methods that were evidently used by the pyramid builders to hollow out the inside of the granite coffers are similar to the methods which would be used to machine out the inside of components today."

Inside the King's Chamber

Dunn says that tool marks on the inside of the granite coffer in the King's Chamber indicate that when the granite was hollowed out, preliminary roughing cuts were made by drilling holes into the granite around the area which was to be removed. According to Petrie, these drill holes were made with tube-drills, which left a central core that had to be knocked away after the hole had been cut. After all the holes had been drilled, and all the cores removed, Petrie surmises that the coffer was then handworked to its desired dimension. The machinists on this particular piece of granite once again let their tools get the better of them, and the resulting errors are still to be found on the inside of the coffer in the King's Chamber:

"On the E. inside is a portion of a tube-drill hole remaining, where they had tilted the drill over into the side by not working it vertically. They tried hard to polish away all that part, and took off about 1/10 inch thickness all around it; but still they had to leave the side of the hole 1/10 deep, 3 long, and 1.3 wide; the bottom of it is 8 or 9 below the original top of the coffer. They made a similar error on the N. inside, but of a much less extent. There are traces of horizontal grinding lines on the W. inside."

Says Dunn, "The errors noted by Petrie are not uncommon in modern machine shops, and I must confess to having made them myself on occasion. Several factors could be involved in creating this condition, although I cannot visualize any one of them being a hand operation. Once again, while working their drill into the granite, the machinists had made a mistake before they had time to correct it.

"Let us speculate for a moment that the drill was being worked by hand. How far into the granite would they be able to cut before the drill had to be removed to permit cleaning the waste out of the hole? Would they be able to drill 8 or 9 inches into the granite without having to remove their drill? It is inconceivable to me that such a depth could be achieved with a hand-operated drill without the frequent withdrawal of the drill to clean out the hole, or provisions being made for the removal of the waste while the drill was still cutting. It is possible, therefore, that frequent withdrawals of the drill would expose their error, and that they would have noticed the direction their drill was taking before it had cut a .200 inch gouge into the side of the coffer, and before it had reached a depth of 8 or 9 inches.

Can't we see the same situation with the drill as with the saw? Here we have two high speed operations where errors are made before the operators have time to correct them.

"Although the ancient Egyptians are not given credit for having a simple wheel, the evidence proves that they not only had the wheel, they had a more sophisticated use for it. The evidence of lathe work is markedly distinct on some of the artifacts housed in the Cairo Museum, as well as those that were studied by Petrie. Two pieces of diorite in Petrie's collection he identified as being the result of true turning on a lathe."

Dunn notes that Petrie did not identify the means by which he inspected the core, whether he used metrology instruments, a microscope or the naked eye. He also mentions that all Egyptologists do not universally accept Petrie's conclusions. For instance, in *Ancient Egyptian Materials and Industries,* author Lucas takes issue with Petrie's conclusion that the grooves were the result of fixed jewel points. He states:

"In my opinion, to suppose the knowledge of cutting these gem stones to form teeth and of setting them in the metal in such a manner that they would bear the strain of hard use, and to do this at the early period assigned to them, would present greater difficulties than those explained by the assumption of their employment. But were there indeed teeth such as postulated by Petrie? The evidence to prove their presence is as follows.

(a) A cylindrical core of granite grooved round and round by a graving point, the grooves being continuous and forming a spiral, with in one part a single groove that may be traced five rotations round the core.

(b) Part of a drill hole in diorite with seventeen equidistant grooves due to the successive rotation of the same cutting point.

(c) Another piece of diorite with a series of grooves ploughed out to a depth of over one-hundredth of an inch at a single cut.

(d) Other pieces of diorite showing the regular equidistant grooves of a saw.

(e) Two pieces of diorite bowls with hieroglyphs incised with a very free-cutting point and neither scraped nor ground out.

But if an abrasive powder had been used with soft copper saws and drills, it is highly probable that pieces of the abrasive would have been forced into the metal, where they might have remained for some time, and any such accidental and temporary teeth would have produced the same effect as intentional and permanent ones..."[98]

Lucas goes on to speculate that

withdrawing the tube-drill in order to remove waste and insert fresh grit into the hole created the grooves. There are problems with this theory. Dunn says that it is doubtful that a simple tool that is being turned by hand will remain turning while the artisans draw it out of the hole. Likewise, placing the tool back into a clean hole with fresh grit would not require that the tool rotate until it was at the workface. There is also the question of the taper on both the hole and the core. Both would effectively provide clearance between the tool and the granite, thereby making sufficient contact to create the grooves impossible under these conditions.

Says Dunn, "The method I propose explains how the holes and cores found at Giza could have been cut. It is capable of creating all the details that Petrie and myself puzzled over. Unfortunately for Petrie, the method was unknown at the time he made his studies, so it is not surprising that he could not find any satisfactory answers.

"The application of ultrasonic machining is the only method that completely satisfies logic, from a technical viewpoint, and explains all noted phenomena. Ultrasonic machining is the oscillatory motion of a tool that chips away material, like a jackhammer chipping away at a piece of concrete pavement, except much faster and not as measurable in its reciprocation. The ultrasonic tool-bit, vibrating at 19,000 to 25,000 cycles per second (Hertz) has found unique application in the precision machining of odd-shaped holes in hard, brittle material such as hardened steels, carbides, ceramics and semiconductors. An abrasive slurry or paste is used to accelerate the cutting action."[98]

Ultrasonic Machining the Granite Core

Says Dunn, "The most significant detail of the drilled holes and cores studied by Petrie is that the groove is cut deeper through the quartz than the feldspar. Quartz crystals are employed in the production of ultrasonic sound and, conversely, are responsive to the influence of vibration in the ultrasonic ranges and can be induced to vibrate at high frequency. In machining granite, using ultrasonics, the harder material (quartz) would not necessarily offer more resistance, as it would during conventional machining practices. An ultrasonically vibrating tool-bit would find numerous sympathetic partners while cutting through granite, embedded right in the granite itself! Instead of resisting the cutting action, the quartz would be induced to respond and vibrate in sympathy with the high frequency waves and amplify the abrasive action as the tool cut through it.

"The fact that there is a groove may be explained several ways. An uneven flow of energy may have caused the tool to oscillate more on one side than the other. The tool may have been improperly mounted. A buildup of abrasive on one side of the tool may have cut the groove as the tool spiraled into the granite.

"The tapering sides of the hole and the core are perfectly normal when we consider the basic requirements for all types of cutting tools. This requirement is that clearance be provided between the tool's non-machining surfaces and the workpiece. Instead of having a straight tube, therefore, we would have a tube with a wall thickness that gradually became thinner along its length. The outside diameter getting gradually smaller, creating clearance between the tool and the hole, and the inside diameter getting larger, creating clearance between the tool and the central core. This would allow a free flow of abrasive slurry to reach the cutting area.

"A tube drill of this design would also explain the tapering of the sides of the hole and the core. By using a tube-drill made of softer material than the abrasive, the cutting edge would gradually wear away. The dimensions of the hole, therefore, would correspond to the dimensions of the tool at the cutting edge. As the tool became worn, the hole and the core would reflect this wear in the form of a taper."[98]

Dunn claims that with ultrasonic machining, the tool can plunge straight down into the workpiece. It can also be screwed into the workpiece. The spiral groove can be explained if we consider one of the methods that is predominantly used to uniformly advance machine components. The rotational speed of the drill is not a factor in this cutting method. The rotation of the drill is merely a means to advance the drill into the workpiece. Using a screw and nut method the tube drill could be efficiently advanced into the workpiece by turning in a clockwise direction. The screw would gradually thread through the nut, forcing the oscillating drill into the granite. It would be the ultrasonically induced motion of the drill that would do the cutting and not the rotation. The latter would only be needed to sustain a cutting action at the workface. By definition, the process is not a drilling process by conventional standards, but a grinding process in which abrasives are caused to impact the material in such a way that a controlled amount of material is removed.

Says Dunn, "Another method by which the grooves could have been created is through the use of a spinning trepanning tool that has been mounted off-centered to its rotational axis. Clyde Treadwell of Sonic Mill Inc., Albuquerque, NM, explained to me that when an off-centered drill rotated into the granite, it would gradually be forced into alignment with the rotational axis of the drilling machine's axis. The grooves, he claims, could be created as the drill was rapidly withdrawn from the hole.

"If Treadwell's theory is the correct one, it still requires a level of technology that is far more developed and sophisticated than what the ancient pyramid builders are given credit for. This method may be a valid alternative to the theory of ultrasonic machining, even though ultrasonics resolves all the unanswered questions where other theories have fallen short. Methods may have been proposed that might cover a singular aspect of the machine marks and not progress to the method described here. It is when we search for a single method that provides an answer for all data that we move away from primitive, and even conventional machining, and are forced to consider methods that are somewhat anomalous for that period in history."[98]

Granite Boxes in Rock Tunnels

In February 1995 Dunn joined Graham Hancock and Robert Bauval in Cairo to participate in a documentary. While there, he came across and measured some artifacts produced by the ancient pyramid builders, which prove beyond a shadow of a doubt that highly advanced and sophisticated tools and methods were employed by this ancient civilization. The group were examining artifacts found in the rock tunnels at the temple of Serapeum at Saqqarra, the site of the step pyramid and Zoser's tomb. Says Dunn, "We were in the stifling atmosphere of the tunnels, where the dust kicked up by tourists lay heavily in the still air. These tunnels contain 21 huge granite boxes. Each box weighs an estimated 65 tons, and, together with the huge lid that sits on top of them, the total weight of the assembly is around 100 tons. Just inside the entrance of the tunnels there is a lid that had not been finished and beyond this lid, barely fitting within the confines of one of the tunnels, is a granite box that had also been rough hewn.

"The granite boxes are approximately 13 ft. long, 7 1/2 ft. wide and 11 ft. high. They are installed in 'crypts' that were cut out of the limestone bedrock at staggered intervals along the tunnels. The floors of the crypts were about 4 ft. below the tunnel floor, and the boxes were set into a recess in the center. Bauval was addressing the engineering aspects of installing such huge boxes within a confined space where the last crypt was located near the end of the tunnel. With no room for the hundreds of slaves pulling on ropes to position these boxes, how were they moved into place?

"While Hancock and Bauval were filming, I jumped down into a crypt and placed my parallel against the outside surface of the box. It was perfectly flat. I shone the flashlight and found no deviation from a perfectly flat surface. I clambered through a broken out edge into the inside of another giant box and again, I was astonished to find it astoundingly flat. I looked for errors and couldn't find any. I wished at that time that I had the proper equipment to scan the entire surface and ascertain the full scope of the work. Nonetheless, I was perfectly happy to use my flashlight and straight edge and stand in awe of this incredibly precise and incredibly huge artifact. Checking the lid and the surface on which it sat, I found them both to be perfectly flat. It occurred to me that this gave the manufacturers of this piece a perfect seal. Two perfectly flat surfaces pressed

Christopher Dunn

together, with the weight of one pushing out the air between the two surfaces. The technical difficulties in finishing the inside of this piece made the sarcophagus in Khafra's pyramid seem simple in comparison. Canadian researcher Robert McKenty accompanied me at this time. He saw the significance of the discovery and was filming with his camera. At that moment I knew how Howard Carter must have felt when he discovered Tutankhamen's tomb.

"The dust-filled atmosphere in the tunnels made breathing uncomfortable. I could only imagine what it would be like if I was finishing off a piece of granite, regardless of the method used, how unhealthy it would be. Surely it would have been better to finish the work in the open air? I was so astonished by this find that it didn't occur to me until later that the builders of these relics, for some esoteric reason, intended for them to be ultra precise. They had taken the trouble to bring into the tunnel the unfinished product and finish it underground for a good reason! It is the logical thing to do if you require a high degree of precision in the piece that you are working. To finish it with such precision at a site that maintained a different atmosphere and a different temperature, such as in the open under the hot sun, would mean that when it was finally installed in the cool, cave-like temperatures of the tunnel, you would lose that precision. The granite would change its shape through thermal expansion and contraction. The solution then as it is now, of course, is to prepare precision surfaces in the location in which they were going to be housed.

"This discovery, and the realization of its critical importance to the artisans that built it, went beyond my wildest dreams of discoveries to be made in Egypt. For a man of my inclination, this was better than King Tut's tomb. The Egyptians' intentions with respect to precision are perfectly clear, but to what end? Further studies of these artifacts should include thorough mapping and inspection with the following tools. A laser interferometer with surface flatness checking capabilities. An ultrasonic thickness gauge to check the thickness of the walls to determine their consistency to uniform thickness. An optical flat with monochromatic light source. Are the surfaces really finished to optical precision?"[98]

Dunn contacted four precision granite manufacturers in the US and hasn't been able to find one who can do this kind of work. He received a letter from Eric Leither of Tru-Stone Corp about the technical feasibility of creating several Egyptian artifacts, including the giant granite boxes found in the bedrock tunnels the temple of Serapeum at Saqqarra. The letter read as follows:

"Dear Christopher,

First I would like to thank you for providing me with all the fascinating information. Most people never get the opportunity to take part in something like this. You mentioned to me that the box was derived from one solid block of granite. A piece of granite of that size is estimated to weigh 200,000 pounds if it was Sierra White granite which weighs approximately 175 lbs. per cubic foot. If a piece of that size was available, the cost would be enormous. Just the raw piece of rock would cost somewhere in the area of $115,000.00. This price does not include cutting the block to size or any freight charges. The next obvious problem would be the transportation. There would be many special permits issued by the D.O.T. and would cost thousands of dollars. From the information that I gathered from your fax, the Egyptians moved this piece of granite nearly 500 miles. That is an incredible achievement for a society that existed hundreds of years ago."

Says Dunn, "Eric went on to say that his company did not have the equipment or capabilities to produce the boxes in this manner. He said that his company would create the boxes in 5 pieces, ship them to the customer and bolt them together on site.

"Another artifact I inspected was a piece of granite that I, quite literally, stumbled across while strolling around the Giza Plateau later that day. I concluded, after doing a preliminary check of this piece, that the ancient pyramid builders had to have used a machine with three axes of movement (X-Y-Z) to guide the tool in three-dimensional space to create it. Outside of being incredibly precise, normal flat surfaces, being simple geometry, can justifiably be explained away by simple methods. This piece, though, drives us beyond the question, "What tools were used to cut it?" to a more far reaching question, "What guided the cutting tool?" In addressing this question and being comfortable with the answer, it is helpful to have a working knowledge of contour machining.

"Many of the artifacts that modern civilization creates would be impossible to produce using simple handwork. We are surrounded by artifacts that are the result of men and women employing their minds to create tools which overcome physical limitations. We have developed machine tools to create the dies that produce the aesthetic contours on the cars that we drive, the radios we listen to and the appliances we use. To create the dies to produce these items, a cutting tool has to accurately follow a predetermined contoured path in three dimensions. In some ap-

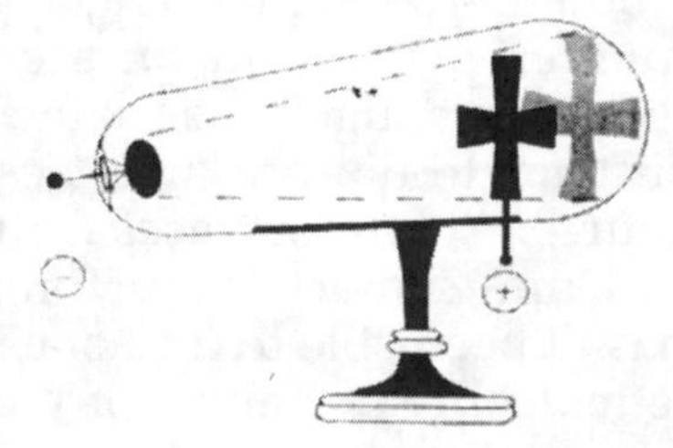

plications it will move in three dimensions, simultaneously using three or more axes of movement. The artifact that I was looking at required a minimum of three axes of motion to machine it. When the machine tool industry was relatively young, techniques were employed where the final shape was finished by hand, using templates as a guide. Today, with the use of precision computer numerical control machines, there is little call for handwork. A little polishing to remove unwanted tool marks may be the only handwork required. To know that an artifact has been produced on such a machine, therefore, one would expect to see a precise surface with indications of tool marks that show the path of the tool. This is what I found on the Giza Plateau, laying out in the open south of the Great Pyramid about 100 yards east of the second pyramid.

"There are so many rocks of all shapes and sizes lying around this area that to the untrained eye, this one could easily be overlooked. To a trained eye, it may attract some cursory attention and a brief muse. I was fortunate that it both caught my attention, and that I had some tools with which to inspect it. There were two pieces laying close together, one larger than the other. They had originally been one piece and had been broken. I found I needed every tool that I had brought with me to inspect it. I was most interested in the accuracy of the contour and its symmetry.

"What we have is an object that, three dimensionally as one piece, could be compared in shape to a small sofa. The seat is a contour that blends into the walls of the arms and the back. I checked the contour using the profile gauge along three axes of its length, starting at the blend radius near the back, and ending near the tangency point, which blended smoothly where the contour radius meets the front. The wire radius gauge is not the best way to determine the accuracy of this piece. When adjusting the wires at one position on the block and moving to another position, the gauge could be re-seated on the contour, but questions could be raised as to whether the hand that positioned it compensated for some inaccuracy in the contour. However, placing the parallel at several points along and around the axes of the contour, I found the surface to be extremely precise. At one point near a crack in the piece, there was light showing through, but the rest of the piece allowed very little to show.

"During this time, I had attracted quite a crowd. It's difficult to traverse the Giza Plateau at the best of times without getting attention from the camel drivers, donkey riders and purveyors of trinkets. It wasn't long after I had pulled the tools out of my backpack that I had two willing helpers, Mohammed and Mustapha, who weren't at all interested in compensation. At least that's what they told me, but I can honestly say that I lost my shirt on that adventure. I had cleaned sand and dirt out of the corner of the larger block and washed it out with water. I used a white T-shirt that I was carrying in my backpack to wipe the corner out so I could get an impression of it with forming wax. Mustapha talked me into giving him the shirt before I left. I was so inspired by what I had found I tossed it to him. Mohammed held the wire gauge at different points along the contour while I took photographs of it. I then took the forming wax and heated it with a match, kindly provided by the Movenpick hotel, then pressed it into the corner blend radius. I shaved off the splayed part and positioned it at different points around. Mohammed held the wax still while I took photographs. By this time there was an old camel driver and a policeman on a horse looking on.

"What I discovered with the wax was a uniform radius, tangential with the contour, the back and the side wall. When I returned to the US, I measured the wax using a radius gauge and found that it was a true radius measuring 7/16 inch. The side (arm) blend radius has a design feature that is a common engineering practice today. By cutting a relief at the corner, a mating part that is to match or butt up against the surface with the large blend radius may have a smaller radius. This feature provides for a more efficient machining operation, because it allows a cutting tool with a large diameter, and, therefore, a large radius to be used. With greater rigidity in the tool, more material can be removed when making a cut. I believe there is more, much more, that can be gleaned using these methods of study. I believe the Cairo Museum contains many artifacts that when properly analyzed, will lead to the same conclusion that I have drawn from this piece."[98]

High-Speed, Motorized Machinery Must Have Been Used

Says Dunn in conclusion, "The use of high-speed motorized machinery, and what we might call modern techniques in non-conventional machining, in manufacturing the granite artifacts found at Giza and other locations in Egypt warrants serious study by qualified, open-minded people who could approach the subject without preconceived notions.

"In terms of a more thorough understanding of the level of technology employed by the ancient pyramid builders, the implications of these discoveries are tremendous. We are not only presented with hard evidence that seems to have eluded us for decades, and which provide further evidence proving the ancients to be advanced, we are also provided with an opportunity to re-analyze everything from a different perspective. Understanding how something is made opens up a different dimension when trying to determine why it was made.

"The precision in these artifacts is irrefutable. Even if we ignore the question of how they were produced, we are still faced with the question of why such precision was needed. Revelation of new data invariably raises new questions. In this case it's understandable to hear, 'Where are

the machines?' Machines are tools. The question should be applied universally and can be asked of anyone who believes other methods may have been used. The truth is that no tools have been found to explain any theory on how the pyramids were built or granite boxes were cut! More than eighty pyramids have been discovered in Egypt, and the tools that built them have never been found. Even if we accepted the notion that copper tools are capable of producing these incredible artifacts, the few copper implements that have been uncovered do not represent the number of such tools that would have been used if every stonemason who worked on the pyramids at just the Giza site owned one or two. In the Great Pyramid alone, there are an estimated 2,300,000 blocks of stone, both limestone and granite, weighing between 2 tons and 70 tons each. That is a mountain of evidence, and there are no tools surviving to explain its creation.

"The principle of 'Occams Razor,' where the simplest means of manufacturing holds force until proven inadequate, has guided my attempt to understand the pyramid builders' methods. With Egyptologists, there is one component of this principle that has been lacking. The simplest methods do not satisfy the evidence, and they have been reluctant to consider other less simple methods. There is little doubt that the capabilities of the ancient pyramid builders have been seriously underestimated. The most distinct evidence that I can relate is the precision and mastery of machining technologies that have only been recognized in recent years.

"Some technologies the Egyptians possessed still astound modern artisans and engineers primarily for this reason. The development of machine tools has been intrinsically linked with the availability of consumer goods and the desire to find a customer. One reference point for judging a civilization to be advanced has been our current state of manufacturing evolution. Manufacturing is the manifestation of all scientific and engineering effort. For over a hundred years industry has progressed exponentially. Since Petrie first made his critical observations between 1880 and 1882, our civilization has leapt forward at breakneck speed to provide the consumer with goods, all created by artisans, and still, over a hundred years after Petrie, these artisans are utterly astounded by the achievements of the ancient pyramid builders. They are astounded not so much by what they perceive a society is capable of using primitive tools, but by comparing these prehistoric artifacts with their own current level of expertise and technological advancement.

"The interpretation and understanding of a civilization's level of technology should not hinge on the preservation of a written record of every technique that they had developed. The 'nuts and bolts' of our society do not always make good copy, and a stone mural will more than likely be cut to convey an ideological message rather than the technique used to inscribe it. Records of the technology developed by our modern civilization rest in media that is vulnerable and could conceivably cease to exist in the event of a worldwide catastrophe, such as a nuclear war or another ice age. Consequently, after several thousand years, an interpretation of an artisan's methods may be more accurate than an interpretation of his language. The language of science and technology doesn't have the same freedom as speech. So even though the tools and machines have not survived the thousands of years since their use, we have to assume, by objective analysis of the evidence, that they did exist.

"There is much to be learned from our distant ancestors, if only we can open our minds and accept that another civilization from a distant epoch may have developed manufacturing techniques that are as great or perhaps even greater than our own. As we assimilate new data and new views of old data, it is wise to heed the advice Petrie gave to an American who visited him during his research at Giza. The American expressed a feeling that he had been to a funeral after hearing Petrie's findings, which had evidently shattered some favorite pyramid theory of the time. Petrie said, 'By all means let the old theories have a decent burial; though we should take care that in our haste none of the wounded ones are buried alive.'

"With such a convincing collection of artifacts that prove the existence of precision machinery in ancient Egypt, the idea that the Great Pyramid was built by an advanced civilization that inhabited the Earth thousands of years ago becomes more admissible. I am not proposing that this civilization was more advanced technologically than ours on all levels, but it does appear that, as far as masonry work and construction are concerned, they were exceeding current capabilities and specifications. Making routine work of precision machining huge pieces of extremely hard igneous rock is astonishingly impressive.

"Considered logically, the pyramid builders' civilization must have developed their knowledge in the same manner any civilization would and had reached their "state of the art" through technological progress over many years. As of this writing, there is much research being conducted by many professionals throughout the world. These people are determined to find answers to the many unsolved mysteries indicating that our planet Earth has supported other advanced societies in the distant past. Perhaps when this new knowledge and insight is assimilated, the history books will be rewritten and, if mankind is able to learn from historical events, then perhaps the greatest lesson we can learn is now being formulated for the benefit of future generations. New technology and advances in the sciences are enabling us to take a closer look at the foundations upon which world history has been built, and these foundations seem to be crumbling. It would be illogical, therefore, to dogmatically adhere to any theoretical point concerning ancient civilizations."[98]

The Great Pyramid and the Mighty Crystal

How then can we can design an object to respond sympathetically with the earth's vibration? How do we utilize that energy? How can we turn it into usable electricity? If we could somehow utilize this energy, it would possibly be the greatest invention in the world.

Says Dunn, "We must, first of all, understand what a transducer is. Early on we discussed the piezoelectric effect vibration has on quartz crystal. Alternately compressing and releasing the quartz produces electricity. Microphones and other modern electronic devices work on this principle. Speak into a microphone and the sound of your voice (mechanical vibration) is converted into electrical impulses. The reverse happens with a speaker where electrical impulses are converted into mechanical vibrations. It has also been speculated that quartz-bearing rock creates the phenomenon known as ball lightning. The quartz crystal is the transducer. It transforms one form of energy into another. Understanding the source of the energy and having the means to tap into it, all we need to do is convert the unlimited mechanical stresses therein into usable electricity utilizing quartz crystals!

"The Great Pyramid was a geomechanical power plant that responded sympathetically with the earth's vibrations and converted that energy into electricity! They used the electricity to power their civilization, which included machine tools with which they shaped hard igneous rock.

"Ok, you may say, how does this power plant work? It's all very well to throw out a broad statement like that which rationalizes your own theory on machining, but we need more facts and proof that what has been stated is more than an interesting and radical theory. It has to have more proof based on truth and fact!

"Well let's start with the power crystal, or transducers. It so happens that the transducers for this power plant are an integral part of the construction that is designed to resonate in harmony with the pyramid itself, and also the earth. The King's Chamber, in which a procession of visitors have noted unusual effects, and in which Tom Danley detected the infrasonic vibrations of the earth, is, in itself, a mighty transducer.

"In any machine there are devices that function to make the machine work. This machine was no different. Though the inner chambers and passages of the Great Pyramid seem to be devoid of what we would consider to be mechanical or electrical devices, there are devices still housed there that are similar in nature to mechanical devices created today.

Dunn's Egyptian satellite.

"These devices could also be considered to be electrical devices in that they have the ability to convert or transduce mechanical energy into electrical energy. You might think of other examples, as the evidence becomes more apparent. The devices, which have resided inside the Great Pyramid since it was built, have not been recognized for what they truly were. Nevertheless, they were an integral part of this machine's function.

"The granite out of which this chamber is constructed is an igneous rock containing silicon quartz crystals. This particular granite, which was brought from the Aswan Quarries, contains 55% or more quartz crystal."

Work by two scientists, Dee Jay Nelson and David H. Coville, shows a special significance in the stone the builders chose in building the King's Chamber. They write:

"This means that lining the King's Chamber, for instance, are literally hundreds of tons of microscopic quartz particles. The particles are hexagonal, by-pyramidal or rhombohedral in shape. Rhomboid crystals are six-sided prisms with quadrangle sides that present a parallelogram on any of the six facets. This guarantees that embedded within the granite rock is a high percentage of quartz fragments whose surfaces, by the law of natural averages, are parallel on the upper and lower sides. Additionally, any slight plasticity of the granite aggregate would allow a 'piezotension' upon these parallel surfaces and cause an electromotive flow. The great mass of stone above the pyramid chambers presses downward by gravitational force upon the granite walls thereby converting them into perpetual electric generators.

"...The inner chambers of the Great Pyramid have been generating electrical energy since their construction 46 centuries ago. A man within the King's Chamber would thus come within a weak but definite induction field."[98]

Dunn comments that "While Nelson and Coville have made an interesting observation and speculation regarding the granite inside the pyramid, I am not sure that they are correct in stating that the pressure of thousands of tons of masonry would create an electromotive flow in the granite. The pressure on the quartz would need to be alternatively pressed and released in order for electricity to flow. The pressure they are describing would be static and, while it would undoubtedly squeeze the quartz to some degree, the electron flow would cease after the pressure came to rest. Quartz crystal does not create energy; it just converts one kind of energy into another. Needless to say, this point in itself leads to some interesting observations regarding the characteristics of the granite complex."[98]

The Acoustics of the Great Pyramid

One key to Dunn's theory of the Giza Power Plant is the acoustics of the Great Pyramid. Above the King's Chamber are five rows of granite beams, making a total of 43 beams weighing up to 70 tons each. Each layer is separated by a space large enough to crawl into. The red granite beams are cut square and parallel on three sides but were left seemingly untouched on the top surface, which was rough and uneven. Some of them even had holes gouged into the top of them.

In cutting these giant monoliths, the builders evidently found it necessary to treat the beams destined for the uppermost chamber with the same respect as those intended for the ceiling directly above the King's Chamber. Each beam was cut flat and square on three sides, with the topside seemingly untouched. This is interesting, considering that the ones directly above the King's Chamber would be the only ones visible to those entering the pyramid. Even so, the attention these granite-ceiling beams received was nonetheless inferior to the attention commanded by the granite out of which the walls were constructed.

William Flinders Petrie writes: "The roofing beams are not of 'polished granite,' as they have been described; on the contrary, they have rough-dressed surfaces, very fair and true so far as they go, but without any pretense to polish."

From his observations of the granite inside the King's Chamber, Petrie continues with those of upper chambers: "All the chambers over the King's Chamber are floored with horizontal beams of granite, rough dressed on the under sides which form the ceilings, but wholly unwrought above."

Says Dunn, "It is remarkable that the builders would exert the same amount of effort in finishing the 34 beams which would not be seen once the pyramid was built, as they did nine beams forming the ceiling of the King's Chamber which would be seen. Even if these beams were imperative to the strength of the complex, deviations in accuracy would surely be allowed, making the cutting of the blocks less time consuming. Unless, of course, they were either using these upper beams for a specific purpose, and/or were using standardized machinery methods that produced parts with little variation.

"Traditional theory has it that the granite beams served to relieve pressure on the chamber and allow this chamber to be built with a flat ceiling. I disagree. The pyramid builders knew about and were already utilizing a design feature that was structurally sound on a lower level inside the pyramid. If we look at the cantilevered arched ceiling of the Queen's Chamber, we can see that it has more masonry piled on top of it than the King's Chamber. The question could be asked, therefore, that if the builders had wanted to put a flat ceiling in this chamber, wouldn't they have only needed to add one layer of beams? For the distance between the walls, a single layer of beams in the Queen's Chamber, like the 43 granite beams above the King's Chamber, would be supporting no more than their own weight.

"This leads me to ask, "Why five layers of these beams?" To include so many monolithic blocks of granite in the structure is redundant. Especially when we consider the amount of incredibly difficult work that must have been invested in quarrying, cutting, transporting them 500 miles from the Aswan quarries, and then raising them to the 175 foot level of the pyramid. There is surely another reason for such an enormous effort and investment of time."

Continues Dunn, "The 43 giant beams above the King's Chamber were not included in the structure to relieve the King's Chamber from excessive pressure from above, but were included to fulfill a more advanced purpose. A simple yet refined technology can be discerned in the granite complex at the heart of the Great Pyramid, and with this technology the ancient power plant operated.

"The giant granite beams above the King's Chamber could be considered to be 43 individual bridges. Like the Tacoma Narrows bridge, each one is capable of vibrating if a suitable type and amount of energy is introduced. If we were to concentrate on forcing just one of the beams to oscillate, with each of the other beams tuned to that frequency or a harmonic of that frequency, the other beams would be forced to vibrate at the same frequency or a harmonic. If the energy contained within the forcing frequency was great enough, this transfer of energy from one beam to the next could affect the entire series of beams. A situation could exist, therefore, in which one individual beam in the ceiling directly above the King's Cham-

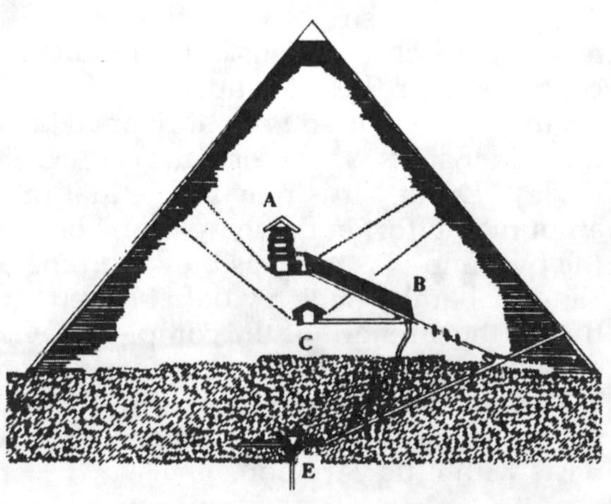

ber could indirectly influence another beam in the uppermost chamber by forcing it to vibrate at the same frequency as the original forcing frequency or one of its harmonic frequencies. The amount of energy absorbed by these beams from the source, would depend on the natural resonant frequency of the beam.

"The ability of the beams to dissipate the energy they are subject to would have to be considered, as well as the natural resonating frequency of the granite beam. If the forcing frequency (sound input) coincided with the natural frequency of the beam, and there was little damping (the beams were not restrained from vibrating), then the transfer of energy would be maximized. Consequently, so would the vibration of the beams.

"It is quite clear that the giant granite beams above the King's Chamber have a length of 17 feet (the width of the Chamber) in which they can react to induced motion and vibrate without restraint. Some damping may occur if the beams adjacent faces are so close that they rub together. However, if the beams vibrate in unison, it is possible that such damping would not happen. To perfect the ability of the 43 granite beams to resonate with the forcing frequency, the natural frequency of each beam would have to be of the same frequency as the forcing frequency, or be in harmony with it."

Tuned Granite Beams

Dunn asserts, "It would be possible to tune a length of granite, such as those found in the Great Pyramid, by altering its physical dimensions. A precise frequency could be attained by either altering the length of the beam which is allowed to vibrate (as in the playing of stringed instruments), or by removing material from the beam's mass, as in the tuning of bells. (A bell is tuned to a fundamental hum and its harmonics by removing metal from critical areas.) Striking it while it was being held in a position similar to that of the beams above the King's Chamber, as one would strike a tuning fork, could induce oscillation of the beam. The frequency of the vibration would be sampled and more material removed until the correct frequency had been reached.

"Rather than a lack of attention, therefore, the top surfaces of these granite beams may have arrived at their present shape through the application of more careful attention and work than the sides or the bottom. Before being placed inside the Great Pyramid, each beam may have been suspended on each end in the same position that it would hold once placed inside the Great Pyramid, and a considerable amount of attention paid to the upper surface. Each granite beam was shaped and gouged on the topside as it was tuned! Thousands of tons of granite were actually tuned to resonate in harmony with the fundamental frequency of the earth and the pyramid!

"The granite beams above the King's Chamber resemble what a granite beam might look like after it has been tuned in such a manner. After cutting three sides square and true to each other, the remaining side could have been cut and shaped until it reached a specific resonating frequency. The removal of material on the upper side of the beam would take into consideration the elasticity of the beam, as a variation of elasticity in the beam might result in more material being removed at one point along the beam's length than another. The fact that the beams above the King's Chamber are all shapes and sizes would support this speculation. In some of the granite beams, it wouldn't be surprising to find holes gouged out of the granite as the tuners worked on trouble spots."[98]

Piazzi Smyth also comments on these markings in his classic book *The Great Pyramid,*[110] "These markings, moreover, have only been discovered in those dark holes or hollows, the so-called 'chambers,' but much rather 'hollows of construction,' broken into by Colonel Howard Vyse above the 'King's Chamber' of the Great Pyramid. There, also, you see other traces of the steps of mere practical work, such as the 'Bat-holes' in the stones, by which the heavy blocks were doubtless lifted to their places, and everything is left perfectly rough."[110]

Rather than holes used for lifting the blocks into place, William Flinders Petrie speculates on another reason for Smyth's so-called "bat-holes:"

"The flooring of the top chamber has large holes in it, evidently to hold the butt ends of beams which supported the sloping roof-blocks during the building."[98]

Dunn comments, "Another reason for the holes gouged in the beams near the end of the beams may have been to provide feedback into the center of the beam, instead of transferring vibration into the core masonry. Although we must consider that both reasons given for the "bat-holes" may be possible explanations for their existence, it does not preclude other possibilities, which have yet to be considered.

"According to Boris Said, who was with Tom Danley when he conducted his tests, the King's Chamber resonated at a fundamental frequency and the entire structure of the King's Chamber reinforced this frequency by producing dominant frequencies that created an F sharp chord. Using large amplifiers F sharp is the frequency that is in harmony with the earth. Said claimed that the Indian Shamans tuned their ceremonial flutes to F sharp because it is a frequency that is sacred to mother earth.

"Testing for frequency, Tom Danley placed accelerometers in the spaces above the King's Chamber, but I don't know whether he went as far as checking the frequency of each beam. Said said something in his interview with Art Bell that may be some indication of where Danley was heading with his research, he said that the beams above the King's Chamber were, 'like baffles in a speaker.' Further research would need to be conducted before any assertion could be made as to the relationship these holes may have with tuning these beams to a specific frequency. However, when we consider the characteristics of the entire granite complex, along with other features found in the Great Pyramid, it seems clear that the results of this research will be along the lines of what I am theorizing.

"Without confirmation that the granite beams were carefully tuned to respond to a precise frequency, I will infer that such a condition exists in light of what is found in the area. While I have not found any specific record of anyone striking the beams above the King's Chamber and measuring their resonant frequencies, there has been quite a lot written about the resonating qualities of the coffer inside the chamber itself. The coffer is said to resonate at 438 hertz and is at resonance with the resonant frequency of the chamber. This is easily tested and has been noted by numerous visitors to the Great Pyramid, including myself."[98]

Another interesting discovery was made by the Schor expedition. This is a preliminary report, told to Art Bell by Boris Said, but it was noted that the floor of the King's Chamber does not sit on solid rock. Not only is the entire granite complex surrounded by massive limestone walls with a space between the granite and the limestone, the floor itself sits on what is characterized as "corrugated" shaped rock. It's no wonder the entire chamber "rings" while walking around inside!

Says Dunn concerning this, "Note, also, that the walls of the chamber do not sit on the granite floor, but are supported outside and 5-inches below the floor level. The granite complex inside the Great Pyramid, therefore, is poised ready to convert vibrations from the earth into electricity. What is lacking is a sufficient amount of energy to drive the beams and activate the piezoelectric properties within. The ancients, though, had anticipated the need for more energy than what would be collected only within the King's Chamber. They had determined that they needed to tap into the vibrations of the earth over a larger area inside the pyramid and deliver that energy to the power center—the King's Chamber—thereby substantially increasing the amplitude of the oscillations of the granite.

"While modern research into architectural acoustics might predominantly focus upon minimizing the reverberation effects of sound in enclosed spaces, there is reason to believe that the ancient pyramid builders were attempting to achieve the opposite. The Grand Gallery, which is considered to be an architectural masterpiece, is an enclosed space in which resonators were installed in the slots along the ledge that runs the length of the Gallery. As the earth's vibration flowed through the Great Pyramid, the resonators converted the energy to airborne sound. By design, the angles and surfaces of the walls and ceiling of the Grand Gal-

lery, caused reflection of the sound and its focus into the King's Chamber. Although the King's Chamber was also responding to the energy flowing through the pyramid, much of the energy would flow past it. The design and utility of the Grand Gallery was to transfer the energy flowing through a large area of the pyramid into the resonant King's Chamber. This sound was then focused into the granite resonating cavity at sufficient amplitude to drive the granite ceiling beams to oscillation. These beams, in turn, compelled the beams above them to resonate in harmonic sympathy. Thus, the input of sound and the maximization of resonance, the entire granite complex, in effect, became a vibrating mass of energy.

"The acoustic qualities of the design of the upper chambers of the Great Pyramid have been referenced and confirmed by numerous visitors since the time of Napoleon, whose men discharged their pistols at the top of the Grand Gallery and noted that the explosion reverberated into the distance like rolling thunder.

"Striking the coffer inside the King's Chamber results in a deep bell-like sound of incredible and eerie beauty, and it has been a practice over the years for the Arab guides to demonstrate this resonating sound to the tourists they guide through the pyramid. This sound was included on Paul Horn's album, (*Inside The Great Pyramid,* Mushroom Records, Inc., L.A., CA). After being advised of the significant pitch produced by the coffer when it has been struck, and the response of the chamber to this pitch, Horn brought along a device which would give him the exact pitch and frequency. Horn tuned his flute to this tone which was emitted, which turned out to be 'A' 438 cycles per second."

In a fascinating booklet about his experiences at the Great Pyramid, Horn describes phenomena concerning the acoustic qualities of the inner chambers, "The moment had arrived. It was time to play my flute. I thought of Ben Peitcsh from Santa Rose, California (a man who had told Mr. Horn about the pitch of the coffer) and his suggestions to strike the coffer. I leaned over and hit the inside with the fleshy part of the side of my fist. A beautiful round tone was immediately produced. What a resonance! I remember him also saying when you hear that tone you will be 'poised in history that is ever present.' I took the electronic tuning device I had brought along in one hand and struck the coffer again with the other and there is was—'A' 438, just as Ben predicted. I tuned up to this pitch and was ready to begin. (The album opens with these events so that you can hear all of these things for yourselves.)"

After noting the eerie qualities of the King's and Queen's Chambers, Paul Horn went out onto the Great Step at the top of the Grand Gallery to continue his sound test. The Grand Gallery, he reported, sounded rather flat compared with the other Chambers. He heard something remarkable at this time. He heard the music he was playing coming back to him clearly and distinctly from the King's Chamber. The sound was going out into the Grand Gallery and was being reflected through the passageway and reverberating inside the King's Chamber!

Dunn says that it appears that the coffer inside the King's Chamber was specifically tuned to a precise frequency, and that the room itself was scientifically engineered to be a resonator of that frequency. Perhaps these observations will finally provide an answer to a mystery that William Flinders Petrie had puzzled over at great length. His discovery of a flint pebble under the coffer, after he raised it, did not strike him as being unimportant for reasons he describes in *The Pyramids And Temples Of Gizeh*:

"The flint pebble that had been put under the coffer is important. If any person wished at present to prop the coffer up, there are multitudes of stone chips in the pyramid ready to hand. Therefore, fetching a pebble from the outside seems to show that the coffer was first lifted at a time when no breakages had been made in the pyramid, and there were no chips lying about. This suggests that there was some means of access to the upper chambers, which are always available by removing loose blocks without any forcing. If the stones at the top of the shaft leading from the subterranean part to the gallery had been cemented in place, they must have been smashed to break through them, or if there were granite portcullises in the Antechamber, they must also have been destroyed; and it is not likely that any person would take the trouble to fetch a large flint pebble into the innermost part of the Pyramid, if there were stone chips lying in his path."

Says Dunn, "Is it possible that the flint pebble was placed underneath the coffer at the time of the building? And that the pebble served a purpose for those who placed it there? The alternative answer—that there was free access to the upper chambers—cannot be supported by fact, and even if it was, we are still faced with the question of why someone found it necessary to prop up the coffer. However, if we had just manufactured an object like the coffer and had it tuned to vibrate at a precise frequency, we would know that to sit flat on the floor would dampen the vibrations somewhat. So, by raising one end of the coffer onto the pebble, it could vibrate at peak efficiency.

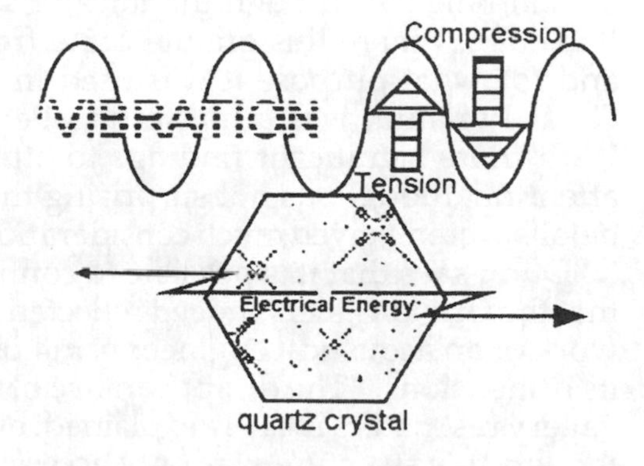

"Another unique feature, which needs to be confirmed by on-site inspection, is the ratchet style roof-line. The problem with coming up with an accurate calculation of the true angle of the overlapping stones is that there is conflicting data from the only two researchers that I have found paying these overlaps any close attention. However, preliminary calculations are interesting to say the least. The angle of the Grand Gallery is 26.3 degrees. Smyth measured the height of the Grand Gallery and found that it varied between 333.9 inches and 346.0 inches. The overlaps are estimated to have approximately a 12-inch tilt. Smyth counted 36 overlaps in the 1844.5 inches length of the roof. The surface of the overlapping stones in the roof line is close to a 45 degree angle from a vertical plane (135 degrees polar coordinates, given that the ends of the gallery are 90 degrees). With this tilt of the roof tiles, a sound wave traveling vertically to the roof would be reflected off the tiles at a 90 degree angle and travel in the direction of the King's Chamber.

"This gives another report, which didn't receive much attention, more pertinence. It has been reported that Al Mamun's men had to break a false floor out of the gallery, and as they broke one stone out, another slid down in its place. It's a sketchy bit of information that would require further research. Al Mamun's men were tearing out so much limestone that little attention was given to this. However, it should be kept in mind that there may have been a ratchet-style tiled floor in this gallery that matched the roof. Much of the stone that Al Mamun cut out of the Ascending Passage was dropped down the Descending Passage. Later explorers, such as Caviglia, Davison and Petrie, eventually cleared this passage of all debris, and most of this debris was dumped on the traditional rubbish pit on the North and East side of the Great Pyramid. Petrie reports finding inside the Great Pyramid a prism shaped stone that had a half round groove running its length. He also found in the Descending Passage a block of granite that was 20.6 inches thick with a section of tubular drilled hole cut through the thickness on one edge. Where this granite came from, and for what purpose it was used in the Great Pyramid, was a mystery to Petrie. With more significant findings to attract attention, though, its not surprising these details weren't given much consideration."

Dunn says that it is possible to confirm that the Grand Gallery indeed reflected the work of an acoustical engineer using only its dimensions, "The disappearance of the gallery resonators is easily explained, even though this structure was only accessible through a tortuously constricted shaft. The original design of the resonators will always be open to question; however, there is one device that performs in a manner that is necessary to respond sympathetically with vibrations. There is no reason that similar devices cannot be created today. There are many individuals who possess the necessary skills to recreate this equipment."[98]

The Helmholtz Resonator and the Grand Gallery Resonators

According to Dunn, a Helmholtz resonator would respond to vibrations coming from within the earth, and actually maximize the transfer of energy! The Helmholtz resonator is made of a round hollow sphere with a round opening that is 1/10—1/5 the diameter of the sphere. The size of the sphere determines the frequency at which it will resonate. If the resonant frequency of the resonator is in harmony with a vibrating source, such as a tuning fork, it will draw energy from the fork and resonate at greater amplitude than the fork will without its presence. It forces the fork to greater energy output than what is normal. Unless the energy of the fork is replenished, the fork will lose its energy quicker than it normally would without the Helmholtz resonator. But as long as the source continues to vibrate, the resonator will continue to draw energy from it at a greater rate.

Dunn says that the Helmholtz resonator is normally made out of metal, but can be made out of other materials. Holding these resonators in place inside the Gallery are members that are "keyed" into the structure by first being installed into the slots, and then held in the vertical position with "shot" pins that locate in the groove that runs the length of the Gallery.

Dunn now thinks that "The material for these members could have been wood, as trees are probably the most efficient responders to natural Earth sounds. There are trees that, by virtue of their internal structure, such as cavities, are known to emit sounds or hum. Modern concert halls are designed and built to interact with the instruments performing within. They are huge musical instruments in themselves. The Great Pyramid can be seen as a huge musical instrument with each element designed to enhance the performance of the other. To choose natural materials, especially in the function of resonating devices, would be a natural and logical decision to make. The qualities of wood cannot be synthesized."

The strange basalt "vases" of the Cairo museum may be the resonators that Dunn is looking for: "One of the most remarkable feats of machining can be found inside the Cairo Museum. I have stood in awe before the stone jars and bowls that are finely machined and perfectly balanced. The schist bowl with three lobes folded toward the center hub is an incredible piece of work. With the application of ultrasonics and sophisticated machinery, I can understand how they could be made, but the purpose for doing so has long escaped me. It seems like a tremendous amount of work to go to just to create a domestic vessel! Perhaps these stone artifacts, of which there were over a thousand found at Saqqarra, were used in some way to convert vibration into airborne sound. Are these vessels the Helmholtz resonators we are looking for?

"The enigmatic Ante Chamber has been the subject of much consternation and discussion. Ludwig Borchardt, Director of the German Institute in Cairo, forwarded one proposal for its use (circa 1925). Borchardt's theory proposed that a series of stone slabs were slid into place after Khufu had been entombed. He theorized that the half-round grooves in the granite wainscoting supported wooden beams that served as windlasses to lower the blocks.

"Borchardt may not have been far off with his analysis of the mechanism that was contained with the antechamber. After building the resonators and installing them inside the Grand Gallery, we would want to focus into the King's Chamber sound of a specific frequency, i.e., a pure tone or harmonic chord. We would be assured of doing so if we installed an acoustic filter between the Grand Gallery and the King's Chamber. By installing baffles inside the antechamber, sound waves travelling from the Grand Gallery through the passageway into the King's Chamber would be filtered as they passed through, allowing only a single frequency or harmonic of that frequency to enter the resonant King's Chamber. Sound wave lengths not coinciding with the dimensions between the baffles are filtered out, thereby ensuring that only no-interference sound waves enter the resonant King's Chamber, a condition that would reduce the output of the system.

"To explain the half-round grooves on one side of the chamber, and the flat surface on the other, we could speculate that when the installation of these baffles took place, they received a final tuning or 'tweaking.' This may have been accomplished by using cams. By rotating these cams, the off-centered shaft would raise or lower the baffles until the throughput of sound was maximized. A slight movement may have been all that was necessary. Maximum throughput is accomplished when the ceiling of the first part of the passageway (from the Grand Gallery), the ceiling of the passageway leading from the acoustic filter to the resonant King's Chamber and the bottom surface of each baffle are in alignment. The shaft suspending the baffles would have then been locked into place in a pillar block located on the flat surface of the wainscoting on the opposite wall."

Knowing that a vibrating system can eventually destroy itself, Dunn says that if there is no means to draw off or dampen the energy, there would have to be some way to control the level of energy at which the system operates. As the output of the resonant cavity would only draw off the energy up to a certain level, that being the maximum amount the granite complex could process, there would have to be some means of controlling the energy as it built up inside the Grand Gallery.

Dunn says that normally there would be three ways to prevent a vibrating system from running out of control:

1. Shut off the source of the vibration. (Can't do that.)
2. Reverse the process that was used to couple the vibration of the pyramid with the Earth.
3. Contrive a means to keep the vibration at a safe level.

Says Dunn, "With the source of vibration being the earth, obviously, numbers 2 and 3 are our best options. There are two ways to eliminate constant vibration, one is to dampen it and the other is to counteract the vibration with an interference wave that cancels it out. Physically dampening the vibration would be impractical, considering the function of the machine. The dampening wouldn't always be necessary, unlike the dampening needs of a bridge, and indeed would have an adverse effect on the efficiency of the machine. Consequently it would involve moving parts—like those in a piano. Faced with this consideration I immediately started to look closer at the Ascending Passage. It is the only feature inside the Great Pyramid that contains 'devices' that are directly accessible from the outside. I call the granite plugs inside this passage 'devices' in the same context that I called the granite beams above the King's Chamber devices because it wasn't necessary to use granite to block this passage and limestone would have been sufficient. It is obvious that their effectiveness at securing the inner chambers from robbers had the reverse effect. They drew attention to the existence of the Ascending Passage and subsequently the entire internal arrangement of passages and chambers. The granite plugs had to have another reason for being there!

"Possibly, they were built into the structure to allow or facilitate interference sound waves being introduced into the Grand Gallery and prevent the build-up of vibration within from reaching destructive levels. It may be the reason that the builders selected granite instead of limestone to plug the Ascending Passage."

Concludes Dunn, "The 3 plugs and their spacing within the passage may have, in fact, provided feedback to signal when the energy was reaching a dangerous level. By directing in- or out-of-phase sound waves up the Ascending Passage, they may have been able to control the energy level of the system. By directing a signal of the correct frequency, they may have also been able to prime the system in this manner also. In other words, the entire system would be forced to vibrate, and once in motion, it would draw energy from the earth with no further input.

"Sir William Flinders Petrie examined these blocks and described them in *Pyramids and Temples of Gizeh.* He remarked that the adjoining faces of the block were not flat but had a wavy finish plus or minus .3 inches. I was unable to confirm this when I was in Egypt, because the blocks, exposed by Al Mamun's tunnel, had slipped since Petrie's day and are now resting against each other. However, it does make for interesting speculation. Were the faces of the blocks cut specifically to modify sound waves? Could the Ascending Passage serve to direct an interference out-of-phase sound wave into the Grand Gallery, thereby controlling the level of energy in the system? There are mysteries still yet to be answered. But, we are not finished yet!"[98]

Those who would take over the earth
And shape it to their will
Never, I notice, succeed.
—Lao Tzu, *Tao Te Ching*

The Great Crystal of Edgar Cayce

In a similar manner to Dunn's theories on the Great Pyramid is the "psychic" information from Edgar Cayce and the Association for Research and Enlightenment in Virginia Beach, Virginia. Known as the "sleeping clairvoyant," Edgar Cayce was born on March 18, 1877 on a farm near Hopkinsville, Kentucky. Even as a child he displayed powers of perception which seem to extend beyond the normal range. In 1898 at the age of twenty-one he became a salesman for a wholesale stationery company and developed a gradual paralysis of the throat muscles which threatened the loss of his voice. When doctors were unable to find a cause for the strange paralysis, he began to see a hypnotist. During a trance, the first of many for Cayce, he recommended medication and manipulative therapy which successfully restored his voice and cured his throat trouble.

He began doing readings for people, mostly of a medical nature, and on October 9, 1910, *The New York Times* carried two pages of headlines and pictures on the Cayce phenomenon. By the time Edgar Cayce died on January 3, 1945, in Virginia Beach, Virginia, he left well over 14,000 documented stenographic records of the telepathic-clairvoyant statements he had given for more than eight thousand different people over a period of 43 years. These typewritten documents are referred to as "readings." Important to our discussion in this book is that many of these "readings" concern Atlantis, persons' former lives in Atlantis, and the airships and motive power used in Atlantis.[120]

In reading 2437-1; Jan. 23, 1941, Cayce told his subject: "...[I]n Atlantean land during those periods of greater expansion as to ways, means and manners of applying greater conveniences for the people of the land—things of transportation, the aeroplane as called today, but then as ships of the air, for they sailed not only in the air but in other elements also."

A number of persons who came to Cayce for individual life readings were, according to Cayce's reading, once navigators or engineers on these aircraft: "[I]n Atlantean land when there were the developments of those things as made for motivative forces as carried the peoples into the various portions of the land and to other lands. Entity a navigator of note then." (2124-3, Oct. 2, 1931)

"...[I]n Atlantean land when peoples understood the law of universal forces entity able to carry messages through space to the other lands, guided crafts of that period." (2494-1; Feb. 76, 1930)

Cayce called the motive power used in these vessels the "nightside of life." "[I]n Atlantean land or Poseidia—entity ruled in pomp and power and in understanding of the mysteries of the application of that often termed the nightside of life, or in applying the universal forces as understood in that period." (2897-1; Dec. 15, 1929)

"...[I]n Atlantean period of those peoples that gained much in understanding of mechanical laws and application of nightside of life for destruction." (2896-1; May 2, 1930)

Cayce speaks of the use of crystals or "firestones" for energy and related applications. He also speaks of the misuse of power and warnings of destruction to come: "[I]n Atlantean land during the periods of exodus due to foretelling or foreordination of activities which were bringing about destructive forces. Among those who were not only in Yucatan but in the Pyrenees and Egyptian land, for the manners of transportation and communications through airships of that period were such as Ezekiel described at a much later date." (4353-4; Nov. 26, 1939. See *Ezekiel* 1:15-25, 10:9-17 RSV.)

"...[I]n Atlantis when there were activities that brought about the second upheaval in the land. Entity was what would be in the present the electrical engineer—applied those forces or influences for airplanes, ships, and what you would today call radio for constructive or destructive purposes." (1574-1; April 19, 1938)

"...[I]n Atlantean land before the second destruction when there was the dividing of islands, when the temptations were begun in activities of Sons of Belial and children of the Law of One. Entity among those that interpreted the messages received through the crystals and the fires that were to be the eternal fires of nature. New developments in air and water travel are no surprise to this entity as these were beginning development at that period for escape." (3004-1; May 15, 1943)

"...[I]n Atlantean land at time of development of electrical forces that dealt with transportation of craft from place to place, photographing at a distance, overcoming gravity itself, preparation of the crystal, the terrible mighty crystal; much of this brought destruction." (519-1; Feb. 20, 1934)

"...[I]n city of Peos in Atlantis—among people who gained understanding of application of nightside of life or negative influences in the earth's spheres, of those who gave much understanding to the manner of sound, voice and picture and such to peoples of that period." (2856-1; June 7, 1930)

"...[I]n Poseidia the entity dwelt among those that had charge of the storage of the motivative forces from the great crystals that so condensed the lights, the forms of the activities, as to guide the ships in the sea and in the air and in conveniences of the body as television and recording voice." (813-1; Feb. 5, 1935)

The use of crystals as an important part of the technology is mentioned in a very long reading from Dec. 29, 1933: "About the firestone—the entity's activities then made such applications as dealt both with the constructive as well as destructive forces in that period. It would be well that there be given something of a description of this so that it may be understood better by the entity in the present.

"In the center of a building which would today be said to be lined with nonconductive stone—something akin to asbestos, with ...other nonconductors such as are now being manufactured in England under a name which is well known to many of those who deal in such things.

"The building above the stone was oval; or a dome wherein there could be ...a portion for rolling back, so that the activity of the stars—the concentration of energies that emanate from bodies that are on fire themselves, along with elements that are found and not found in the earth's atmosphere.

"The concentration through the prisms or glass (as would be called in the present) was in such manner that it acted upon the instruments which were connected with the various modes of travel through induction methods which made much the [same] character of control as would in the present day be termed remote control through radio vibrations or directions; though the kind of force impelled from the stone acted upon the motivation forces in the crafts themselves.

"The building was constructed so that when the dome was rolled back there might be little or no hindrance in the direct application of power to various crafts that were to be impelled through space—whether within the radius of vision or whether directed under water or under other elements, or through other elements.

"The preparation of this stone was solely in the hands of the initiates at the time; and the entity was among those who directed the influences of radiation which arose, in the form of rays that were invisible to the eye but acted upon the stones themselves as set in the motivating forces—whether the aircraft were lifted by the gases of the period; or whether for guiding the more-of-pleasure vehicles that might pass along close to the earth, or crafts on the water or under the water.

"These, then, were impelled by the concentration of rays from the stone which was centered in the middle of the power station, or powerhouse (as would be the term in the present).

"In the active forces of these, the entity brought destructive forces by setting up—in various portions of the land—the kind that was to act in producing powers for the various forms of the people's activities in the same cities, the towns, and the countries surrounding same. These, not intentionally, were tuned too high; and brought the second period of destructive forces to the people of the land—and broke up the land into those isles which later became the scene of further destructive forces in the land.

"Through the same form of fire the bodies of individuals were regenerated; by burning—through application of rays from the stone—the influences that brought destructive forces to an animal organism. Hence the body often rejuvenated itself; and it remained in that land until the eventual destruction; joining with the peoples who made for the breaking up of the land—or joining with Belial, at the final destruction of the land. In this, the entity lost. At first it was not the intention nor desire for destructive forces. Later it was for ascension of power itself.

"As for a description of the manner of construction of the stone: we find it was a large cylindrical glass (as would be termed today); cut with facets in such manner that the capstone on top of it made for centralizing the power or force that concentrated between the end of the cylinder and the capstone itself. As indicated, the records as to ways of constructing same are in three places in the earth, as it stands

Cleopatra's needle

today: in the sunken portion of Atlantis, or Poseidia, where a portion of the temples may yet be discovered under the slime of ages of sea water—near what is known as Bimini, off the coast of Florida. And (secondly) in the temple records that were in Egypt, where the entity acted later in co-operation with others towards preserving the records that came from the land where these had been kept. Also (thirdly) in records that were carried to what is now Yucatan, in America, where these stones (which they know so little about) are now—during the last few months—being uncovered." (440-5; Dec. 20, 1933)[98]

A Giant Pyramid Underwater at Bimini?

Bimini Island is a small member of the Bahamas, located about fifty miles east of Miami. Besides having sandy beaches, coral reefs, a variety of sunken ships, and some excellent fishing areas, Bimini is also the site for a series of very unusual underwater stone formations. This assemblage of huge blocks, many existing in straight patterns, are submerged under only 20 or 30 feet of water. There may also be a large pyramid in the vicinity of Bimini as well—underwater.

The Bimini Wall was first discovered in 1968 by Dr. J. Manson Valentine, Florida's maverick archaeologist. Valentine first saw the wall from the surface of the water when the sea was especially clear. He was with three other divers at the time, Jacques Mayol, Harold Climo, and Robert Angove. Said Valentine in an interview, "An extensive pavement of rectangular and polygonal flat stones of varying size and thickness, obviously shaped and accurately aligned to form a convincingly artifactual arrangement. These stones had evidently lain submerged over a long span of time, for the edges of the biggest ones had become rounded off, giving the blocks the domed appearance of giant loaves of bread or pillows. Some were absolutely rectangular, sometimes approaching perfect squares. (One remembers that absolutely straight lines are never present in natural formations.) The larger pieces, at least ten to fifteen feet in length, often ran the width of parallel-sided avenues, while the small ones formed mosaic-like pavements covering broader sections... The avenues of apparently fitted stones are straight-sided and parallel; the long one is a clear-cut double series interrupted by two expanses containing very large, flat stones propped up at the corners by vertical members (like the ancient dolmens of Europe); and the southeast end of this great roadway terminates in a beautifully curved corner; the three short causeways of accurately aligned large stones are of uniform width and end in *corner* stones..."[44, 61]

Dr. David Zink of the U.S. Air Force Academy in Colorado began doing research around Bimini that continues to this day. His book, *The Stones of Atlantis,*[121] chronicles his many adventures, with many good photographs, in the waters around Bimini. Dr. Zink firmly believes that the

Bimini Road is a man-made structure, but he has debunked a few of the other structures in the area, including a rectangular structure off Andros Island that was once believed to have been a temple site, but is now believed to be a sponge pen built in the 1930s. In 1974 he even photographed an unusual stone column standing upright that was believed to be the point of an obelisk that stood 40 to 50 feet tall, though most of it was buried in the ocean floor mud.

Many believers in Atlantis have been very excited by these finds just opposite Miami, as Edgar Cayce had predicted that the first portion of the lost continent to be discovered would be located in this general area. Cayce, who died in 1945, had predicted that the first portion of Atlantis to rise would be found in 1968 or 1969. Aerial reconnaissance of this region in 1968, and subsequent dives, did indeed reveal these stone structures breaking the surface of the ocean floor.

Another possibility is that these massive blocks are more likely the product of an early indigenous Indian civilization. Such a development could have served as the mother culture of the Olmecs and Maya in nearby Mesoamerica, as well as influenced the high centers of South America. At present, however, the weight of evidence is that these rocks may just be unique natural formations. Geologists and archaeologist have not found enough evidence to convince them to change their point of view. They contend that the area is simply composed of a rather unusual type of fractured beach rock.

Countless books have been written on the Bermuda Triangle, sometimes called the Devil's Triangle, of which Bimini is a part. Most books claim that some sort of vortex, or time warp, is responsible for missing ships, vanished airplanes, instruments that go haywire and weird magnetic and atmospheric phenomena.

There is good evidence that an energy vortex, or "gravitational anomaly" as they are sometimes called, is operating in the heavily trafficked waters off of Florida. In this area between Miami, Bermuda and Puerto Rico, literally hundreds of ships and planes have vanished. In a few odd cases, ships have been found derelict without crews. Very little wreckage has been found.

In 1990 it was announced that the five Navy torpedo bombers that had vanished in the Bermuda Triangle on Dec. 54, 1945 had been discovered off the waters of Fort Lauderdale. Later it was announced that these were not the missing planes, but a different set of crashed airplanes, with two of the craft having the same identification number.

According to Charles Berlitz, the grandson of the founder of the Berlitz Language schools, and the author of the worldwide bestseller *The Bermuda Triangle*[69] (plus other books on Atlantis and mysteries of the world), there are quite a few strange instances that have been recorded concerning the bizarre and life-threatening effects that occur in the Bermuda Triangle.

According to Berlitz:

• An oceanic investigative party on the yacht *New Freedom,* in July 1975, passed through an intense but rainless electromagnetic storm. During one tremendous burst of energy, Dr. Jim Thorpe photographed the exploding sky. The photograph when developed showed the burst in the sky, but it showed, too, a square-rigged ship on the sea about one hundred feet away from the *New Freedom,* although a moment before, the sea had been empty.

• John Sander, a steward on the *Queen Elizabeth-I* saw a small plane silently flying alongside his ship at deck level. He alerted another steward and the officer of the watch while the plane silently splashed into the ocean only seventy-five yards from the ship. The *QE-I* turned around and sent a boat over, but no indication of anything was found.

• Another "phantom plane" silently crashed into the ocean at Daytona Beach on February 17, 1935, in front of hundreds of witnesses, but an immediate search revealed nothing at all in the shallow water by the beach.

• A Cessna 172, piloted by Helen Cascio, took off for Turks Island, Bahamas, with a single passenger. About the time she should have arrived, a Cessna 172 was seen by the tower circling the island but not landing. Voices from the plane could be heard by the tower, but landing instructions from the tower evidently could not be heard by the pilot. A woman's voice was heard saying, "I must have made a wrong turn. That should be Turks, but there's nothing down there. No airport. No houses.

In the meantime, the tower attempted to give landing to the unresponsive Cessna. Finally the woman's voice said, "Is there no way out of this?" and the Cessna, watched by hundreds of people, flew away from Turks into a cloud bank from which it apparently never exited, since the plane, the pilot, and the passenger were never found.[69]

As Berlitz points out, the plane had been visible to the people on Turks, but when the pilot looked down, apparently she saw only an undeveloped island. Had she been seeing the island at a point in time before the airport and the houses were built? Where did this plane finally land? Did it land on the beach of some past or future world?

Various theories have been put forth to explain the Bermuda Triangle mystery. Sudden giant waves or eruptions of underwater volcanoes, whirlpools and "holes in the ocean" have all been

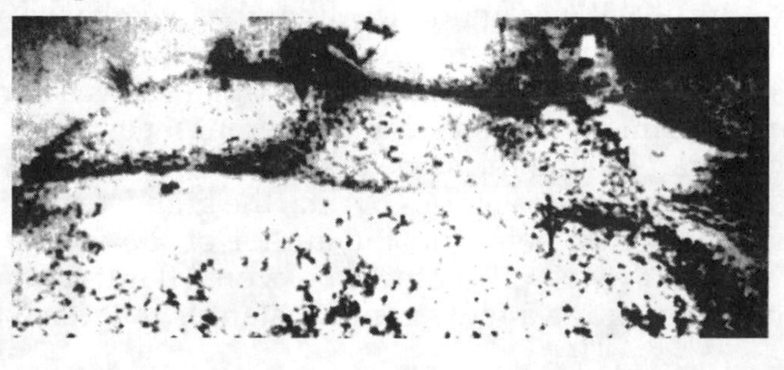

used as possible causes. Most researchers are willing to admit, though, that some sort of electromagnetic disturbance that causes instruments to malfunction is operating in the area.

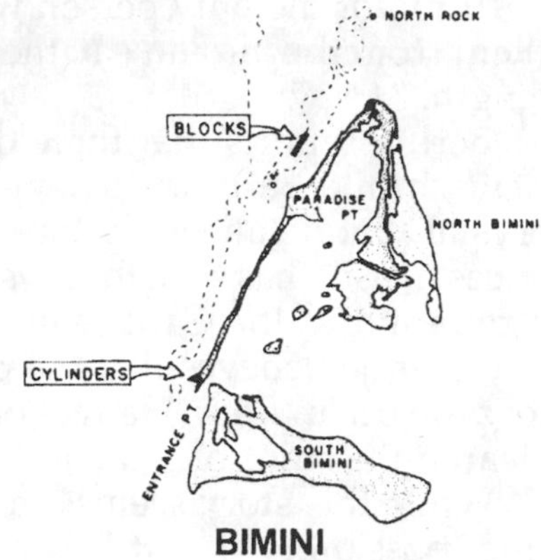

There are local stories of strange dense compact fogs on the surface of the water or in the sky. According to local belief, ships or aircraft that enter these odd clouds do not emerge.

Berlitz gives us the theory of Tom Gary, author of *Adventures of an Amateur Psychic*, claims that the Bermuda Triangle's destructive force comes from energy emanating from beneath the sea. "There is speculation that a power structure is still underwater in the Bermuda area," wrote Gary. According to him, the structure sits atop a large core that extends down through the crust of the earth. "When conditions are right the power structure works intermittently, causing ship and plane captains to lose control of their craft."[44, 69]

According to Gary, streaming ions form an electric current that produces a magnetic field and this causes instrument failure in craft in the vicinity. Magnetic compasses, fuel gauges, altitude indicators, and all electrically operated instruments are affected. Pilots who have survived such activity have reported battery drainage as well.

An incredible story is told by Ray Brown of Mesa, Arizona, that concerns an ancient pyramid off the Berry Islands in the Bahamas. In 1970, Brown claims to have been in a big storm while on the Berry Islands, having been looking for sunken galleons. In the morning after the storm, he says, their compasses were spinning and their magnometers were not giving any readings. "We took off north-east from the island. It was murky but suddenly we could see outlines of buildings under the water. It seemed to be a large exposed area of an underwater city. We were five divers and we all jumped in and dove down, looking for anything we could find," said Brown in an interview with Charles Berlitz.

"As we swam on, the water became clearer. I was close to the bottom at 135 feet and was trying to keep up with the diver ahead of me. I turned to look toward the sun through the murky water and saw a pyramid shape shining like a mirror. About thirty-five to forty feet from the top was an opening. I was reluctant to go inside... but I swam in anyway. The opening was like a shaft debouching into an inner room. I saw something shining. It was a crystal, held by two metallic hands. I had on my gloves and I tried to loosen it. It became loose. As soon as I grabbed it I felt this was the time to get out and not come back.

"I'm not the only person who has seen these ruins—others have seen them from the air and say they are five miles wide and more than that in length."[44, 61, 69]

Berlitz reports that three of the other divers have since died in accidents in the Bermuda Triangle and that Brown occasionally shows the crystal that he allegedly took from the sunken pyramid to lecture audiences. Berlitz has seen the crystal himself, though it is not necessarily from a pyramid in the Caribbean. Brown will not reveal the exact site of the city, but he believes that the pyramid and other buildings extend far below the ocean floor. He was only lucky that the storm the day before had cleared the ruins of sand and silt.

While this story seems almost too fantastic to be true, there is the distinct possibility that it is somehow based on fact—the "fact" that there may be a giant pyramid somewhere near Florida that is causing powerful electromagnetic effect.

This giant pyramid might be another of the giant generating stations that once existed around the world, similar to Christopher Dunn's theories on the Great Pyramid.

The Giza Power Plant.

An old print of initiates by an Egyptian obelisk.

The Sphinx with the Great Pyramid in the Background.

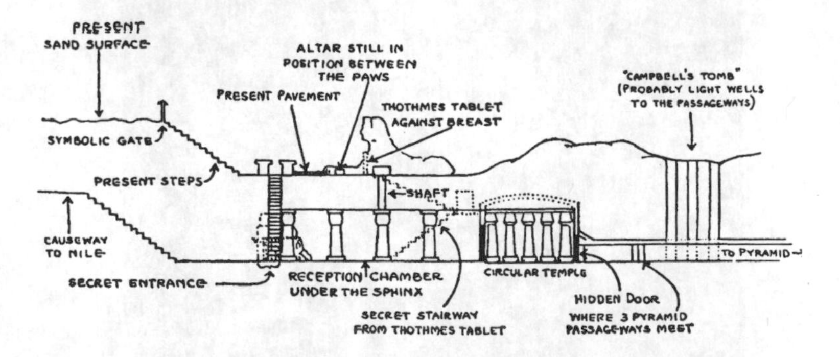

Possible secret chambers beneath the Sphinx.

Possible secret chambers beneath the Sphinx.

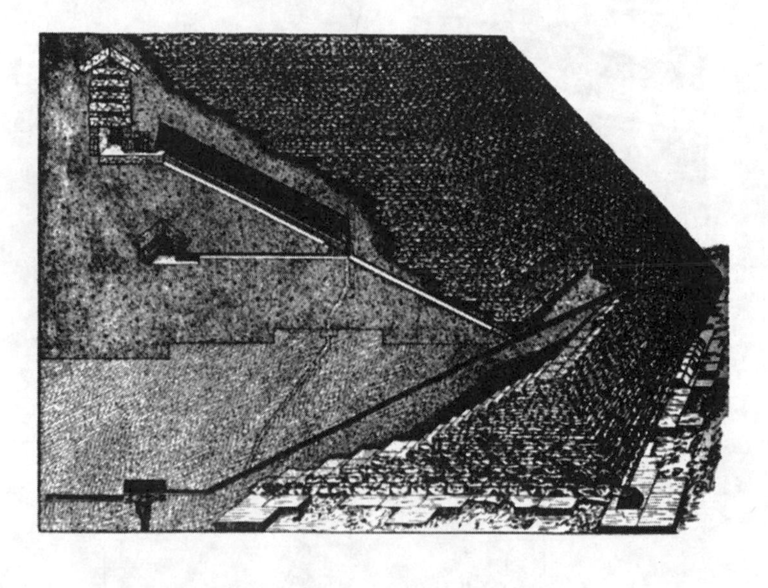

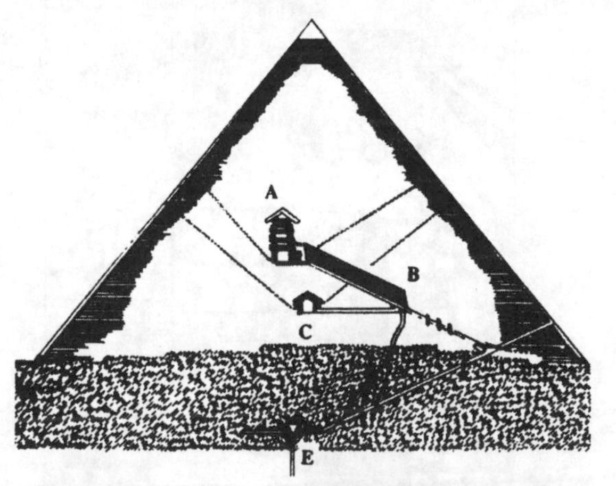

Two views of the interior of the Great Pyramid.

Two views of the Grand Gallery going into the King's Chamber. It appears as if some kind of machinery originally was placed into the strange ascending passage.

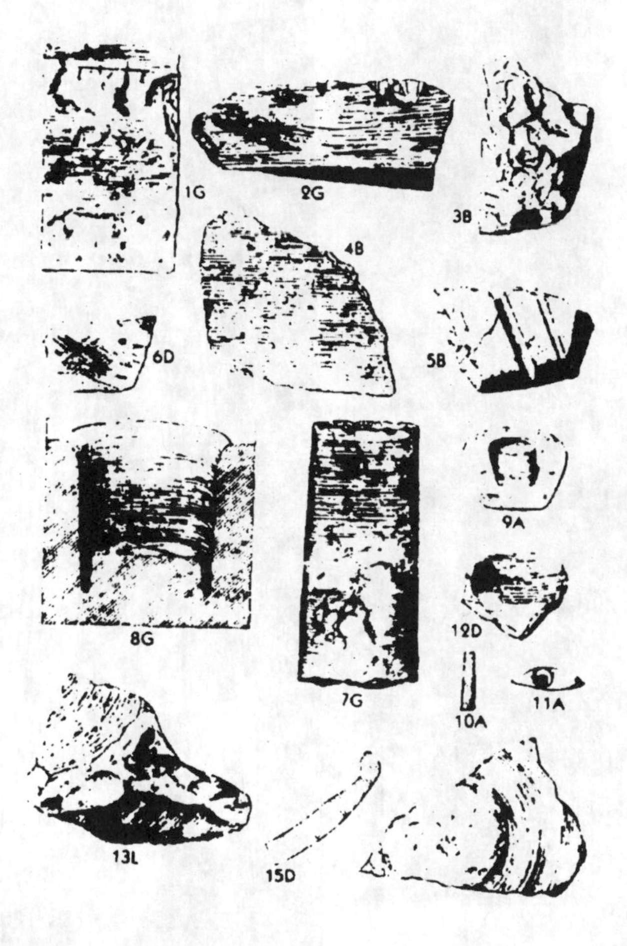

Petrie's examples of machining.

Christopher Dunn's examples of machining on Egyptian granite.

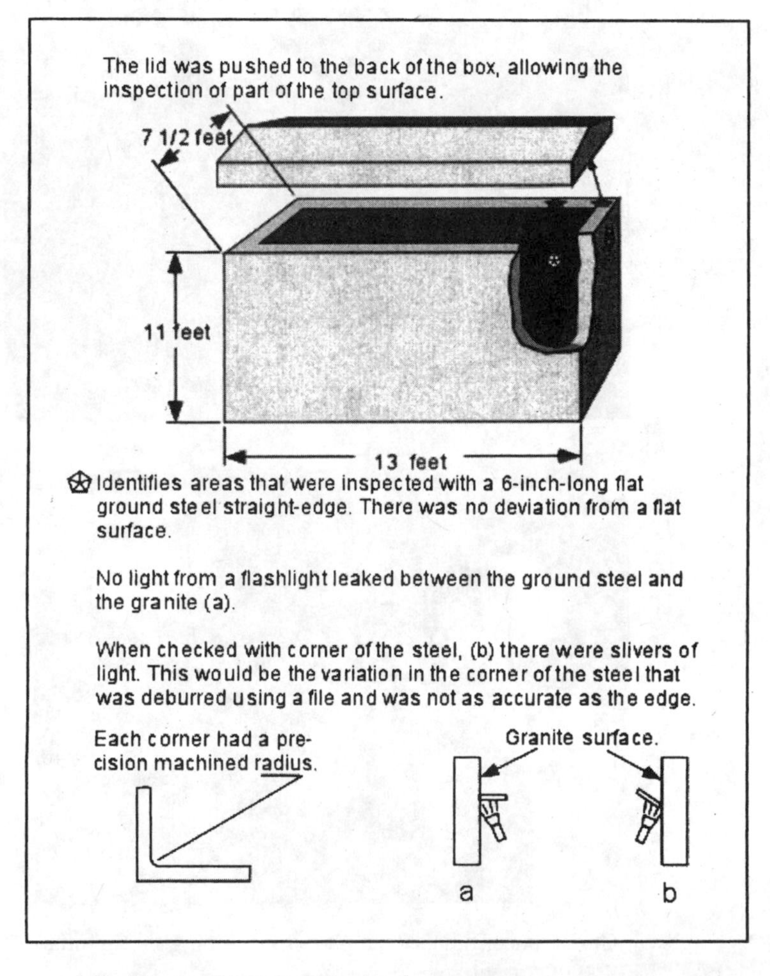

Christopher Dunn's diagram of a machined granite box.

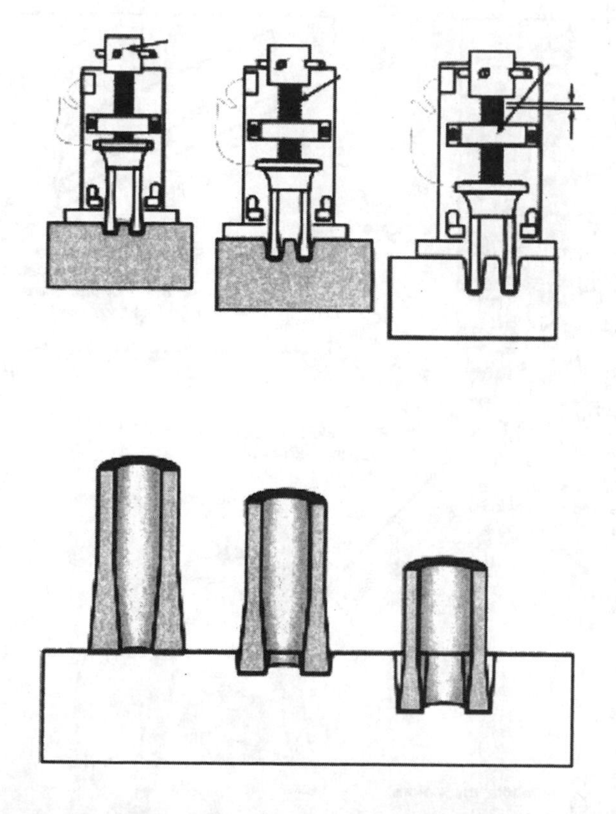

Christopher Dunn's diagram of ultrasonic drilling of granite. Top: The progression of drilling in granite using ultrasonic (vibratory) drill. The drill advances .100 inch plus tool wear for every rotation of the handle. Below: A cross section of the drill showing how the abrasive slurry wears the tool as well as the granite. The length of the tool diminishes as the cut deepens, resulting in a taper on the core and the hole.

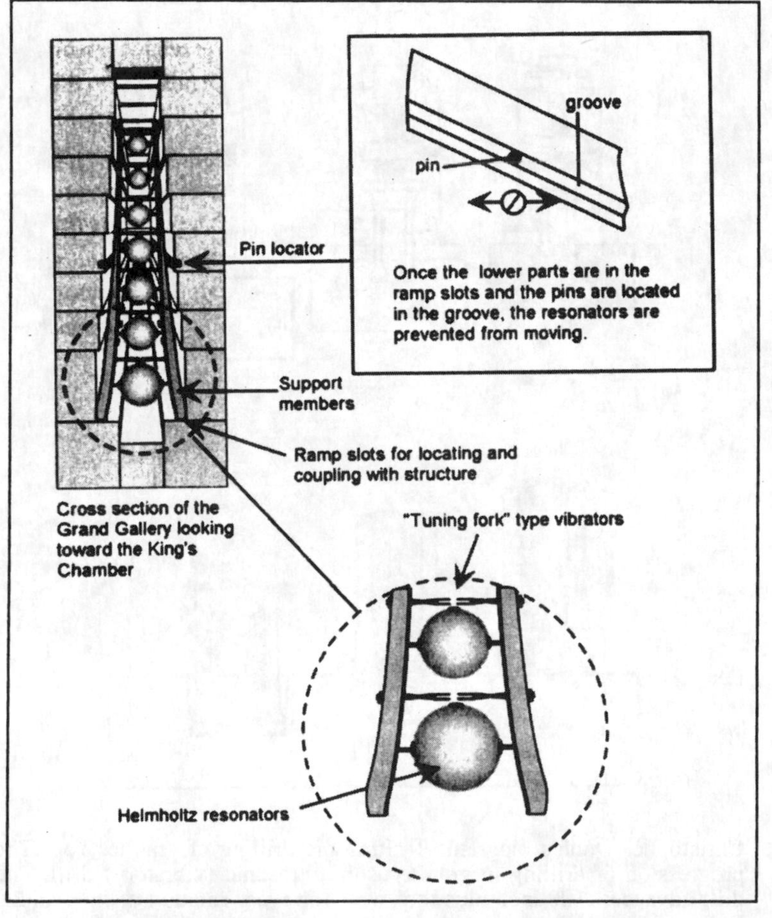

Christopher Dunn's diagram of the design and installation of the Helmholtz resonators.

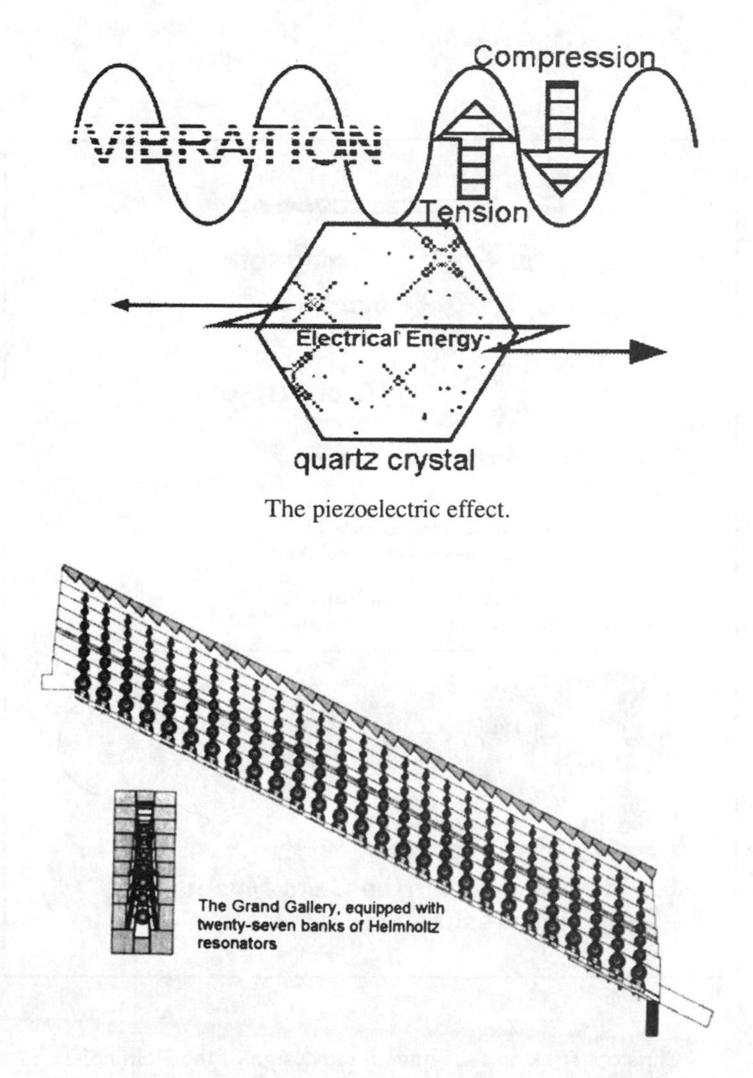

The piezoelectric effect.

Christopher Dunn's diagram of the Grand Gallery Resonators.

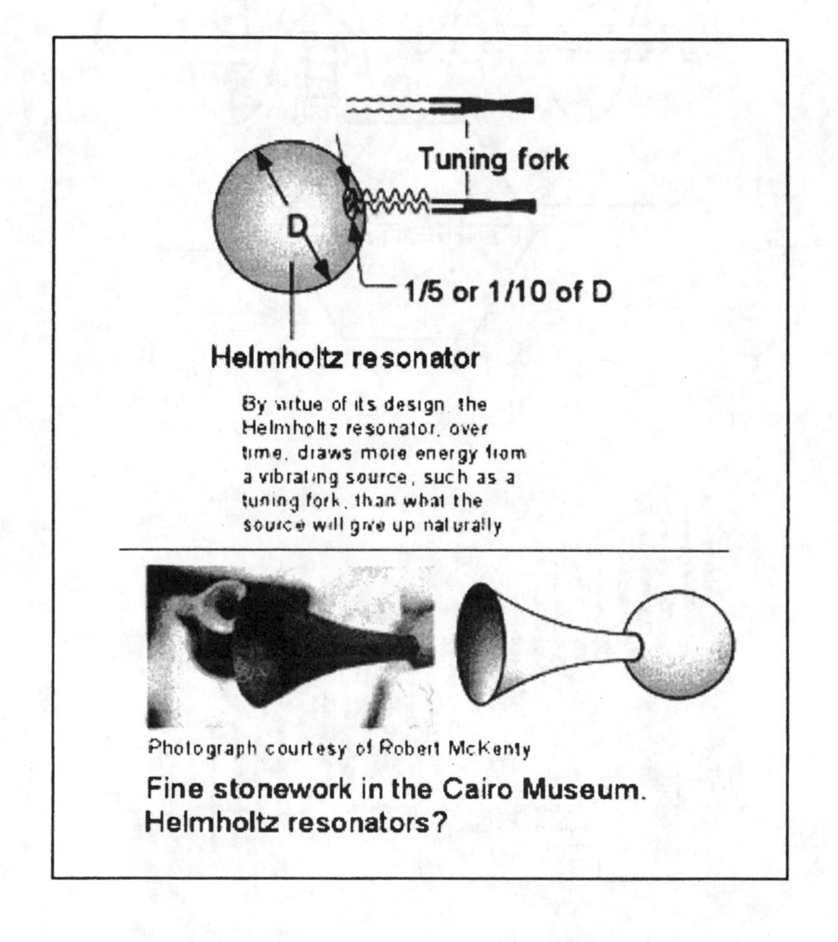

Christopher Dunn's diagram of the design of the Helmholtz resonators.

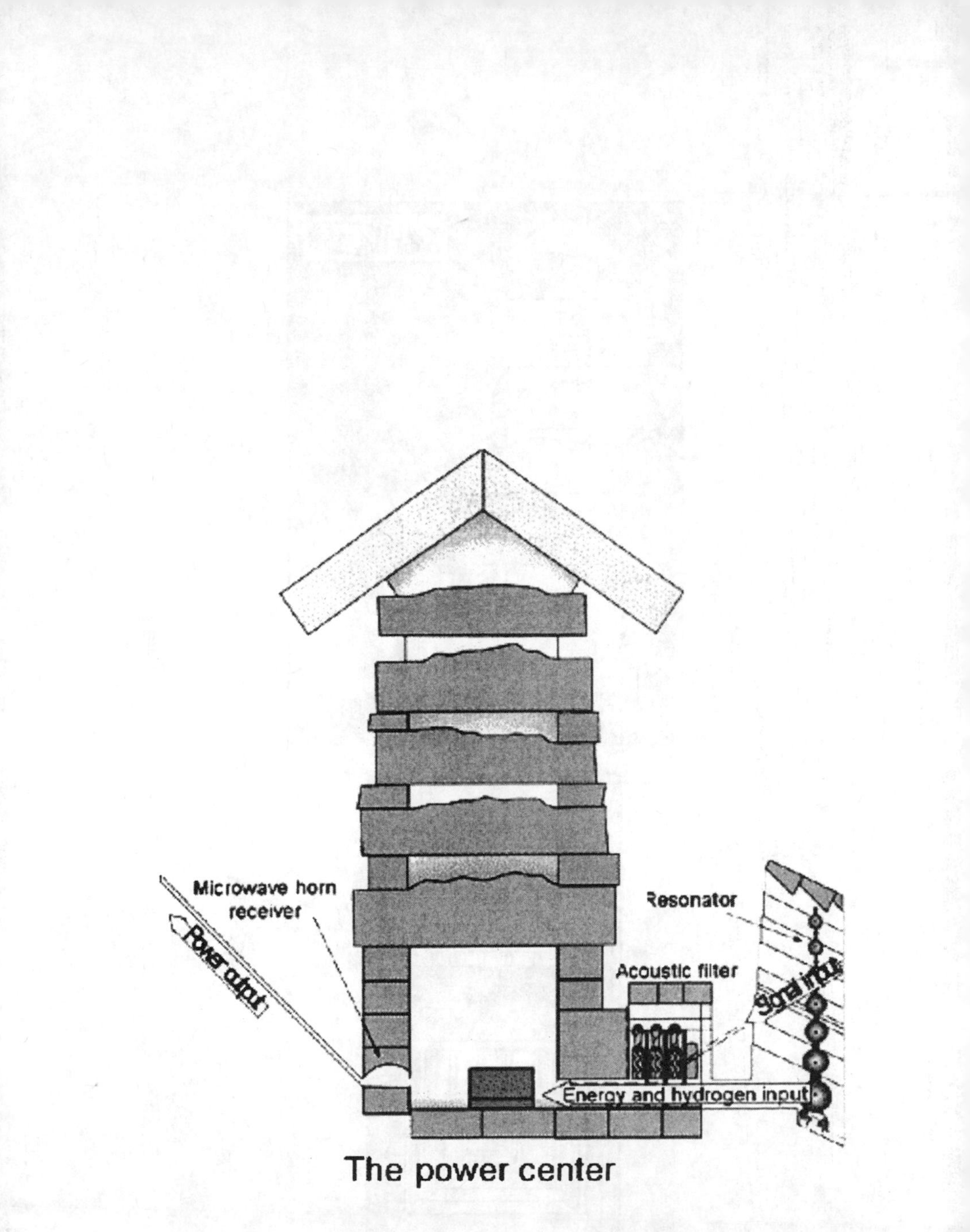

Christopher Dunn's diagram of the King's Chamber as the power center for the huge microwave-Maser power station.

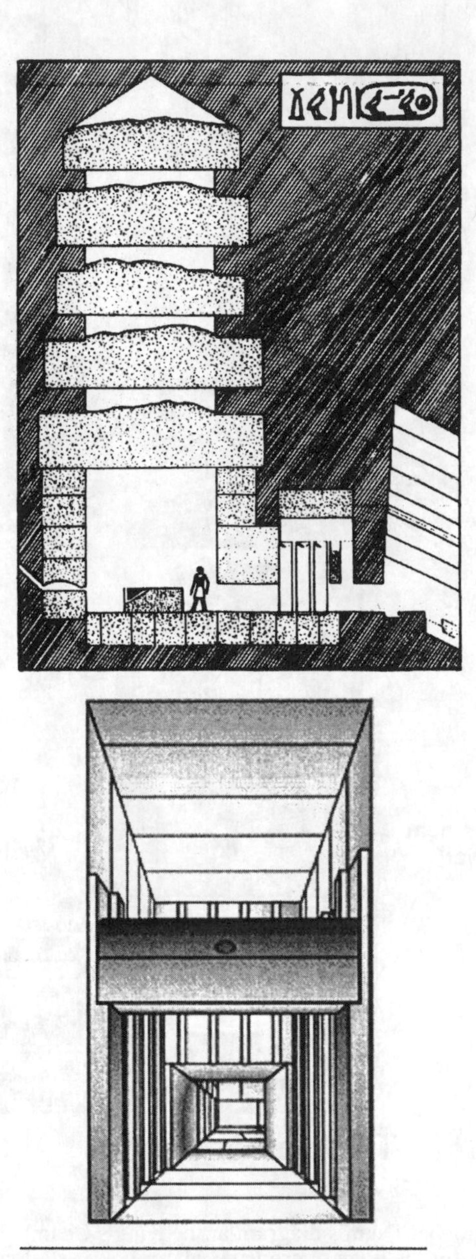

Top: The King's Chamber. Bottom: Christopher Dunn's diagram of the Antechamber between the Grand Gallery and the King's Chamber.

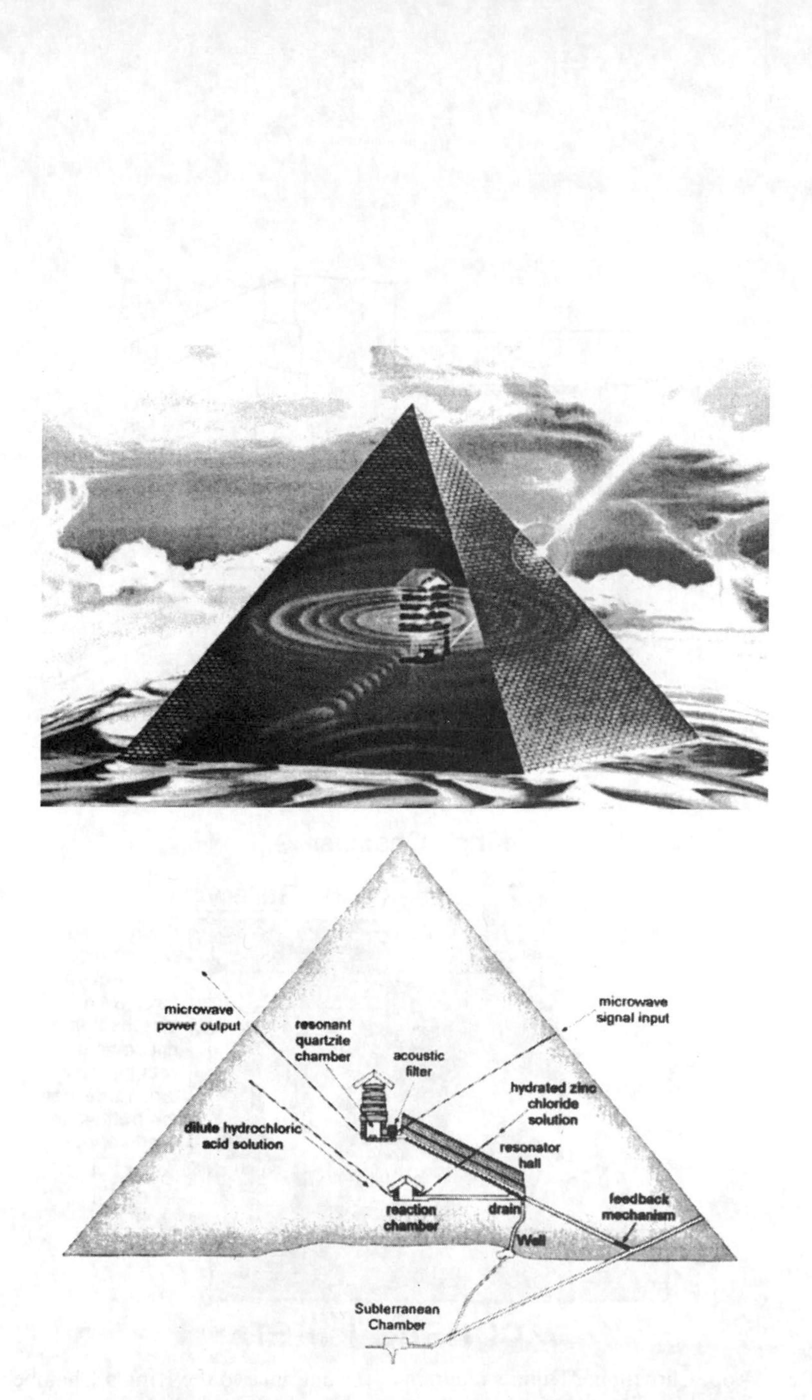

Christopher Dunn's concept of the Giza Power Plant.

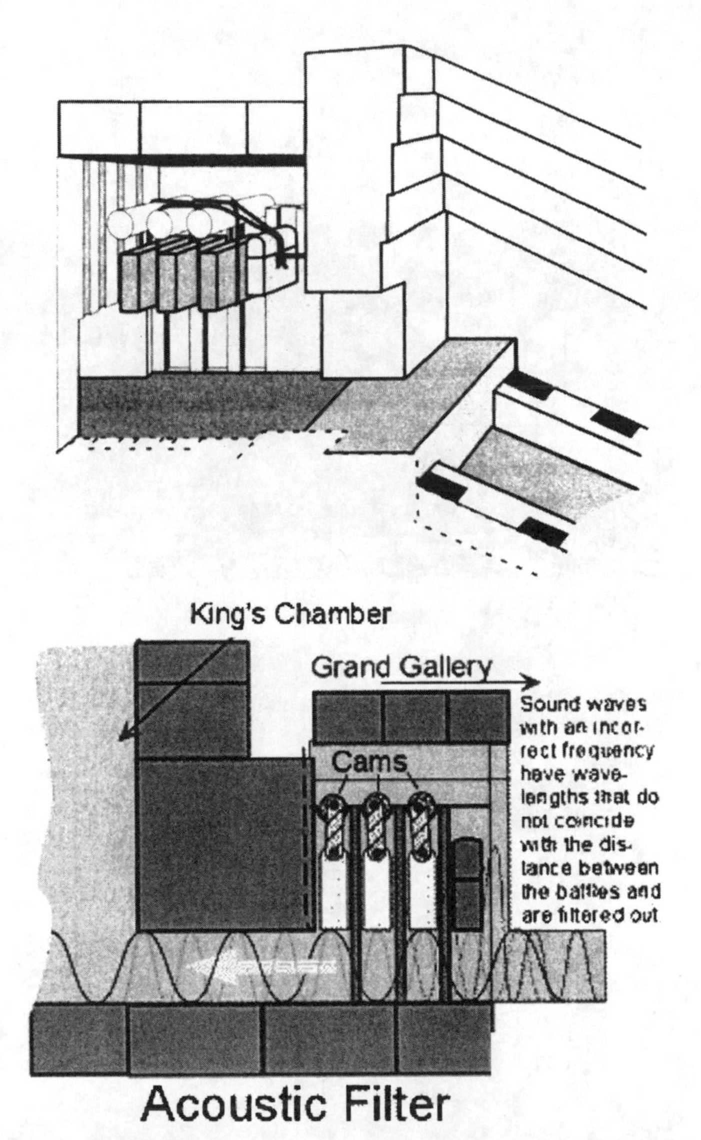

Top: Christopher Dunn's diagram of the entrance to the King's Chamber from the Grand Gallery showing the granite slabs that may have been used to seal the chamber according to the archeologist Borchardt. Below: The Antechamber serving as an acoustic filter. The baffles are raised or lowered to "tweak" system and to maximize its throughput.

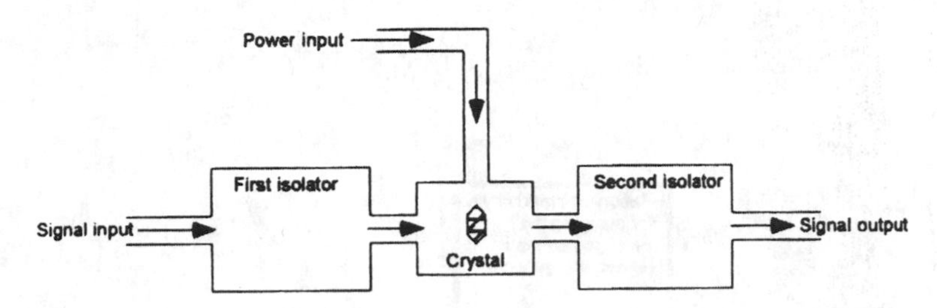

A microwave amplifier.

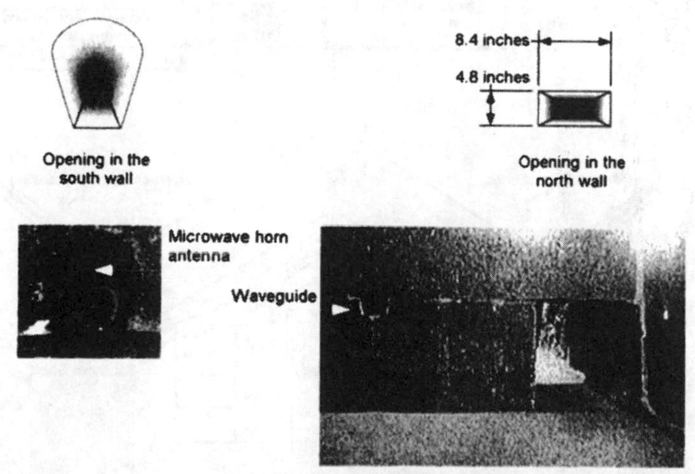

Christopher Dunn's diagram of the microwave horn antenna and waveguide inside the King's Chamber.

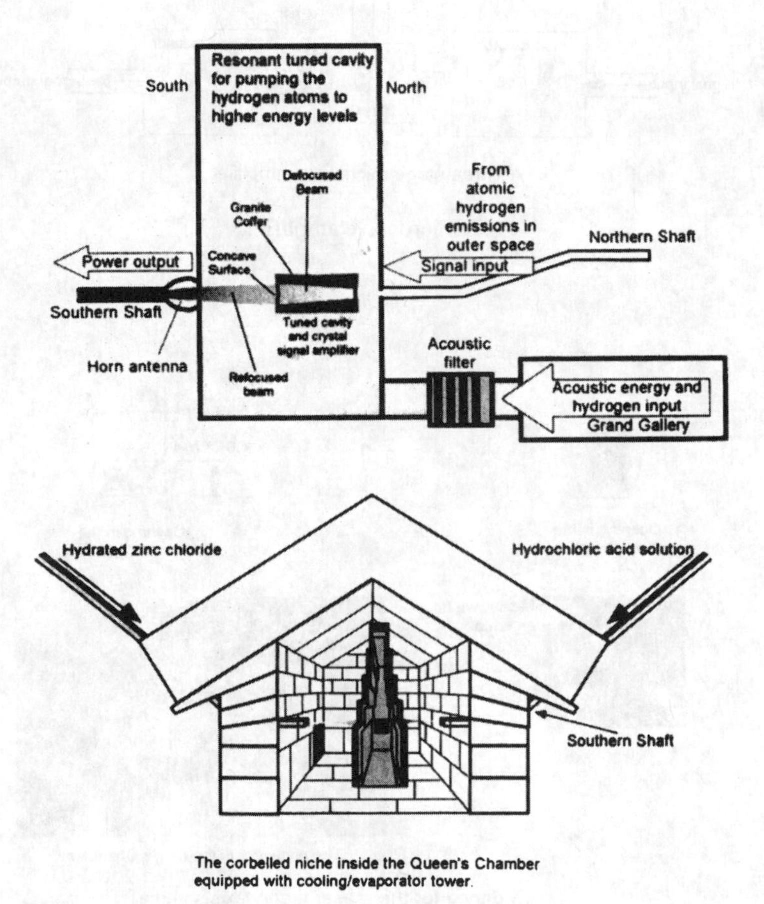

Top: Christopher Dunn's diagram of the MASER system for the pyramid.
Bottom: The mixing chamber and cooling/evaporation tower.

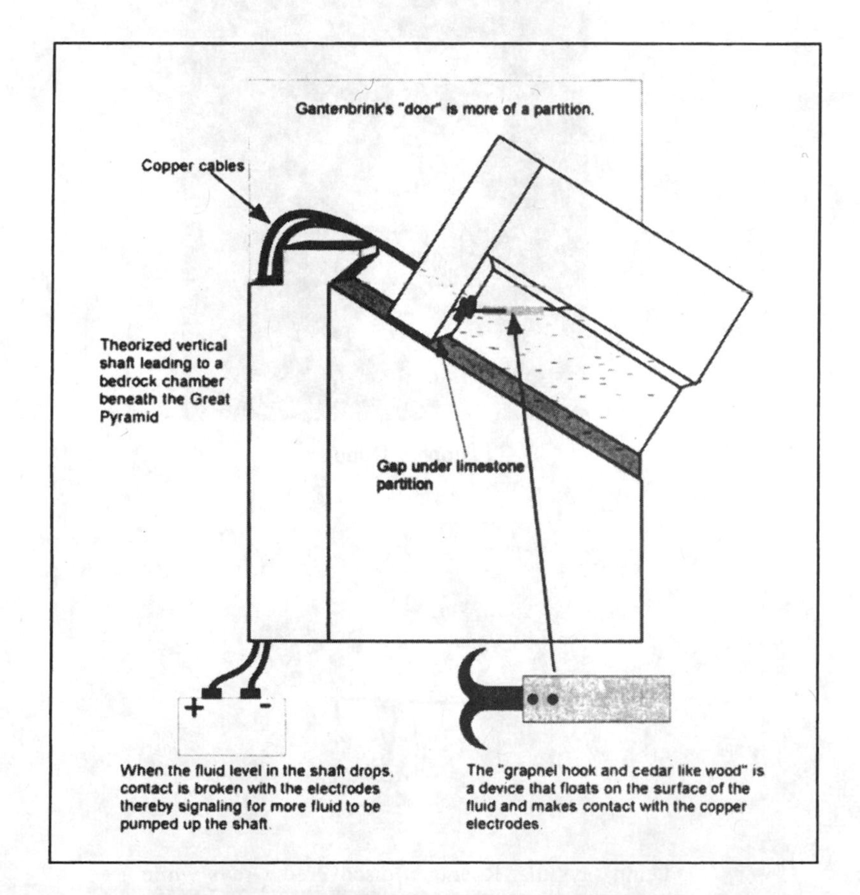

Christopher Dunn's diagram of the "Fluid Switch" for the hydrogen reaction inside the pyramid.

Christopher Dunn

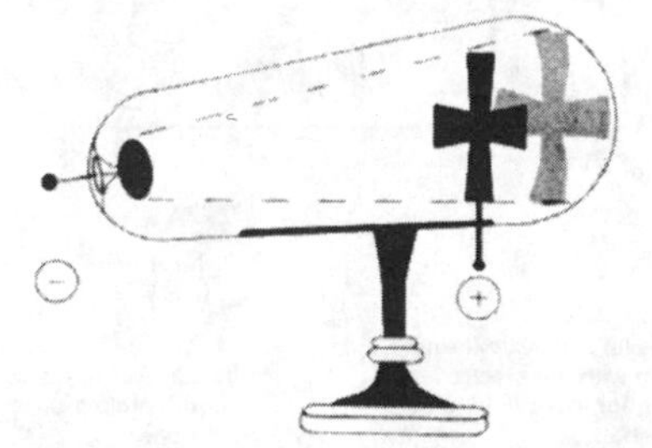

Dunn says that Roentgen discovered x-rays while experimenting with a Crookes Tube. In a Crookes Tube accelerated electrons fly past the anode and form a shadow on the wall of the tube. The Dendera wall carving may be a Crookes Tube.

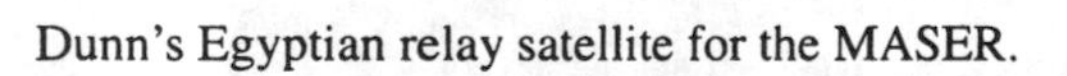

Dunn's Egyptian relay satellite for the MASER.

The Giza Power Plant.

Cleopatra's needle

Above: The crystal power tower of Atlantis, similar to Edgar Cayce's "terrible crystal," as depicted by the Lemurian Fellowship in 1944. Right: Cleopatra's Needle.

1922 illustration of Nikola Tesla's power broadcasting station while anti-gravity ships battle in the waters around it.

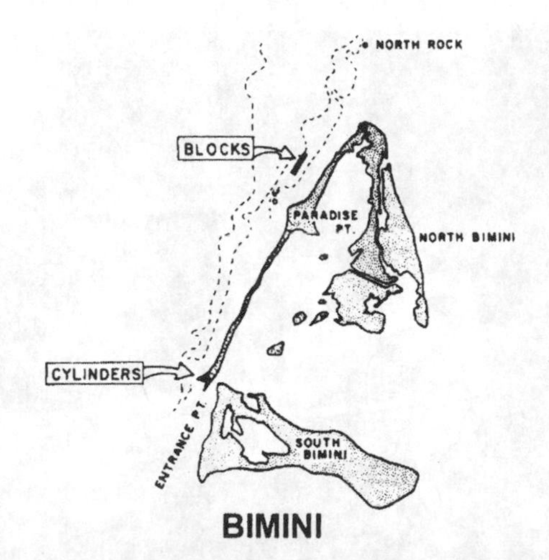

The unusual stones of the Bimini Road.

8.
The Cyclical Nature of History

The world is a dangerous place to live;
not because of the people who are evil,
but because of the people who don't do anything about it.
—Albert Einstein

Life is either a daring adventure or it is nothing.
—Helen Keller

The Cyclical Nature of History

History is cyclical in nature, the evidence shows us. What is today, was before. What was yesterday, will be tomorrow. We need to learn from our mistakes, so that instead of travelling endlessly in a repetitious cycle, we move in an upward spiral toward perfection and utopia.

We are, today, as the gods of yesterday: we fly through the air, we communicate with magic mirrors and talking boxes, we have tremendous war machines, and move things in a way that seems like magic. The cyclical nature of history brings us around to ages of great technology as well as to dark ages of ignorance and scientific repression. To get humankind through the dark ages, secret societies and secret libraries have been created to protect important knowledge, such as the fact that the world is a sphere, that electricity can be used for lighting, etc. Things that are part of everyday life today were yesterday's secrets. How many people have been tortured and killed in the effort to prevent technological and scientific advancement? The list would be a long one.

The Book of Enoch

The *Bible* is important, not only in a religious context, but as a historical document. The *Bible* gives us tales, in many cases derived from ancient Sumeria and Egypt, that would otherwise be lost to us because of the destruction of knowledge

throughout history. Fourteen ancient texts considered for inclusion in the *Bible* were ultimately left out of most versions. These books are collectively called the *Apocrypha.* The apocryphal *Book of Enoch the Prophet* was first discovered in Abyssinia in the year 1773 by a Scottish explorer named James Bruce. Bruce, a sort of 18th century Indiana Jones, may have seen the Ark of the Covenant at Axum (or its copy, as we surmise), and was able to obtain the ancient Coptic Christian text, approximately 2,000 years old. In 1821 *The Book of Enoch* was translated by Richard Laurence and published in a number of successive editions, culminating in the 1883 edition.

Says *The Book of Enoch,* in Chapter XIII, "Moreover Azayel taught men to make swords, knives, shields, breastplates, the fabrications of mirrors, and the workmanship of bracelets and ornaments, the use of paint, the beautifying of the eyebrow, the use of stones of every valuable and select kind, and of all sorts of dyes, so that the world became altered." Here we have yet another example of technology being handed over to mankind in ancient times by a helping "god" or superman, reminiscent of the stories of Osiris, Quetzlcoatl and Tubal Cain. This was world-altering technology, and it is interesting to note that the first items mentioned have to do with waging war. Who was Azayel, and where did he get the knowledge that he imparted?

The Cave of the Ancients

The curious (and rather prolific) British author T. Lobsang Rampa wrote a popular book on the theme of the cyclical nature of history in 1963 called *The Cave of the Ancients.*[122] In the book (reportedly non-fiction) the young Rampa, a boy-monk in Tibet, is taken to a remote spot by his teacher-guru to see the amazing "Cave of the Ancients"—a repository of ancient machines and devices.

After they entered the remote, hidden cave, says Rampa, "the four of us stood silent and frightened gazing at the fantastic sight before us. A sight which would have made any one of us alone think that he had taken leave of his senses. The cave was more like an immense hall, it stretched away in the distance as if the mountain itself was hollow. The light was everywhere, beating down upon us from a number of globes which appeared to be suspended from the the darkness of the roof. Strange machines crammed the place, machines such as we could not have imagined. Even from the high roof hung suspended apparatus and mechanisms. Some, I saw with great amazement, were covered by what appeared to be the clearest of glass…

"Slowly, almost imperceptibly, a misty glow formed in the darkness

before us. At first it was just a suspicion of blue-pink light, almost as if a ghost were materializing before our gaze. The mist-light spread, becoming brighter so that we could see the outlines of incredible machines filling this large hall, all except the center of the floor upon which we sat. The light drew in upon itself, swirling, fading, and becoming brighter and then it formed and remained in spherical shape. I had the strange and unexplainable impression of age-old machinery creaking slowly into motion after eons of time."

The Lama Mingyar Dondup tells young Rampa, "Thousands and thousands of years ago there was a high civilization upon this world. Men could fly through the air in machines which defied gravity; men were able to make machines which would impress thoughts upon the minds of others—thoughts that would appear as pictures. They had nuclear fission, and at last they detonated a bomb which all but wrecked the world, causing continents to sink below the oceans and others to rise. The world was decimated, and so, throughout the religions of this Earth, we now have the story of the Flood."

The lama continues, "There is a similar chamber at a certain place in the country of Egypt. There is another chamber with identical machines located in a place called South America. I have seen them, I know where they are. These secret chambers were concealed by the peoples of old so that their artifacts would be found by a later generation when the time was ready."

The group moves through the galleries inside the mountain, "We moved to the panel which the Lama Mingyar Dondup had told me about previously, and at our approach it opened with a grating creak, so loud in the silence of the place that I think we all jumped up with alarm. Inside was the darkness, profound, almost as if we had clouds of blackness swirling about us. Our feet were guided by shallow channels in the floor. We shuffled along and when the channels ended, we sat down. As we did so, there came a series of clicks, like metal scraping against metal, and almost imperceptibly light stole across the darkness and pushed it aside. We looked about us and saw more machines, strange machines. There were statues here, and pictures carved in metal. Before we had time to more than glance, the light drew in upon itself and formed a glowing globe in the center of the Hall. Colors flickered aimlessly, and bands of light without apparent meaning swirled round the globe. Pictures formed, at first blurred and indistinct, then growing vivid and real and with three-dimensional effect. We watched intently...

"This was the world of Long Long Ago.

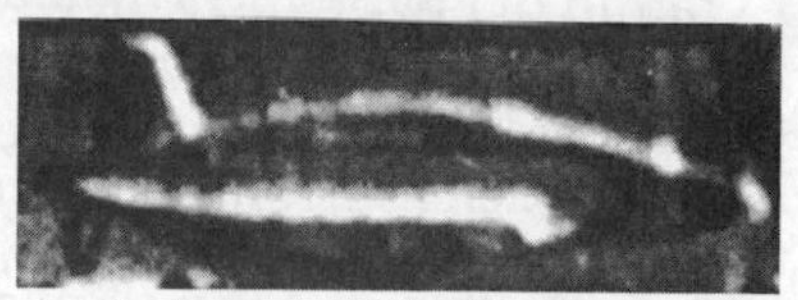

A close-up of the Abydos jet.

When the world was very young. Mountains stood where now there are seas, and the pleasant seaside resorts are now mountain tops. The weather was warmer and strange creatures roamed afield. This was a world of scientific progress. Strange machines rolled along, flew inches from the surface of the Earth, or flew miles up in the air. Great temples reared their pinnacles skywards, as if in challenge to the clouds. Animals and Man talked telepathically together. But all was not bliss; politicians fought against politicians. The world was a divided camp in which each side coveted the lands of the other. Suspicion and fear were the clouds under which the ordinary man lived. Priests of *both* sides proclaimed that they alone were the favored of the gods. In the pictures before us we saw ranting priests—as now—purveying their own brand of salvation. At a price! Priests of each sect taught that it was a "holy duty" to kill the enemy. Almost in the same breath they preached that mankind throughout world were brothers. The illogicality of brother killing brother did not occur to them.

"We saw great wars fought, with most of the casualties being civilians. The armed forces, behind their armor, were mostly safe. The aged, the women and children, those who did not fight, were the ones to suffer. We saw glimpses of scientists working in laboratories, working to produce even deadlier weapons, working to produce bigger and better bugs to drop on the enemy. One sequence of pictures showed a group of thoughtful men planning what they termed a 'Time Capsule' (what we called 'The Cave of the Ancients'), wherein they could store for later generations working models of their machines and a complete, pictorial record of their culture and lack of it. Immense machines excavated the living rock. Hordes of men installed the models and the machines. We saw the cold-light spheres hoisted in place, inert radio-active substances giving off light for millions of years. Inert in that it could not harm humans, active in that the light would continue almost until the end of Time itself."[122]

A Hollow Mountain as an Atomic Refuge

Fantastic as Lobsang Rampa's tale may seem, it is billed as non-fiction, and there are other sources which support the idea of secret repositories of knowledge—also, the existence of caves full of high-tech equipment. Today, we do not make giant pyramids like the Great Pyramid of Egypt. Instead, we make giant underground military bases, such as Area 51 in Nevada. We even hollow out entire mountains! The NORAD defense command at Cheyenne Mountain in Colorado Springs is a hollow mountain with an entire city inside it. The normal citizen

walking inside Cheyenne Mountain would be absolutely astonished at the level of the technology within the facility. These high-tech bases are housed underground to protect them in the event of catastrophe, including nuclear war. Sound similar to the Cave of the Ancients?

Other hollow mountains were said to have existed in ancient times as well. Mount Shasta in northern California is one such mountain. Mount Shasta is sometimes said to have a "Lemurian" city inside of it. Strange lights are said to be seen from time to time on the mountain, and UFOs are reported to appear there.

The great Chinese philosopher, Lao Tzu, often talked of the "Ancient Ones" in his writings, much as Confucius did. They were wise and knowledgeable, human beings that were as gods—powerful, good, loving, and all-knowing. These ancients apparently lived in a secret, remote area of China or Tibet, guarding the wisdom of the ages.

Born around 604 BC, Lao Tzu wrote the book which is still perhaps the most famous Chinese classic of all time, the *Tao Te Ching*. When he finally left China, at the close of his very long life, he journeyed to the west, to the legendary land of Hsi Wang Mu, to seek the headquarters of the Ancient Ones, the Great White Brotherhood. It was as he was leaving, at one of the border posts of western China, that a guard persuaded him to write down the *Tao Te Ching* , so that Lao Tzu's wisdom would not be lost.

The *Ancient Masters* were subtle,
mysterious, profound, responsive.
The depth of their knowledge is unfathomable.
Because it is unfathomable,
all we can do is to describe their appearance.
Watchful, like men crossing a winter stream.
Alert, like men aware of danger.
Courteous, like visiting guests.
Yielding, like ice about to melt.
Simple, like uncarved blocks of wood.

—Lao Tzu,
Tao Te Ching (Chapter 15)

No one ever heard of Lao Tzu again, and it is presumed that he made it to the Land of Hsi Wang Mu. Hsi Wang Mu is another name for the popular Chinese Goddess Kuan Yin, the "Merciful Guardian" and "Queen Mother of the West." Her land, traditionally located in the Kun Lun mountains, was known as the "Abode of the Immortals" and "The Western Paradise."

In *Myths and Legends of China*,[78] a collection published in 1922, Hsi Wang

Mu is connected to a lost continent: "Hsi Wang Mu was formed of the pure quintessence of the Western Air, in the legendary continent of Shen Chou. ...As Mu Kung, formed of the Eastern Air, is the active principle of the male air and sovereign of the Eastern Air, so Hsi Wang Mu, born of the Western Air, is the passive or female principle (yin) and sovereign of the Western Air. These two principles, co-operating, engender heaven and earth and all the beings of the universe, and thus become the two principles of life and of the subsistence of all that exists. She is the head of the troop of genii dwelling on the K'un-lun Mountains (the Taoist equivalent of the Buddhist Sumeru), and from time to time holds intercourse with favored imperial votaries.

"Hsi Wang Mu's palace is situated in the high mountains of the snowy K'un-lun. It is 100 *li* (about 333 miles) in circuit; a rampart of massive gold surrounds its battlements of precious stones. Its right wing rises on the edge of the Kingfishers' River. It is the usual abode of the *Immortals*, who are divided into seven special categories according to the color of their garments—red, blue, black, violet, yellow, green, and 'nature color.' There is a marvelous fountain built of precious stones, where the periodical banquet of the Immortals is held. This feast is called P'an-t'ao Hui, 'the feast of the Peaches.' It takes place on the borders of Yao Ch'ih, Lake of Gems, and is attended by both male and female immortals."[78]

Over the years of Chinese history, many expeditions were sent out to the Kun Lun mountains, the "Mount Olympus" of ancient China, in an attempt to contact the Ancient Ones.

In the Chin Dynasty (265-420 AD) the Emperor Wu-ti ordered the scholar Hsu to re-edit the "bamboo books" found in the tomb of an ancient king named Ling-Wang, the son of Hui-che'ng-wang, ruler of Wei State, circa 245 BC. The books recorded the travels of the Chou-Dynasty emperor "Mu" (1001-946 BC) who journeyed to the Kun Lun mountains to "pay a visit the Royal Mother of the West." The emperor met with Hsi Wang Mu on the auspicious day *chia-tzu.*"The ancient Chinese counted days and years in a special, cyclical fashion, similar to the ancient Mayans of Central America. There are ten characters known as the ten stems of heaven and another twelve characters known as the twelve branches of earth. The combinations of these two sets of characters give names to the sixty years of of the Chinese cycle. They named and counted days in the same way.

Emperor Mu had an audience with Hsi Wang Mu on the bank of Jasper Lake in the Kun Lun range. She blessed him and sang for him, and the emperor promised to return in three years after bringing peace and prosperity to his millions of subjects. He then had rocks engraved as a record of his visit and departed eastward across the desert back to his kingdom.[146]

Not all were so lucky as to meet the goddess, however. While travelling just north of the Kun Lun mountains, in Sinkiang, the famous Russian artist, explorer and mystic Nicholas Roerich first heard of the Valley of the Immortals, located just over the mountains. "Behind that mountain live holy men who are saving humanity through wisdom; many tried to see them but failed—somehow as soon as they go over the ridge, they lose their way," he was told. A native guide told him of huge vaults inside the mountains where treasures had been stored from the beginning of history. He also indicated that tall white people had been disappearing into those rock galleries.[102]

Nicholas Roerich was at one time in the possession of a fragment of "a magical stone from another world," called in Sanskrit the Chintamani Stone. Alleged to come from the star system of Sirius, ancient Asian chronicles claim that a divine messenger from the heavens gave a fragment of the stone to Emperor Tazlavoo of Atlantis.[102] According to legend the stone was later sent to King Solomon in Jerusalem (who, you will remember, was said to have travelled all over Asia and Africa in a vimana airship). He split the stone and made a ring out of one piece.

The stone is believed by some people to be moldavite, a magnetic stone sold in crystal shops, said to have fallen to earth in a meteor shower 14.8 million years ago. Moldavite is said to be a spiritual accelerator and has achieved a certain popularity in recent years. It is entirely possible that the Chintamani stone is a special piece of moldavite. It is worth noting here, too, that the sacred black stone kept in the Kabbah of Mecca to which all Muslims pray, is also a piece of meteorite.

Nicholas Roerich

Nicholas Roerich himself saw what was possibly a vimana from the land of Hsi Wang Mu in the Kun Lun. In his travel diary of August 5th, 1926 while in the Kukunor district, he noted that their caravan saw "something big and shiny reflecting the sun, like a huge oval moving at great speed. Crossing our camp this thing changed in this direction from south to southwest. And we saw how it disappeared in the intense blue sky. We even had time to take our field glasses and saw quite distinctly an oval form with a shiny surface, one side of which was brilliant from the sun."[102]

The strong similarity between the legends of Shambala and the secret land of Hsi Wang Mu is easily noticed. Shambala, reputedly located in Tibet, is famous as a repository of ancient wisdom, sheltered from the ages in a secluded valley. An ancient library was also said to be kept underground in Tibet. In some traditions, the library is said to be near Lhasa, possibly connected to the underground tunnels beneath the Potala, the Dalai Lama's fabulous skyscraper.

The stories of hidden archives and centers of learning are too universal and widespread to be easily dismissed. Incredible at it may seem, there may be a repository of ancient Chinese knowledge in the Kun Lun range of northwestern Tibet. Perhaps a hollow mountain filled with relics of ancient technology?

Technology is War Driven

It is my belief that advanced technology was developed over 12,000 years ago. This technology was used in some civilizations around the world, though not all. Just as Stone Age tribes still live in the highlands of New Guinea and other places today, many people were still primitive in those days as well. We call these ancient high-tech civilizations Atlantis, Rama, Osiris, and other names. Atlantis, I believe, is beneath the mid-Atlantic in the vicinity of the Azores and the Bahamas. Although it was a small continent, its influence reached across the waters to the Americas and to what are now Britain, Ireland and the Mediterranean area. This Atlantic island civilization was contemporary with other civilizations such as the Osirian civilization of the Mediterranean, Egypt and North Africa and the Rama Empire of India. In the far east, perhaps in Indonesia and Southeast Asia, was another advanced civilization strongly connected to ancient India and the Rama Empire.

Around 10000 BC, geological upheavals, perhaps both natural and man-made, sank Atlantis and affected the whole world, especially Europe and the Americas. The Mediterranean was apparently flooded at this time, creating the various islands and unique megalithic cultures around this inland sea. Much of the ancient technology was lost to humankind.

A thousand years after the destruction of Atlantis and upheavals in other empires, the Hittites and Egyptians began to explore the newly created Mediterranean Sea area and the Atlantic. In the Americas, early groups such as the Tiahuanaco culture and the Mayans began to recreate their civilizations. Seafarers of the legendary Atlantean League began to cross the Atlantic again by about 6000 BC. These same Mediterraneans colonized areas of northern Europe, including the British Isles as far north as the Shetlands (Set-lands).

North Sea earthquakes finished off the coastal civilization that inhabited much of the Netherlands, Denmark and Sweden. This civilization was much later than Atlantis, probably reaching its height about 1500 BC. At about this time or shortly thereafter, the Sea Peoples with their horned helmets came from Denmark, England, Holland, Germany and France to the Mediterranean to invade Greece, Egypt and the Hittite empire.

Like today, powerful nations fought wars that spanned entire continents. Secret societies like the Knights Templar turned ancient Phoenician ports into their own strongholds. There is an old saying, "what goes around, comes around." Mankind's love for warfare has both fueled his technology and created much destruction and woe. Great teachers incarnate from time to time to try to teach man to love his neighbor and to live in a peaceful, helpful manner with others.

Yet our history is one of nonstop warfare and invasion. Technology is, in many ways, driven by warfare. Man slaughters man, and the gods look down in pain and sorrow at what we have created for ourselves. Plato and the Egyptian priests have given us the story of one ancient civilization that waged war on the rest of the world with disastrous results.

Today's looming wars have their roots in history: the creation of the Christian church, the creation of the Islamic empire, the creation of the Jewish refugee state of Israel, the conflict of ancient foes over energy sources and land control. The current war being fought by the Russians in Chechnya is also one of religious conflict combined with the desire to control oil wealth from the Caspian Sea.

Now that technology has again reached a point of no return, perhaps we are ready to jump to the next level. A level beyond the technology of today. The technology of the gods tomorrow. A technology that allows man to finally learn to live in harmony with nature and his fellow man.

An old print of the standing stones of Carnac, Brittany.

Bibliography & Footnotes

1. *Technology in the Ancient World,* Henry Hodges, 1970, Marboro Books, London.
2. *Engineering in the Ancient World* J. Landels, 1978, University of California Press, Berkeley.
3. *Arthur C. Clarke's Mysterious World,* Simon Welfare & John Fairley, 1980, Wm. Collins & Sons, London.
4. *The Ancient Greek Computer from Rhodes,* Victor J. Kean, 1991, Estathiadis Group, Athens.
5. *Ancient Man: A Handbook of Puzzling Artifacts,* William Corliss, 1978, The Sourcebook Project, Glen Arm, MD.
6. *The Traveler's Key to Ancient Egypt,* John Anthony West, 1985, Alfred Knopf, NYC.
7. *Egyptian Myth and Legend,* Donald Mackenzie, 1907, Bell Publishers, NYC.
8. *Investigating the Unexplained,* Ivan T. Sanderson, 1972, Prentice Hall, Englewood Cliffs, NJ.
9. *The World's Last Mysteries,* Nigel Blundell, 1980, Octopus Books, London.
10. *Vimana Aircraft of Ancient India & Atlantis,* Childress, Sanderson, Josyer, 1991, Adventures Unlimited Press, Kempton, Illinois.
11. *Lemurian Fellowship,* Lesson Material, 1936, Ramona, California.
12. *The Ultimate Frontier,* Eklal Kueshana, 1962, The Stelle Group, Stelle, IL.
13. *Strange Artifacts,* William Corliss, 1974, The Sourcebook Project, Glen Arm, MD.
14. *The World's Last Mysteries,* Reader's Digest, 1976, Reader's Digest Association, Inc., Pleasantville, New York.
15. *Timeless Earth,* Peter Kolosimo, 1974, University Press Seacaucus, NJ.
16. *Mysteries of Time & Space,* Brad Steiger, 1974, Prentice Hall, Englewood Cliffs, NJ.
17. *Strangest of All,* Frank Edwards, 1956, Ace Books, New York.
18. *Legends of the Lost,* Peter Brookesmith, ed, 1984, Orbis Publishing, London.
19. *Stranger Than Science,* Frank Edwards, 1959, Bantam Books, NYC.
20. *A Dweller on Two Planets,* Frederick Spencer Oliver, 1884, Borden Publishing, Alhambra, California.
21. *The Ancient Secret: Fire from the Sun,* Flavia Anderson, 1953, R.I.L.K.O.

Books, Orpington, Kent, England.
22. *YHWH,* Jerry Ziegler, 1985, Star Publishers, Morton, Illinois.
23. *Lost Cities of North & Central America,* D.H. Childress, 1992, Adventures Unlimited Press, Kempton, IL.
24. *We Are Not the First,* Andrew Tomas, 1971, Souvenir Press, London.
25. *The Curse of the Pharaohs,* Philipp Vandenberg, 1975, J.B. Lippincott Co., Philadelphia, PA.
26. *Ancient Astronauts: A Time Reversal?,* Robin Collyns, 1976, Sphere Books, London.
27. *The Ancient Engineers,* L. Sprague de Camp, 1960, Ballentine Books, New York.
28. *War in Ancient India,* V. R. Dikshitar, 1944, Oxford University Press (1987 edition published by Motilal Banarsidass, Delhi).
29. *The Bible As History,* Werner Keller, 1965, Hodder & Stoughton, London.
30. *Footprints on the Sands of Time,* L.M. Lewis, 1975, Signet Books, New York
31. *Lost Worlds,* Alistair Service, 1981, Arco Publishing, New York.
32. *Secrets of the Lost Races,* Rene Noorbergen, 1977, Barnes & Noble Publishers, NYC.
33. *The Vymanika Shastra,* Maharishi Bharadwaaja, translated and published 1979 by G.R. Josyer, Mysore, India.
34. *Weird America,* Jim Brandon, 1978, E.P. Dutton, New York.
35. *Cataclysms of the Earth,* Hugh A. Brown, 1967, Twayne Pubs., NYC.
36. *The Path of the Pole,* Charles Hapgood, 1970, Adventures Unlimited, Kempton, IL.
37. *Geological Anomalies,* William Corliss, 1974, The Sourcebook Project, Glen Arm, MD.
38. *Strange Artifacts,* William Corliss, 1974, The Sourcebook Project, Glen Arm, MD.
39. *The World's Last Mysteries,* Reader's Digest, 1976, Reader's Digest Association, Inc., Pleasantville, New York.
40. *Laserbeams From Star Cities,* Robyn Collins, 1971, Sphere Books, London.
41. *Arthur C. Clarke's Mysterious World,* Simon Welfare & John Fairley, 1980, Wm. Collins & Sons, London.
42. *Lost Cities of China, Central Asia & India,* David Hatcher Childress, 1991, Adventures Unlimited Press, Stelle, Illinois.
43. *2000 Years of Space Travel,* Russell Freedman, 1963, William Collins & Sons, London.
44. *Mysteries of Forgotten Worlds,* Charles Berlitz, 1972, Doubleday, NYC..

45. *Riddles of Ancient History*, A. Gorbovsky, 1966, Soviet Publishers, Moscow.
46. *Mysterious Britain*, Janet & Colin Bord, 1972, Granada Publishing, London.
47. *The Mysterious Past*, Robert Charroux,1973, Robert Laffont, NYC.
48. *Living Wonders*, John Mitchell & Robert Rickard, 1982, Thames & Hudson, NYC.
49. *Enigmas*, Rupert Gould,1945, University Books, NYC.
50. *Lost Outpost of Atlantis*, Richard Wingate, 1980, Everest House, NY.
51. *Strange World*, Frank Edwards, 1964, Bantam Books, NYC.
52. *Stranger Than Science*, Frank Edwards, 1959, Bantam Books, NYC.
53. *Strangest of All*, Frank Edwards, 1956, Ace Books, NYC.
54. *Technology In the Ancient World*, Henry Hodges, 1970, Marboro Books, London.
55. *Unearthing Atlantis*, Charles Pellegrino, 1991, Random House, NYC.
56. *Along Civilization's Trail*, Ralph M. Lewis, 1940, AMORC, San José, CA.
57. *Lost Cities & Ancient Mysteries of South America*, David Hatcher Childress, 1987, AUP, Stelle, Illinois.
58. *Lost Cities & Ancient Mysteries of Africa & Arabia*, David Hatcher Childress, 1990, AUP, Stelle, Illinois.
59. *Lost Cities of Ancient Lemuria & the Pacific*, David Hatcher Childress, 1988, AUP, Stelle, Illinois.
60. *The Chronicle of Akakor*, Karl Brugger, 1977, Delacorte Press, NYC.
61. *Atlantis, The Lost Continent Revealed*, Charles Berlitz, 1984, Macmillan,London.
62. *Timeless Earth*, Peter Kolosimo, 1974, University Press Seacaucus, NJ.
63. *Extraterrestrial Intervention: The Evidence*, Jacques Bergier, 1974, Henry Regnery, Chicago.
64. *Arthur C. Clarke's Mysterious World*, Simon Welfare & John Fairley, 1980, Wm. Collins & Sons, London.
65. *Mysterious Britain*, Janet & Colin Bord, 1972, Granada Publishing, London.
66. *Riddles of Ancient History*, A. Gorbovsky, 1966, Soviet Publishers, Moscow.
67. *Megaliths and Masterminds*, Peter Lancaster Brown, 1979, Charles Scribner's Sons, New York.
68. *The God-Kings & the Titans*, James Bailey, 1973, St. Martin's Press, NYC.
69. *The Bermuda Triangle*, Charles Berlitz, 1974, Doubleday, NYC.
70. *Lost Worlds*, Robert Charroux, 1973, Collins, Glasgow, Great Britain.
71. *Chariots of the Gods*, Erich Von Daniken, 1969, Putnam, NYC.
72. *The History of Atlantis*, Lewis Spence, 1926, London, reprinted 1995, Adventures Unlimited Press, Kempton, Illinois.

73. *The View Over Atlantis,* John Michell, 1969, Ballantine Books, NYC.
74. *The Occult Sciences In Atlantis,* Lewis Spence, 1943, Rider & Co., London.
75. *Lost Atlantis,* James Bramwell, 1938, Harper & Brothers, New York.
76. *Lost Cities of China, Central Asia & India,* David Hatcher Childress, 1991, Adventures Unlimited Press, Stelle, Illinois.
77. *Megalithomania,* John Michell, 1982, Thames & Hudson, London.
78. *Shambala, Oasis of Light,* Andrew Tomas, 1978, Souvenir Press, London..
79. *Ice, The Ultimate Disaster,* R. Noone, 1982, Crown Publishers, NYC.
80. *Atlantis In Andalucia,* E.M. Whishaw, 1928, Rider, London. Reprinted as *Atlantis In Spain* by Adventures Unlimited Press, Kempton, Illinois.
81. *Atlantis & the Giants,* Denis Saurat, 1957, Faber & Faber, London.
82. *The Problem of Atlantis,* Lewis Spence, 1924, Rider & Co., London.
83. *Edgar Cayce on Atlantis,* Edgar Evans Cayce, 1968, Warner Books, NYC.
84. *Atlantis: From Legend to Discovery,* Andrew Tomas, 1972, Robert Laffont, Paris. (Sphere Books, 1973, London).
85. *The Shadow of Atlantis,* Colonel A. Braghine, 1940, London reprinted 1996, Adventures Unlimited Press, Kempton, Illinois.
86. *Mysteries of Ancient South America,* Harold Wilkins, 1946, Citadel Press, NYC.
87. *Secret Cities of Old South America,* Harold Wilkins, 1952, London, reprinted 1998, Adventures Unlimited Press, Kempton, Illinois.
88. *Exploration Fawcett (Lost Trails, Lost Cities),* Brian Fawcett, 1953, Hutchinson & Co., London.
89. *Indra Girt by Maruts,* Jerry Ziegler, 1994, Next Millennium Publishers, Stamford, CT.
90. *KA: A Handbook of Mythology, Sacred Practices, Electrical Phenomena, and their Linguistic Connections in the Ancient World,* H. Crosthwaite, 1992, Metron Publlications, Princeton, NJ.
91. *100 Tons of Gold,* David Chandler, 1978, Doubleday, Garden City, NY.
92. *The Birth & Death of the Sun,* George Gamow, 1940, Viking Press, NYC.
93. *Danger My Ally,* F.A. Mitchel-Hedges, 1954, Elek Books, London.
94. *The Crystal Skull,* Richard Garviin, 1973, Doubleday, NY.
95. *Stonehenge Decoded,* Gerald Hawkins, 1965, Doubleday, NY.
96. *The Sphinx and the Megaliths,* John Ivimy, 1974, Sphere Books, London.
97. *The Carbon-14 Dating of Iron,* Nikolass van der Merwe, 1969, University of Chicago Press, Chicago and London.
98. *The Giza Power Plant: Technologies of Ancient Egypt,* Christopher Dunn, 1998, Bear & Company, Sante Fe, New Mexico.
99. *The Genius of China: 3,000 Years of Science, Discovery and Invention,* Robert Temple,1987, Simon & Schuster, New York.
100. *Warfare In Ancient India,* Ramachandra Dikshitar, 1944, University of

Madras/Oxford University.
101. *The Queen of Sheba and Her Only Son Menyelek (Kebra Nagast)*, translated by Sir E.A. Wallis Budge, 1932, Dover, London.
102. *Shambala*, Nicholas Roerich, 1930, Roerich Museum, New York.
103. *The Five Sons of King Pandu, The Story of the Mahabharata*, retold by Elizabeth Seeger, 1970, Dent & Sons, London.
104. *With Mystics and Magicians In Tibet*, Alexandra David-Neel, 1931, Dover, New York.
105. *Merury: UFO Messenger of the Gods*, William Clendenon, 1990, Adventures Unlimited Press, Kempton, IL.
106. *Anti-Gravity & the Unified Field*, Edited by David Hatcher Childress, 1990, Adventures Unlimited Press, Kempton, Illinois.
107. *The Magic of Obelisks*, Peter Tompkins, 1981, Harper & Row, NYC.
108. *The Serpent in the Sky*, John Anthony West, 1984. Harper & Row, NYC.
109. *Migdar–The Secret of the Sphinx*, F.L. Oscott, Neville Spearman, Suffolk, UK
110. *The Great Pyramid*, Piazzi Smyth, 1880, Bell Publishing Co. NYC.
111. *The Riddle of the Pyramids*, Kurt Mendelssohn, Thames & Hudson, London.
112. *A Traveller's Key to Ancient Egypt*, John Anthony West, 1988,. Quest Books, Wheaton, Illinois.
113. *Forbidden Archeology*, Michael Cremo and Richard Thompson, 1993, Bhativedanta Book Trust, Los Angeles, CA.
114. *The Pyramids: An Enigma Solved*, Joseph Davidovits & Margie Morris, 1988, Hippocrene Books, New York.
115. *Secrets of the Great Pyramid*, Peter Tompkins, 1971, Harper & Row, NYC.
116 *Egypt Before the Pharaohs*, Michael A. Hoffman, 1979, Alfred Knopf, NYC.
117. *The Sphinx and the Megaliths*, John Ivimy, 1974, Sphere Books, London.
118. *The Great Pyramid: Man's Monument to Man*, Tom Valentine, 1975, Pinnacle, NYC.
119. *The Sirius Mystery*, Robert Temple, 1976, Harper & Row, NYC.
120. *Edgar Cayce on Atlantis*, Edgar Evans Cayce, 1968, Warner Books, NYC.
121. *The Stones Of Atlantis*, David Zink, 1978, Prentice-Hall, Englewood Cliffs, NJ.
122. *The Cave of the Ancients*, T. Lobsang Rampa, 1963, Ballantine Books, NYC.
123. *Extraterrestrial Intervention: The Evidence*, Jacques Bergier, 1974, Henry Regnery, Chicago.
124. *Mysteries of Ancient South America*, Harold Wilkins, 1946, Citadel Press, NYC.

125. *Secret Cities of Old South America,* Harold Wilkins, 1952, Library Publications, Inc., NYC.
126. *Exploration Fawcett (Lost Trails, Lost Cities),* Brian Fawcett, 1953, Hutchinson & Co., London.
127. *The Bridge To Infinity,* Bruce Cathie, 1989, Adventures Unlimited Press, Kempton, IL.
128. *The Lost Realms,* Zechariah Sitchin, 1990, Avon Books, NYC.
129. *Nazca: Journey to the Sun,* Jim Woodman, 1977, Simon & Schuster, New York.
130. *Cleanliness and Godliness,* Reginald Reynolds, 1946, Doubleday, Garden City, New York.
131. *Arthur C. Clarke's Mysterious World,* Simon Welfare & John Fairley, 1980, Wm. Collins & Sons, London.
132. *The Rediscovery of Lost America,* Arlington Mallery & Mary Harrison, 1951,1979, E.P. Dutton, NY.
133. *The Ancient Greek Computer From Rhodes,* Victor J. Kean, 1991, Estathiadis Group, Athens.
134. *Mysterious Britain,* Janet & Colin Bord, 1972, Granada Publishing, London.
135. *Wonders of Ancient Chinese Science,* Robert Silverburg, 1972, Ballentine Books, NY.
136. *On the Trail of the Sun Gods,* Marcel Homet, 1965, Neville Spearman, London.
137. *Preliminary Catalogue of the Comalcalco Bricks,* Neil Steede, 1984, Centro de Investigacion Precolombina, Tabasco, Mexico.
138. *The Sirius Mystery,* Robert Temple, 1976, Harper & Row, NYC.
139. *Men Who Dared the Sea,* Gardner Soule, 1976, Thomas Crowell Co., NYC.
140. *Michigan Prehistory Mysteries,* Betty Sodders, 1990, Avery Studios, Au Clair, Michigan.
141. *Michigan Prehistory Mysteries II,* Betty Sodders, 1991, Avery Studios, Au Clair, Michigan.
142. *The Unexplained,* William Corliss, 1976, Bantam Books, New York.
143. *Riddles in History,* Cyrus H. Gordon, 1974, Crown Publishers, NYC.
144. *The Maldive Mystery,* Thor Heyerdahl, 1986, Adler-Adler, Bethesda, MD.
145. *The Incas,* Garcilaso de la Vega, 1961 (first published in 1608), Orion Press, NY.
146. *Daily Life In Carthage,* Gilbert Charles-Picard, 1961, Macmillan Co. NY.
147. *The Traveler's Key to Ancient Egypt,* John Anthony West, 1985, Alfred Knopf, NYC.
148. *Egyptian Myth & Legend,* Donald Mackenzie, 1907, Bell Publishers,

NYC.
149. *The Phoenicians,* Donald Harden, 1962, Praeger Publishers, New York.
150. *Cities In the Sea,* Nicholas C. Flemming, 1971, Doubleday, New York.
151. *The Alexandria Project,* Stephen Schwartz, 1983, Dell Books, New York.
152. *Alexandria: A History & a Guide,* E.M. Forster, 1922, Morris, Alexandria.
153. *The Phoenicians,* Gerhard Herm, 1975, William Morrow & Co. New York.
154. *Baalbek,* Friedrich Ragette, 1980, Chatto & Windus, London.
155. *Mystery Religions In the Ancient World,* Joscelyn Godwin, 1981, Thames & Hudson, London.
156. *The Sea Peoples,* N.K. Sandars, 1978, Thames & Hudson, London.
157. *Atlantis Illustrated,* H.R. Stahel, 1982, Grosset & Dunlap, New York.
158. *The World of Megaliths,* Jean-Pierre Mohen, 1989, Facts on File, NYC.
159. *The Standing Stones of Europe,* Alastair Service & Jean Bradbery,1979, Orion Publishing, London.
160. *Secrets of the Great Pyramid,* Peter Thompkins, 1971, Harper & Row, New York.
161. *Fingerprints Of the Gods,* Graham Hancock, 1995, Crown Publishers, New York.
162. *The Sign and the Seal,* Graham Hancock, 1992, Crown Publishers, New York.
164. *The Great Pyramid Speaks,* Joseph B. Gill, 1984, Barnes & Noble Books, New York.

Get these fascinating books from your nearest bookstore or directly from: Adventures Unlimited Press

www.adventuresunlimitedpress.com

VIMANA:
Flying Machines of the Ancients
by David Hatcher Childress

According to early Sanskrit texts the ancients had several types of airships called vimanas. Like aircraft of today, vimanas were used to fly through the air from city to city; to conduct aerial surveys of uncharted lands; and as delivery vehicles for awesome weapons. David Hatcher Childress, popular *Lost Cities* author and star of the History Channel's long-running show Ancient Aliens, takes us on an astounding investigation into tales of ancient flying machines. In his new book, packed with photos and diagrams, he consults ancient texts and modern stories and presents astonishing evidence that aircraft, similar to the ones we use today, were used thousands of years ago in India, Sumeria, China and other countries. Includes a 24-page color section.

408 Pages. 6x9 Paperback. Illustrated. $22.95. Code: VMA

THE ENIGMA OF CRANIAL DEFORMATION
Elongated Skulls of the Ancients
By David Hatcher Childress and Brien Foerster

In a book filled with over a hundred astonishing photos and a color photo section, Childress and Foerster take us to Peru, Bolivia, Egypt, Malta, China, Mexico and other places in search of strange elongated skulls and other cranial deformation. The puzzle of why diverse ancient people—even on remote Pacific Islands—would use head-binding to create elongated heads is mystifying. Where did they even get this idea? Did some people naturally look this way—with long narrow heads? Were they some alien race? Were they an elite race that roamed the entire planet? Why do anthropologists rarely talk about cranial deformation and know so little about it? Color Section.

250 Pages. 6x9 Paperback. Illustrated. $19.95. Code: ECD

ARK OF GOD
The Incredible Power of the Ark of the Covenant
By David Hatcher Childress

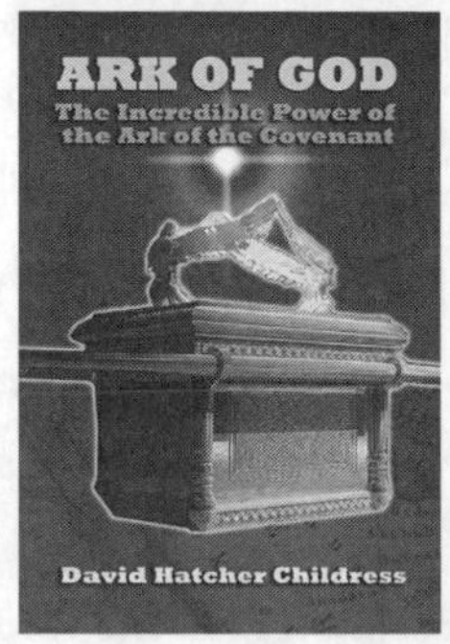

Childress takes us on an incredible journey in search of the truth about (and science behind) the fantastic biblical artifact known as the Ark of the Covenant. This object made by Moses at Mount Sinai—part wooden-metal box and part golden statue—had the power to create "lightning" to kill people, and also to fly and lead people through the wilderness. The Ark of the Covenant suddenly disappears from the Bible record and what happened to it is not mentioned. Was it hidden in the underground passages of King Solomon's temple and later discovered by the Knights Templar? Was it taken through Egypt to Ethiopia as many Coptic Christians believe? Childress looks into hidden history, astonishing ancient technology, and a 3,000-year-old mystery that continues to fascinate millions of people today. Color section.

420 Pages. 6x9 Paperback. Illustrated. $22.00 Code: AOG

THE LOST WORLD OF CHAM
The Trans-Pacific Voyages of the Champa
By David Hatcher Childress

The mysterious Cham, or Champa, peoples of Southeast Asia formed a megalith-building, seagoing empire that extended into Indonesia, Tonga, and beyond—a transoceanic power that reached Mexico and South America. The Champa maintained many ports in what is today Vietnam, Cambodia, and Indonesia and their ships plied the Indian Ocean and the Pacific, bringing Chinese, African and Indian traders to far off lands, including Olmec ports on the Pacific Coast of Central America. Topics include: Cham and Khem: Egyptian Influence on Cham; The Search for Metals; The Basalt City of Nan Madol; Elephants and Buddhists in North America; The Cham and Lake Titicaca; Easter Island and the Cham; the Magical Technology of the Cham; tons more. 24-page color section.
328 Pages. 6x9 Paperback. Illustrated. $22.00 Code: LPWC

ADVENTURES OF A HASHISH SMUGGLER
by Henri de Monfreid

Nobleman, writer, adventurer and inspiration for the swashbuckling gun runner in the *Adventures of Tintin*, Henri de Monfreid lived by his own account "a rich, restless, magnificent life" as one of the great travelers of his or any age. The son of a French artist who knew Paul Gaugin as a child, de Monfreid sought his fortune by becoming a collector and merchant of the fabled Persian Gulf pearls. He was then drawn into the shadowy world of arms trading, slavery, smuggling and drugs. Infamous as well as famous, his name is inextricably linked to the Red Sea and the raffish ports between Suez and Aden in the early years of the twentieth century. De Monfreid (1879 to 1974) had a long life of many adventures around the Horn of Africa where he dodged pirates as well as the authorities.
284 Pages. 6x9 Paperback. $16.95. Illustrated. Code AHS

NORTH CAUCASUS DOLMENS
In Search of Wonders
By Boris Loza, Ph.D.

Join Boris Loza as he travels to his ancestral homeland to uncover and explore dolmens firsthand. Throughout this journey, you will discover the often hidden, and surprisingly forbidden, perspective about the mysterious dolmens: their ancient powers of fertility, healing and spiritual connection. Chapters include: Ancient Mystic Megaliths; Who Built the Dolmens?; Why the Dolmens were Built; Asian Connection; Indian Connection; Greek Connection; Olmec and Maya Connection; Sun Worshippers; Dolmens and Archeoastronomy; Location of Dolmen Quarries; Hidden Power of Dolmens; and much more! Tons of Illustrations! A fascinating book of little-seen megaliths. Color section.
252 Pages. 5x9 Paperback. Illustrated. $24.00. Code NCD

GIANTS: MEN OF RENOWN
By Denver Michaels

Michaels runs down the many stories of giants around the world and testifies to the reality of their existence in the past. Chapters and subchapters on: Giants in the Bible; Texts; Tales from the Maya; Stories from the South Pacific; Giants of Ancient America; The Stonish Giants; Mescalero Tales; The Nahullo; Mastodons, Mammoths & Mound Builders; Pawnee Giants; The Si-Te-Cah; Tsul 'Kalu; The Titans & Olympians; The Hyperboreans; European Myths; The Giants of Britain & Ireland; Norse Giants; Myths from the Indian Subcontinent; Daityas, Rakshasas, & More; Jainism: Giants & Inconceivable Lifespans; The Conquistadors Meet the Sons of Anak; Cliff-Dwelling Giants; The Giants of the Channel Islands; Strange Tablets & Other Artifacts; more. Tons of illustrations with an 8-page color section.
320 Pages. 6x9 Paperback. Illustrated. $22.00. Code: GMOR

ANTARCTICA AND THE SECRET SPACE PROGRAM
By David Hatcher Childress

David Childress, popular author and star of the History Channel's show *Ancient Aliens*, brings us the incredible tale of Nazi submarines and secret weapons in Antarctica and elsewhere. He then examines Operation High-Jump with Admiral Richard Byrd in 1947 and the battle that he apparently had in Antarctica with flying saucers. Through "Operation Paperclip," the Nazis infiltrated aerospace companies, banking, media, and the US government, including NASA and the CIA after WWII. Does the US Navy have a secret space program that includes huge ships and hundreds of astronauts?

392 Pages. 6x9 Paperback. Illustrated. $22.00 Code: ASSP

HAUNEBU: THE SECRET FILES
The Greatest UFO Secret of All Time
By David Hatcher Childress

Childress brings us the incredible tale of the German flying disk known as the Haunebu. Although rumors of German flying disks have been around since the late years of WWII it was not until 1989 when a German researcher named Ralf Ettl living in London received an anonymous packet of photographs and documents concerning the planning and development of at least three types of unusual craft. Chapters include: A Saucer Full of Secrets; WWII as an Oil War; A Saucer Called Vril; Secret Cities of the Black Sun; The Strange World of Miguel Serrano; Set the Controls for the Heart of the Sun; Dark Side of the Moon: more. Includes a 16-page color section. Over 120 photographs and diagrams.

352 Pages. 6x9 Paperback. Illustrated. $22.00 Code: HBU

ANDROMEDA: THE SECRET FILES
The Flying Submarines of the SS
By David Hatcher Childress

Childress brings us the amazing story of the Andromeda craft, designed and built during WWII. Along with flying discs, the Germans were making long, cylindrical airships that are commonly called motherships—large craft that house several smaller disc craft. Chapters include: Gravity's Rainbow; The Motherships; The MJ-12, UFOs and the Korean War; The Strange Case of Reinhold Schmidt; Secret Cities of the Winged Serpent; The Green Fireballs; Submarines That Can Fly; more. Includes a 16-page color section.

382 Pages. 6x9 Paperback. Illustrated. $22.00 Code: ASF

VRIL: SECRETS OF THE BLACK SUN
By David Hatcher Childress

Childress unveils the amazing story of the Vril, Haunebu and Andromeda. This volume closes with how the SS operates today in the Ukraine and how the Wagner second in command, Dimitry Utkin, killed in the fiery crash of Yevgeny Prigozhin's private jet between Moscow and St. Petersburg in August of 2023, had SS tattoos on his shoulders and often signed his name with the SS runes. Chapters include: Secrets of the Black Sun; The Extra-Territorial Reich; The Rise of the SS; The Marconi Connection; Yellow Submarine; Ukraine and the Battalion of the Black Sun; more. Includes an 8-page color section. Over 120 photographs and diagrams.

382 Pages. 6x9 Paperback. Illustrated. $22.00 Code: VSBS

OBELISKS: TOWERS OF POWER
The Mysterious Purpose of Obelisks
By David Hatcher Childress

Some obelisks weigh over 500 tons and are massive blocks of polished granite that would be extremely difficult to quarry and erect even with modern equipment. Why did ancient civilizations in Egypt, Ethiopia and elsewhere undertake the massive enterprise it would have been to erect a single obelisk, much less dozens of them? Were they energy towers that could receive or transmit energy? With discussions on Tesla's wireless power, and the use of obelisks as gigantic acupuncture needles for earth, Chapters include: The Crystal Towers of Egypt; The Obelisks of Ethiopia; Obelisks in Europe and Asia; Mysterious Obelisks in the Americas; The Terrible Crystal Towers of Atlantis; Tesla's Wireless Power System; Obelisks on the Moon; more. 8-page color section.

336 Pages. 6x9 Paperback. Illustrated. $22.00 Code: OBK

THE ANTI-GRAVITY HANDBOOK
edited by David Hatcher Childress

The new expanded compilation of material on Anti-Gravity, Free Energy, Flying Saucer Propulsion, UFOs, Suppressed Technology, NASA Cover-ups and more. Highly illustrated with patents, technical illustrations and photos. This revised and expanded edition has more material, including photos of Area 51, Nevada, the government's secret testing facility. This classic on weird science is back in a new format!

230 PAGES. 7x10 PAPERBACK. ILLUSTRATED. $16.95. CODE: AGH

ANTI–GRAVITY & THE WORLD GRID

Is the earth surrounded by an intricate electromagnetic grid network offering free energy? This compilation of material on ley lines and world power points contains chapters on the geography, mathematics, and light harmonics of the earth grid. Learn the purpose of ley lines and ancient megalithic structures located on the grid. Discover how the grid made the Philadelphia Experiment possible. Explore the Coral Castle and many other mysteries, including acoustic levitation, Tesla Shields and scalar wave weaponry. Browse through the section on anti-gravity patents, and research resources.

274 PAGES. 7x10 PAPERBACK. ILLUSTRATED. $14.95. CODE: AGW

ANTI–GRAVITY & THE UNIFIED FIELD
edited by David Hatcher Childress

Is Einstein's Unified Field Theory the answer to all of our energy problems? Explored in this compilation of material is how gravity, electricity and magnetism manifest from a unified field around us. Why artificial gravity is possible; secrets of UFO propulsion; free energy; Nikola Tesla and anti-gravity airships of the 20s and 30s; flying saucers as superconducting whirls of plasma; anti-mass generators; vortex propulsion; suppressed technology; government cover-ups; gravitational pulse drive; spacecraft & more.

240 PAGES. 7x10 PAPERBACK. ILLUSTRATED. $14.95. CODE: AGU

THE TIME TRAVEL HANDBOOK
A Manual of Practical Teleportation & Time Travel
edited by David Hatcher Childress

The Time Travel Handbook takes the reader beyond the government experiments and deep into the uncharted territory of early time travellers such as Nikola Tesla and Guglielmo Marconi and their alleged time travel experiments, as well as the Wilson Brothers of EMI and their connection to the Philadelphia Experiment—the U.S. Navy's forays into invisibility, time travel, and teleportation. Childress looks into the claims of time travelling individuals, and investigates the unusual claim that the pyramids on Mars were built in the future and sent back in time. A highly visual, large format book, with patents, photos and schematics. Be the first on your block to build your own time travel device!

316 PAGES. 7x10 PAPERBACK. ILLUSTRATED. $16.95. CODE: TTH

ANCIENT ALIENS ON THE MOON
By Mike Bara

What did NASA find in their explorations of the solar system that they may have kept from the general public? How ancient really are these ruins on the Moon? Using official NASA and Russian photos of the Moon, Bara looks at vast cityscapes and domes in the Sinus Medii region as well as glass domes in the Crisium region. Bara also takes a detailed look at the mission of Apollo 17 and the case that this was a salvage mission, primarily concerned with investigating an opening into a massive hexagonal ruin near the landing site. Chapters include: The History of Lunar Anomalies; The Early 20th Century; Sinus Medii; To the Moon Alice!; Mare Crisium; Yes, Virginia, We Really Went to the Moon; Apollo 17; more. Tons of photos of the Moon examined for possible structures and other anomalies.

248 Pages. 6x9 Paperback. Illustrated.. $19.95. Code: AAOM

ANCIENT ALIENS ON MARS
By Mike Bara

Bara brings us this lavishly illustrated volume on alien structures on Mars. Was there once a vast, technologically advanced civilization on Mars, and did it leave evidence of its existence behind for humans to find eons later? Did these advanced extraterrestrial visitors vanish in a solar system wide cataclysm of their own making, only to make their way to Earth and start anew? Was Mars once as lush and green as the Earth, and teeming with life? Chapters include: War of the Worlds; The Mars Tidal Model; The Death of Mars; Cydonia and the Face on Mars; The Monuments of Mars; The Search for Life on Mars; The True Colors of Mars and The Pathfinder Sphinx; more. Color section.

252 Pages. 6x9 Paperback. Illustrated. $19.95. Code: AMAR

ANCIENT ALIENS ON MARS II
By Mike Bara

Using data acquired from sophisticated new scientific instruments like the Mars Odyssey THEMIS infrared imager, Bara shows that the region of Cydonia overlays a vast underground city full of enormous structures and devices that may still be operating. He peels back the layers of mystery to show images of tunnel systems, temples and ruins, and exposes the sophisticated NASA conspiracy designed to hide them. Bara also tackles the enigma of Mars' hollowed out moon Phobos, and exposes evidence that it is artificial. Long-held myths about Mars, including claims that it is protected by a sophisticated UFO defense system, are examined. Data from the Mars rovers Spirit, Opportunity and Curiosity are examined; everything from fossilized plants to mechanical debris is exposed in images taken directly from NASA's own archives.

294 Pages. 6x9 Paperback. Illustrated. $19.95. Code: AAM2

ANCIENT TECHNOLOGY IN PERU & BOLIVIA
By David Hatcher Childress

Childress speculates on the existence of a sunken city in Lake Titicaca and reveals new evidence that the Sumerians may have arrived in South America 4,000 years ago. He demonstrates that the use of "keystone cuts" with metal clamps poured into them to secure megalithic construction was an advanced technology used all over the world, from the Andes to Egypt, Greece and Southeast Asia. He maintains that only power tools could have made the intricate articulation and drill holes found in extremely hard granite and basalt blocks in Bolivia and Peru, and that the megalith builders had to have had advanced methods for moving and stacking gigantic blocks of stone, some weighing over 100 tons.

340 Pages. 6x9 Paperback. Illustrated.. $19.95 Code: ATP

BIGFOOT NATION
A History of Sasquatch In North America
By David Hatcher Childress

Childress takes a deep look at Bigfoot Nation—the real world of bigfoot around us in the United States and Canada. Whether real or imagined, that bigfoot has made his way into the American psyche cannot be denied. He appears in television commercials, movies, and on roadside billboards. Bigfoot is everywhere, with actors portraying him in variously believable performances and it has become the popular notion that bigfoot is both dangerous and horny. Indeed, bigfoot is out there stalking lovers' lanes and is even more lonely than those frightened teenagers that he sometimes interrupts. Bigfoot, tall and strong as he is, makes a poor leading man in the movies with his awkward personality and typically anti-social behavior. Includes 16-pages of color photos that document Bigfoot Nation!

320 Pages. 6x9 Paperback. Illustrated. $22.00. Code: BGN

MEN & GODS IN MONGOLIA
by Henning Haslund

Haslund takes us to the lost city of Karakota in the Gobi desert. We meet the Bodgo Gegen, a god-king in Mongolia similar to the Dalai Lama of Tibet. We meet Dambin Jansang, the dreaded warlord of the "Black Gobi." Haslund and companions journey across the Gobi desert by camel caravan; are kidnapped and held for ransom; withness initiation into Shamanic societies; meet reincarnated warlords; and experience the violent birth of "modern" Mongolia.

358 Pages. 6x9 Paperback. Illustrated. $18.95. Code: MGM

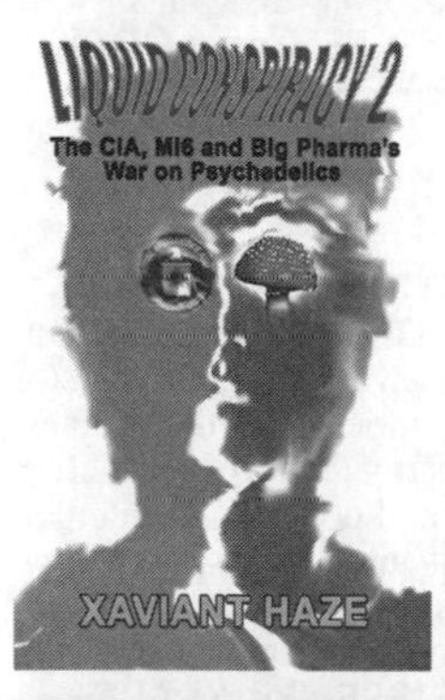

LIQUID CONSPIRACY 2:
The CIA, MI6 & Big Pharma's War on Psychedelics
By Xaviant Haze

Underground author Xaviant Haze looks into the CIA and its use of LSD as a mind control drug; at one point every CIA officer had to take the drug and endure mind control tests and interrogations to see if the drug worked as a "truth serum." Chapters include: The Pioneers of Psychedelia; The United Kingdom Mellows Out: The MI5, MDMA and LSD; Taking it to the Streets: LSD becomes Acid; Great Works of Art Inspired and Influenced by Acid; Scapolamine: The CIA's Ultimate Truth Serum; Mind Control, the Death of Music and the Meltdown of the Masses; Big Pharma's War on Psychedelics; The Healing Powers of Psychedelic Medicine; tons more.

240 pages. 6x9 Paperback. Illustrated. $19.95. Code: LQC2

THE GODS IN THE FIELDS
Michael, Mary and Alice—
Guardians of Enchanted Britain
By Nigel Graddon

This book offers for the first time detailed insights into England's St. Michael leyline, the celebrated "straight track" whose dragon energies (Michael and Mary) travel coast-to-coast from Cornwall to Norfolk. Aspects of these teachings are found all along the St. Michael ley: at Glastonbury, the location of Merlin and Arthur's Avalon; in the design and layout of the extraordinary Somerset Zodiac of which Glastonbury is a major part; in the amazing stone circles and serpentine avenues at Avebury and nearby Silbury Hill: portals to unimaginable worlds of mystery and enchantment.

280 Pages. 6x9 Paperback. Illustrated. Bibliography. $19.95. Code: GIF

ANCIENT ALIENS & SECRET SOCIETIES
By Mike Bara

Did ancient "visitors"—of extraterrestrial origin—come to Earth long, long ago and fashion man in their own image? Were the science and secrets that they taught the ancients intended to be a guide for all humanity to the present era? Bara establishes the reality of the catastrophe that jolted the human race, and traces the history of secret societies from the priesthood of Amun in Egypt to the Templars in Jerusalem and the Scottish Rite Freemasons. Bara also reveals the true origins of NASA and exposes the bizarre triad of secret societies in control of that agency since its inception. Chapters include: Out of the Ashes; From the Sky Down; Ancient Aliens?; The Dawn of the Secret Societies; The Fractures of Time; Into the 20th Century; The Wink of an Eye; more.

288 Pages. 6x9 Paperback. Illustrated. $19.95. Code: AASS

THE CRYSTAL SKULLS
Astonishing Portals to Man's Past
by David Hatcher Childress and Stephen S. Mehler

Childress introduces the technology and lore of crystals, and then plunges into the turbulent times of the Mexican Revolution form the backdrop for the rollicking adventures of Ambrose Bierce, the renowned journalist who went missing in the jungles in 1913, and F.A. Mitchell-Hedges, the notorious adventurer who emerged from the jungles with the most famous of the crystal skulls. Mehler shares his extensive knowledge of and experience with crystal skulls. Having been involved in the field since the 1980s, he has personally examined many of the most influential skulls, and has worked with the leaders in crystal skull research, including the inimitable Nick Nocerino, who developed a meticulous methodology for the purpose of examining the skulls.

294 pages. 6x9 Paperback. Illustrated. $18.95. Code: CRSK

LEY LINE & EARTH ENERGIES
An Extraordinary Journey into the Earth's Natural Energy System
by David Cowan & Chris Arnold

The mysterious standing stones, burial grounds and stone circles that lace Europe, the British Isles and other areas have intrigued scientists, writers, artists and travellers through the centuries. How do ley lines work? How did our ancestors use Earth energy to map their sacred sites and burial grounds? How do ghosts and poltergeists interact with Earth energy? How can Earth spirals and black spots affect our health? This exploration shows how natural forces affect our behavior, how they can be used to enhance our health and well being.

368 PAGES. 6x9 PAPERBACK. ILLUSTRATED. $18.95. CODE: LLEE

THE GIZA DEATH STAR REVISITED
By Joseph P. Farrell

Join revisionist author Joseph P. Farrell for a summary, revision, and update of his original *Giza Death Star* trilogy in this one-volume compendium of the argument, the physics, and the all-important ancient texts, from the Edfu Temple texts to the Lugal-e and the Enuma Elish that he believes may have made the Great Pyramid a tremendously powerful weapon of mass destruction. Those texts, Farrell argues, provide the clues to the powerful physics of longitudinal waves in the medium that only began to be unlocked centuries later by Sir Isaac Newton and his well-known studies of the Great Pyramid, and even later by Nikola Tesla's "electro-acoustic" experiments.

360 Pages. 6x9 Paperback. Illustrated. $19.95. Code: GDSR